INSIGHTS
OF
GENIUS

INSIGHTS OF GENIUS

Imagery and Creativity in Science and Art

Arthur I. Miller

The MIT Press
Cambridge, Massachusetts
London, England

COPERNICUS
An Imprint of Springer-Verlag

First MIT Press paperback edition, 2000

Originally published in 1996 by Copernicus, an imprint of Springer-Verlag, New York, Inc.

Cover art (clockwise from top): Detail from Figure 3, p. 163, showing a bubble chamber
photograph of a collision between a neutrino and a proton; Minkowski's sketches of space-time
diagrams; Pablo Picasso, composition study for *Guernica*, dated "May 1 37 (III)"

Library of Congress Cataloging in Publication Data

Miller, Arthur I.

Insights of genius : imagery and creativity in science and art / Arthur I. Miller.—

1st MIT Press pbk. ed.

p. cm.

Originally published: New York : Copernicus, © 1996.

Includes bibliographical references and index.

ISBN 0-262-63199-7 (pb : alk. paper)

1. Physics—Methodology. 2. Science—Methodology. 3. Creative ability in science. I. Title.

QC6 .M44 2000

530'.01—dc21   99-059185

Text design by Irmgard Lochner

10 9 8 7 6 5 4 3 2

To Norma

PREFACE

"For there are 'made' laws, 'discovered' laws, but also laws—a truth for all time. These are more or less hidden in the reality which surrounds us and do not change. Not only science but art also, shows us that reality, at first incomprehensible, gradually reveals itself, by the mutual relations that are inherent in things."

—Piet Mondrian, *Figurative Art and Nonfigurative Art*
(1937)

Scientists have always expressed a strong urge to think with visual images, especially today with our new and exciting possibilities for the visual display of information. We can "see" elementary particles interacting and "see" images of the brain. But "see" is a complex term.

Artists and scientists alike seek a visual representation of worlds both visible and invisible. They attempt to "read" nature. This book explores their efforts, with emphasis on the late nineteenth and twentieth centuries, the age of modern art and modern physics. Our journey will take us through the philosophy of mind and language and into cognitive science and neurophysiology, as we seek the origins and meaning of visual imagery and, so too, of science.

Asking what visual representation means to science throws fresh light on such problems as

- What is the connection between common sense intuition and scientific intuition?
- By what means does physics progress?
- Are there limits to physics?
- What are the relations between art and science?

Today these problems are no longer solely grist for academic debates. Too often, the media portray science as a godless, dehumanizing exercise whose excesses undermine the very fabric of society. Studying scientists' personal struggles to understand nature, to convince their peers, to inform the public, and to deal with cultural reactions to their research gives us a much different picture.

I focus on physics for a number of reasons. Chief among them are its long history, its foundation rich in philosophical implications, and its interface with so many other scientific disciplines. Another, no less important reason is my familiarity with the subject, which enables me to draw upon significant cases of scientific research for the purposes of motivation and illustration. Toward writing a self-contained book, I have elaborated on a handful of concepts and principles chosen specifically because they pervade all of the physical sciences and so ought to be understood by everyone. Some of these concepts come from other disciplines such as cognitive science and mathematics, which often play a key role in the analyses to follow.

The book concentrates on intuition, aesthetics, realism, representation, and visual imagery. We explore their interrelations and how they became transformed both in order to advance science and in response to another key concept in this book, scientific progress. Allied to these concepts are the principles of causality, relativity, energy conservation, entropy, and the correspondence principle. For example, exploring the origins of the principle of relativity entails investigating the transition from everyday common sense to the intuition of Galilean–Newtonian science with accompanying changes in our notion of cause and effect, or causality. Studying the origins of the principles of energy conservation and entropy leads us to the introduction of probability into the physical sciences with the allied concept of atomism, which opens the subject of scientific realism. The correspondence principle, introduced systematically into physics by Niels Bohr in 1913, turns out to be a key part of scientific progress, a concept whose exploration ties together many themes discussed thus far and introduces others such as metaphorical thought and theory of meaning from philosophy. The concept of aesthetics arises often in scientific creativity and in considerations of the relations between art and science. We will find that scientists' notions of an aesthetic can be articulated and bear similarities with the artists' notions. Larger issues emerge from discussions of these fundamental concepts and principles; these deeper topics include the meaning of science, the origins of scientific concepts and their relation to our cognitive apparatus, the origins of visual imagery and its role in thinking, particularly in creative scientific thought, the architecture of our mind that gives rise to our mental processes, and the relation between art and science.

The Ariadne's thread running throughout the book is that science has developed in a way that extends our intuition from common sense into an understanding of a world beyond our perceptions—a world where heavier objects fall at the same acceleration as lighter ones, where there is a relativity of space and time, and where there is a wave/particle duality. The final chapter extends this analysis to late nineteenth- and early twentieth-century art.

After touching on ancient science, Chapter 1 discusses Galileo and explores his use of thought experiments to abstract to possible worlds which turned out to be our own. This abstraction enabled Galileo to extend our intuition into a world in which there can be vacuums. For such a world he made the dazzling hypothesis that all bodies would fall with the same acceleration regardless of their weight. There were no laboratory vacuums for Galileo to test out this hypothesis, which turned out to be correct.

We then proceed to Isaac Newton's genial elaboration and extensions of Galileo's groundwork into a magisterial theory that unified phenomena on the Earth and in the heavens, yielding as well a new concept of causality strikingly different from the one in Greek physics. With further thought experiments, Albert Einstein realized that our concept of intuition had to be transformed yet again in order to understand a world in which space and time are relative quantities. As in every case in this book, the analysis is carried out by bringing to bear notions from fields other than science, such as from philosophy and psychology, all the while placing figures such as Galileo and Newton in their societal context. In all situations investigated in Chapter 1, the visual representations scientists used are abstracted from the world of sense perceptions. This time-honored method worked again and again, and nothing succeeds like success.

In Chapter 2, however, the situation gets more complex because twentieth-century research into atomic physics revealed that this sort of visual representation is not only inappropriate, but also wrong. This situation led to further transformations in intuition, causality, substance, and visual representations for scientific theories. We are led to investigate how scientists use experiments to probe a world beyond appearances from which, in turn, arises the importance of a form of reasoning fundamentally different from deduction and induction. We delve into the concept of visual imagery that was of importance to physicists in the German-cultural milieu in which atomic physics was primarily developed during 1913 to 1927. This chapter presents a dispute over essentially aesthetic tastes in the polemic between Werner Heisenberg and Erwin Schrödinger over whether atomic physics should be based exclusively on particles or waves. The issue was decided in 1927 by further redefinitions of the concept of intuition brought about by Heisenberg's uncertainty principle and then Bohr's complementarity principle. We realize that we can trace how theories emerge from Newtonian mechanics such as special relativity and quantum mechanics by studying the particular values possessed by universal constants of nature characteristic of these theories: the velocity of light and Planck's constant. As a result of certain limiting cases of these constants, relativity and quantum mechanics revert back to Newtonian mechanics, which has become intuitive to us because it reflects the world of perceptions in which there is no relativity of space and time and no wave/particle duality. These limiting cases are studied with Bohr's correspondence principle.

Having traced transformations in the concepts of intuition and visual imagery, in Chapter 3 we look in more detail as to how this came about. From exploring how scientists use data for constructing theories and for defending them against competing ones, we realize that these procedures are far from straightforward and not amenable to being encapsulated by any one philosophical theory of scientific method. Specific instances from the research of Galileo and Einstein are discussed, in addition to the physics of electrons in the early twentieth century. We then explore the unlikely origins of two guiding principles in the sciences, the principles of relativity and conservation of energy. Chapter 3 concludes with a discussion of the desire for unification of known forces, which goes back to the very beginnings of science in ancient Greece and continues to be a guideline for scientists. The roots of these principles, and of unification, turn out to be entangled with religion and mystical philosophical beliefs.

Chapter 4 descends one level deeper, inquiring whether the universe is indeed an ordered one, as assumed by Galileo and Newton, among others. The first inkling that this is not the case was revealed through the emergence of the concept of entropy along with the entry of probability into science. Some scientists assumed that these two developments required everyone to take atomism seriously. The ramifications of entropy, with its message of a tendency of the universe toward disorder, went beyond science. Nevertheless, most physicists continued to assume that the probabilities entering physics through gas theory, with its apparently attendant atomism, was just a reflection of our ignorance in tracing the paths of individual atoms. They believed that physics would revert to its completely deterministic basis as soon as Newton's mechanics was more finely tuned to apply to the atomic world. But this turned out not to be the case, as Chapter 2 discussed. Chapter 4 ends by examining how a new and intrinsic probability entered atomic physics along with wildly counterintuitive predictions, which have been established experimentally and thus call for yet further transformations of intuition.

Since atoms have played a central role in the transformation of intuition and the appearance of probabilities in physical theory, we must investigate further how we can argue that invisible entities can actually exist. This is the view of scientific realism explored in some detail in Chapters 5, 6, and 7. What emerges is that theories based on scientific realism make far-reaching claims such as the universality of science which can deliver absolute truth about the nature of physical reality. In contrast, antirealists claim that there are no such entities as electrons and no absolute truth, and some maintain that science may well be an artificial social construct. Comparison of these viewpoints opens up such problems as what we mean by the rationality and objectivity of science and so, too, the concept of scientific progress. The analysis in Chapter 5 necessarily takes us into the meaning of science and thus into criticism of science, which stems from the early nineteenth century to current postmodernism.

Weighing the comparative advantages of scientific realism and antirealism, I opt for the former. The burden of argument is heavy for the scientific realist. The reason is that logic is on the side of the antirealist, because of the underdetermination thesis (which is introduced in Chapter 3). According to this thesis, there are in principle an infinite number of scientific theories that can describe any set of experimental data. This holds for any science. So why should a single theory, with a claim to the reality of unobservable entities, be taken as the correct one? The chapter presents arguments from science and philosophy, including experimental evidence for atomism in conjunction with the fertility of this concept, and the need for atoms in order to explain natural phenomena.

A further basis for scientific realism is presented in Chapter 6, which explores the relation between mathematics and physics. This age-old problem has been discussed from Plato, to Newton, to Poincaré, and to Einstein, and it surfaced again at the forefront of physics as a result of Werner Heisenberg's uncertainty principle paper of 1927 wherein he emphasized that mathematics is to be the guide toward understanding atomic physics. The nature of this relation is essential for establishing whether the mathematical formulation of physical theories is a means for understanding worlds beyond sense perceptions. This investigation takes us through Platonism into the cognitive status of science, the origins and testability of geometries, and then into inquiring into whether physics can generate mathematics, which is an important test as to whether mathematics is the fabric of nature.

In analyzing scientific progress and the usefulness of metaphors for extending scientific theories into new domains, Chapter 7 ties together all themes discussed thus far. A model for scientific progress is presented based in scientific realism and with proper philosophical and psychological dimensions in addition to important elements from cognitive psychology as well as Bohr's correspondence principle. The model takes into account transformations in intuition as well as in visual representation that emerged from research in atomic and then nuclear physics, culminating in the so-called Feynman diagrams that are generated by the mathematics of quantum electrodynamics. Specific examples are explored on which the model of scientific progress is based.

A great deal of weight has been placed on the role of visual representations in creative scientific thought as well as on the notion of visual imagery and thought experiments. Chapter 8 explores these issues with results from artificial intelligence, cognitive science, and neurophysiology on how visual images are generated and manipulated. These views are compared with conclusions from previous chapters particularly to ascertain how they match results from the history of scientific thought and what light this subject can shed on general problems of visual imagery.

Thus far we have analyzed and explored the content and meaning of scientific theories, with some attention paid to their discovery. Chapter 9 focuses on creative scientific thought. In so doing it investigates theories of digital and analog thought and presents a model for creative scientific thinking based on results from previous chapters such as guidelines for constructing scientific theories. The model assumes the importance of unconscious parallel processing of information and is compared with the introspections of Henri Poincaré and Einstein, which are the centerpiece of this chapter. Their views on creativity, aesthetics, and intuition are compared and contrasted. This analysis sheds further light on the fascinating problem of why it was Einstein who formulated special relativity in 1905 and not Poincaré, despite both men having at their disposal the same experimental data and having arrived at the same mathematical formulation.

Throughout the book thus far the concept of representing the world about us has been central and we have mentioned problems that scientists and artists share. That there were close relations between art and science in the Renaissance is incontestable. With the appearance of Newtonian science in the late seventeenth century, and with it the onset of the Age of Rationalism, many educated people considered science to be the only pursuit of absolute truth. This view began to change noticeably by the end of the nineteenth century. What I mean by relations between art and science is that at the end of the nineteenth century both subjects happened to be moving toward greater abstraction. Why? Chapter 10 concentrates on this fascinating topic. It returns to concepts discussed with relation only to science, such as intuition and aesthetics, and opens them up here, taking into account the differences and similarities between art and science. The discussion of aesthetics is then expanded by defining it in more detail and comparing how it is used by artists and scientists. The interplay between art and science, which we discuss in some depth, concerns the development of Cubism by Pablo Picasso and Georges Braque, and influences on them, like Paul Cézanne. Then we move to elaborations of Cubism toward nonfigurative art by Juan Gris and Piet Mondrian, among other artists such as the Abstract Expressionist Mark Rothko. In order to seek deeper insights into scientific progress, we compare histories of art with histories of science. We find many similarities between artistic creativity and creative scientific thought, particularly the role of unconscious parallel processing of information.

I wrote *Insights* for lovers of science, be they educated lay readers, students, professional scientists, or artists. Especially important to me are science students and the growing number of lay readers for whom science has become a passion. I hope that *Insights* amplifies their interest by answering some of their questions while whetting their appetite for others. I hope that professional scientists, who often can spare little time for historical and philosophical literature, will find in *Insights* the kind of foundational analysis that gives new awareness into the philosophical underpinnings of their work.

Having based the material in Chapter 10 on firm historical and artistic investigations grounded in the results of previous chapters, I believe that my analyses of aesthetic ideas and the differences and similarities between art and science will interest artists and art historians. I think everyone agrees that part of the intellectual and emotional adventure in art and science resides in the ultimately one-on-one struggle with nature. In this arena barriers between disciplines dissolve.

While the laws of science per se make no statements on ethics, morals, or religion, its practitioners sometimes do. By and large, scientists have been in the vanguard of the fight to defend purely speculative thought against accusations of religious or political sedition. In reply to antiscience currents in France in 1902, the great mathematician–

philosopher–scientist Henri Poincaré rose to science's defense in "Science for its own sake." His words convey the intensely personal endeavor demanded by art and science:

> It is only through science and art that civilization is of value. Some have wondered at the formula: science for its own sake; and yet it is as good as life for its own sake. . . . Thought is only a gleam in the midst of a long night. But it is this gleam which is everything.

With Poincaré's passionate message in mind, let us explore how artists and scientists seek a visual representation of our universe.

—Arthur I. Miller
Department of Science & Technology Studies
University College London

ACKNOWLEDGMENTS

I have thought about this book a great deal for well over a decade, during which I have had the pleasure to teach some of the material and to discuss much of it with numerous colleagues. Let me reserve mention here to those who have been of great help on the manuscript itself. I should like to thank Brian Balmer, Hasok Chang, A. R. Jonkheere, Allistair McClelland, David Miller, Steven Miller, Constantine Moutoussis, Piyo Rattansi, Jon Turney, and Scott Walter. For comments on Chapter 5 it is a particular pleasure to thank Marcello Pera. On Chapter 6 I acknowledge conversations with Marcus Giaquinto, Jeremy Gray, and Andrew Gregory. For sharing their expertise with me on vision which helped to get Chapter 8 into shape I am most grateful to J. Michael Brady, Michael Duff, and Semir Zeki. Chapter 10 was a difficult one and I deeply appreciate criticisms from David Bindman, Bernard Cohen, Nathan Cohen, John Golding, Norma Miller, and Sarah Wilson.

It is more than appropriate for me to express my gratitude to Terry Kornak at Springer-Verlag for invaluable editorial assistance, and to William Frucht for his insightful editing.

CONTENTS

3

SCIENTIFIC METHODS

The Verdict of Experiment or Not? / Data, Data
Everywhere . . . / Unwanted Precision / Picking and
Choosing Data / Sociology Influences Science / The
Principle of Relativity / Conservation of Energy / Heat Is
Energy / Unification of the Sciences

71

4

FAITH IN AN ORDERED UNIVERSE

Certainty and/or Knowledge? / The Sign of the Time / Carnot's
Imaginary Engine / The Birth of Uncertainty / A Mal du Siècle
Over the Fin de Siècle / Causality in Quantum Physics / The
Image of Light / Quantum Effects and the Corsican Brothers

105

5

SPEAKING REALISTICALLY ABOUT SCIENCE

Let's Have a Reality Sandwich / Atoms in Antiquity / Doubts
About Atomic Reality / Positive Thinking / A Mach Attack /
Planck Converts to Atomism / Atomic Vindication / On Mach's
Philosophical Heirs / Boredom Relieved, Momentarily /
Scientific Realism / Scientific Antirealism / The Social Side of
Science / Antirealism and Scientific Practice / The Chic of
Antirealism: Postmodernism / The Two Cultures / Some Very
Modern Postmodernists / Back to Reality

129

6

THE REASONABLE EFFECTIVENESS
OF MATHEMATICS IN PHYSICS

Mathematics and Physics / Pythagoras's Preestablished Harmony /
Newton, Einstein, and Minkowski's Dream / Kant, Geometry, and

1
COMMON SENSE AND SCIENTIFIC INTUITION

Common sense tells us that heavier objects fall faster than lighter ones, that we can catch up with whatever we want by increasing our speed without limit, and that light travels infinitely fast. This chapter explores the power of thought experiments to probe deeper insights into nature than common sense. In this way common sense developed from daily experiences is found to be naive and inappropriate for extending our knowledge. Thought experiments can lead to scientific theories that form the basis for a new intuition of nature. For instance, Galileo used thought experiments to overturn the "intuitive" Aristotelian view of falling bodies.

The history of scientific thought shows that this cycle occurs many times. In this chapter we cover major transformations in the history of scientific thought: Galileo's thought experiments setting a basis for Newton's theory of falling bodies on the Earth and the motion of comets in the heavens, which offered both a new common-sense concept and a new basis for intuition, and with further thought experiments, Einstein showing the limitations of the Newtonian view and proposing a theory that offered a newer common sense, one that encompassed properties of light and offered deeper insights into the nature of time. Intuition turns out to depend heavily on scientific theories that eventually become the customary ways to understand nature.

At the Beginning

What we know as science has roots in places as far afield as China, India, and the Middle East. But the crowning achievement of inventing science falls to the Greeks in the seventh century B.C. A number of contributing factors explain the Greek origin of science. Besides the relative Pax Graeca, which lent a peaceful atmosphere in which these people could travel and study, as well as religious freedom, Greece is strategically situated at the crossroads of great caravan trails. This fortuitous situation provided the likes of Homer, Thales, Democritus, Pythagoras, Plato, and Aristotle peace of mind, with which to contemplate eternity with the best data available.

At the beginning of our story, an almost inconceivable six thousand years ago in the region bounded by the Euphrates and Tigris Rivers known as Babylonia, forgotten men in fabulous cities with exotic names systematically looked at the sky. After observing for over three thousand years, the Babylonians arrived at a world view in which the Earth was at the center of all they perceived. Even now, looking up at the heavens on a clear night, we can imagine what the Babylonians saw: stars that seem to be painted on a hemispherical bowl rotate about an immobile Earth. The Babylonians saw regularity from the vantage point of a stationary platform, gazed upon by gods who were their creators. Except for the theological part, this is a reasonable image. Open-shutter photographs of the night sky display concentric streaks of light with ourselves at the very center. By the Middle Ages, this view of the cosmos became part and parcel of common sense. If the Earth moved, after all, would not hurricane-like winds blow across its surface, with clouds and birds flying haphazardly across the sky?

Other common-sense notions were incorporated into what was turning out to be an organized manner of dealing with phenomena on the Earth and in the heavens. Is it not common sense that heavier objects fall faster than lighter ones? By and by, such commonsensical statements formed the basis of a theory of motion so successful that it stood for over one thousand years. Spun by such giants of Western thought as Pythagoras and Plato, this theory was codified most systematically by Aristotle in 350 B.C.

The Aristotelian cosmos is purposeful because all objects move *in order that* they reach their natural place. In Aristotelian science, everything is composed from four basic elements: earth, water, air, and fire, arranged in ascending order of their ability to permeate Earth. These elements are in motion toward their "natural place." Once there, they remain at rest: Every motion has a goal. Every motion has a beginning and an end, which are preordained from the object's place in the grand order of things. Objects possess a tendency toward their destinies. This purposeful movement toward a final state is the sort of causality philosophers call "teleology." Motions away from a natural place are "unnatural" or "violent." Consequently, a ball thrown upward into the air will at first undergo an unnatural motion and eventually turn around "in order that" it return to its natural place, on the Earth. Violent motions give way to natural ones, never vice versa, because this explanation or principle is confirmed by our observations.

These principles were consistent with our common sense, but they were not correct. It was a start, however, and the only conceivable one. We build frameworks of knowledge based on the world about us and see where the structure of explanation leads. The chief underlying assumption has always been that science would lead somewhere, somehow. By the sixteenth century A.D., Greek science had become entwined with Judeo-Christian dogma. During Galileo's lifetime, the Bible was deemed the last word in matters of religion and science. Within Judeo-Christian tenet, the Babylonian cosmos became God's blueprint: an immobile Earth at the center of a finite universe. After all, if the universe were infinite, it would take a much longer time for souls to journey to heaven, while hell is just below our feet, and this subterranean geography was demonstrated by volcanic action. Unlike the vulgar, irregular Earth, the heavenly bodies are perfect spheres. They move about the Earth in the most perfect of all paths—the circle—which has no beginning or end, with every point equidistant from a single center. From the moon's orbit outwards, everything is perfect and without blemish, and this perfection can be seen with the naked eye.

It took ingenious thought experiments and far-reaching mental abstractions to strike down this view. To set the stage for these developments, let us try to get some inkling of the mindset of Renaissance scientists, which today might seem quite extraordinary to us. As the Newton scholars J. E. McGuire and P. M. Rattansi write:

> It is possible to imagine a multiplicity of Newtons, one engaging in "science" and the others dabbling in theology, Biblical chronology, and other similar disputes. . . . Newton, however, was not a "scientist" but a Philosopher of Nature. In the intellectual environment of his century, it was a legitimate task to use a variety of materials to reconstruct the unified wisdom of Creation.

THE MINDSET OF RENAISSANCE SCIENTISTS

In 1543, some two decades before Galileo's birth, the Polish churchman and astronomer Nicholas Copernicus offered an alternative view to the

Church's cosmology: a sun-centered universe. Not a shred of experimental evidence existed to support it. Copernicus's system rested on elegance and symmetry of presentation and on neo-Platonic philosophy, a version of Platonic thought proposed by the fifth-century A.D. Greek philosopher Proclus. By "elegance of presentation," I mean that in a sun-centered universe the planets' orbits are related to how fast they go around the sun. This is not the case in an Earth-centered system. According to neo-Platonic philosophy, a sun-centered universe better displays the hand of God because the sun is the source of warmth and light. A point in favor of Copernicus's universe is that it offered a calendar superior to the Julian one. With this advantage in hand, the Church authorities could not relegate Copernicus's work to the Index, the list of books the Church condemned. Rather, they interpreted his work as an instrument for calculation. The scientist who would alter this viewpoint forever was Galileo Galilei.

Italians honor their great thinkers by addressing them by their first names. Galileo was born in 1564, the year Michelangelo died and William Shakespeare and Christopher Marlowe were born. Among Galileo's other contemporaries were the philosopher René Descartes; the poet John Milton; the discoverer of the blood circulatory system, William Harvey; and the astronomer Johannes Kepler. With good reason, the era in which Galileo lived is known as the Age of Genius.

Galileo was an urbane man, a master of wit and sarcasm, as well as an eloquent writer. He was also a pious man, who looked upon science as a means to better understand the mind of God. He was fond of saying that the Bible tells one how to go to heaven, but not how the heavens go. This was the mindset of Copernicus, Descartes, Kepler, and Isaac Newton, who must be understood as men of the Renaissance, for whom science was a natural extension of religion. Perhaps the most extreme thinker among this group was Newton. From the austere mathematics in which Newton necessarily formulated his theory, we might project back to its author the personality of a cold calculator, much the Hollywood image of the scientist. Yet one of the fruits of recent Newtonian scholarship is the discovery that chief among Newton's interests were alchemy, magic, and biblical chronology—correlating historical facts with prophecy. In what follows we will come across a mixture of these subjects with scientific pursuits which, while striking the twentieth-century mind as highly peculiar—even outrageous—was quite normal for someone of the seventeenth century.

The very basis of Copernicus's and Galileo's research, which Newton elaborated and deepened in his magisterial 1687 book, *Philosophiae Naturalis Principia Mathematica*, is a universe in which the Earth moves about the sun. There would be no evidence of this until the middle of the nineteenth century and by then it was a moot point. But is this not amazing? If science is a rational enterprise, how could such eminent practitioners as Galileo and Newton have based a spectacular theory of motion on an assumption for which there was no direct evidence? The reasons lay partly in the theory's predictive and explanatory powers: It predicted the existence of unknown planets and explained fearsome comets, whose appearances it also predicted, merely as wayward members of our solar system. Newton succeeded in unifying motions in the heavens and on Earth within a single theory. This was the first great unification in modern science.

But we are running ahead of ourselves. In seventeenth-century Italy, the Earth-centered universe was official Church doctrine and therefore dangerous to oppose. What about terrestrial physics? This, too, had its roots in Aristotle's teachings, which the fourteenth-century churchman Thomas Aquinas had folded into Christian theology. Within this framework, Aristotle became "the Philosopher." For his travail, Aquinas was elevated to sainthood.

GALILEO'S IMAGINATION

Sometimes the early lives of great thinkers are singularly unremarkable, except in retrospect. Galileo and Einstein, for example, both had utterly undistinguished college careers. In 1583 Galileo was forced to leave the university in Pisa. Besides the animosity of the faculty, which he earned mainly by asking questions they could not answer, Galileo had run out of money. After a series of odd jobs, and a brief sojourn, due to interesting results of his in physics, at his old university, in 1592 Galileo accepted a position at Italy's most important university, in Padua. The 18 years he spent there were the happiest and most creative of his life.

From this period in Padua emerged his spectacular results on falling bodies. The results are all based on the extremely counterintuitive assumption that all bodies fall in a vacuum with the same acceleration, regardless of their weight. Galileo reached this conclusion through thought experiments. As the name indicates, scientists perform thought experiments in their "mind's eye." These experiments are run on idealized apparatuses and so require a high degree of abstraction. They are inexpensive to run but require tremendous ingenuity. Galileo was a master at this process. Through them he realized that the proper motions of bodies ought not be imposed on nature—such as the natural and unnatural motions we discussed earlier. Rather, they should be drawn out of nature through mathematics and experiment.

To see what Galileo was up against, here is an example of how common sense can lead one astray. Consider a typical argument from Greek physics against the idea of a vacuum, which may appear reasonable from the viewpoint of common sense. Recall that a vacuum is a space with nothing in it. Somehow if a vacuum were to exist, the Greeks believed, then a natural motion could be transformed into an unnatural one. For example, when you let go of a stone, it falls down in order to reach its natural place, which is on the ground. No one has ever seen the stone's continuous fall become erratic or unnatural. On the basis of many such observations, Aristotle inferred that there are no vacuums.

Another instructive example of how common sense can be misleading is Aristotle's inference from common sense that heavier objects fall faster than lighter ones. They do, but this misses the point of free fall in all its wonderful generality. Galileo wondered about free fall through a vacuum in which there is no air resistance. There is an apocryphal story that Galileo refuted this hypothesis by dropping cannon balls from the leaning tower of Pisa and timing their fall. Considering the state of timing devices, he knew better and therefore used a thought experiment rather than actually dropping the cannon balls. He described one such experiment in his masterpiece of 1632, *Dialog on the Two Chief World Systems.*

Let's follow his experiment: Consider a large stone, stone 2, which falls at, say, eight units of speed and a smaller one, stone 1, which falls at four units (see Figure 1). This difference in speed is expected, according to Aristotle's physics. Now tie the two stones together. What will happen? According to Aristotle, the combination will fall at a speed somewhere between four and eight units, because the lighter one will retard the fall of the heavier one. But how do the stones know they are tied together? Why isn't it possible for the combination to fall at a rate greater than eight? This is one of several thought experiments on the basis of which Galileo demonstrated that Aristotle's physics is riddled with internal inconsistencies.

To get an inkling of the actual experiments carried out by Galileo on falling bodies, one must walk just across from the leaning tower in Pisa to the cathedral, which contains a small bronze lamp that the sacristan set swinging at Sunday services. A plaque in the cathedral tells us that watching the back-and-forth motion of this lamp may have piqued Galileo's fascination with periodic motion, leading him to abstract to the case where bodies fall through a vacuum. The experiments he performed on pendula incorporated material of not drastically differing weights. He found that the time elapsed for one complete motion back and forth—the pendulum's period—depends only on the square root of the length of the pendulum's arm, but *not* on the weight of its bob as would be the case for free fall through a vacuum.*

---

*Most historians of science agree that Galileo actually performed experiments on pendula and inclined planes. A contentious point is how much his experimental data influenced his theorizing.

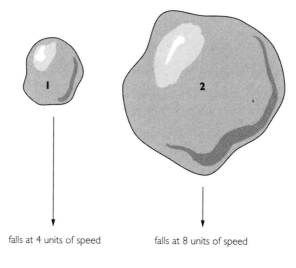

falls at 4 units of speed   falls at 8 units of speed

(a)

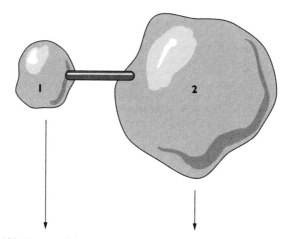

With what speed does the combination of stones 1 and 2 fall?

(b)

FIGURE 1.

Galileo's thought experiment concerning the free fall of two stones. (a) Stone 1 is lighter than stone 2, and so according to Aristotle, stone 1 falls more slowly than stone 2. (b) But what about when the two are bolted together? How fast does the joined combination fall?

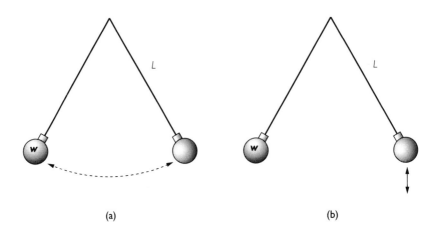

(a)   (b)

FIGURE 2.

Part (a) is a pendulum of length L with a pendulum bob of weight w on its end. *Seeing as* is seeing the pendulum bob going to and fro. Part (b) shows the point of view of seeing the phenomenon in (a) as *seeing that*, where the pendulum bob *falls* and then *rises*. Galileo had already ascertained that the period of the pendulum's motion depends on the square root of L and not on the weight of the pendulum bob.

During the 1620s, when Galileo worked on the *Dialogue*, he recalled this result from Pisa and *saw* it in a unique manner.* There is an interesting distinction between *seeing as* and *seeing that*. Everyone *sees* the pendulum *as* going to and fro periodically [(Figure 2(a)]. Galileo *saw that* the pendulum bob fell and then rose again [Figure 2(b)] and that, within reason, the weight of different bobs did not appreciably affect the pendulum's time of rise and fall. He supported his conclusion with further

*Although Galileo's research in Padua went extremely well, heavy teaching duties and burdensome administration took its toll. In addition, there were financial burdens placed upon him by ne'er-do-well brothers and sisters, coupled with his high lifestyle. Interestingly, in Galileo's time the differential pay scale was tipped in favor of philosophers and not scientists. In 1610 he returned to Tuscany, where he was offered a dual position as Chief Mathematician to the Grand Duke of Tuscany as well as a chair of mathematics at his old undergraduate school in Pisa. The catalyst for the offer was Galileo's book of 1610, *Starry Messenger*, which contains some of his spectacular telescopic observations supporting a sun-centered universe. This book elevated Galileo to the status of the most famous scientist in Italy.

thought experiments that included perfectly round spheres rolling on perfectly smooth inclined planes.

Pushing this conclusion to cover all bodies regardless of their weight required Galileo to abstract to a world in which a vacuum exists. In this way he could propose the astounding hypothesis that all objects fall through a vacuum with the same acceleration regardless of their weight. There would be no laboratory vacuums large enough to demonstrate this spectacular idea until years after Galileo's death. Today, this demonstration is standard fare in many science museums, where there are two evacuated columns in which a brick and feather released at the same time fall side by side and hit the floor together. These experiments are generalizations of actual ones Galileo performed with inclined planes, which he used in order to slow what would have been the object's descent in free fall.

To convince yourself of the counterintuitivity of Galileo's results on falling bodies, you might want to perform the following experiment (Figure 3). Take two stones of different sizes and weights. Hold one over a slot in a table and the other just on the edge. Simultaneously let the first one drop and push the second one off the table. Which one

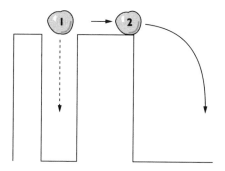

FIGURE 3.

Stone 1 is held over a slot in a table, while stone 2 is placed at the table's edge. At the same time, drop stone 1 and give stone 2 a horizontal push off the table. Which stone lands on the floor first?

lands on the floor first? They hit the ground together. Why? They both are acted on by the same downward force—their weight—while the horizontal portion of stone 2's motion remains unchanged. Thus, despite the fact that their weights, which differ, cause the stones to fall, they hit the ground simultaneously.

Behind the counterintuitive result in Figure 3 is Galileo's hypothesis of the weight independence of free fall through a vacuum, which is reflected in the basic mathematics of Galilean-Newtonian mechanics: The end result of calculating the time of fall for either stone does not contain its weight. We may as well be living in a vacuum. We don't, yet we can get away with assuming this to be true most of the time. In situations where wind effects are important, such as in ballistics tables for long-range artillery, or in objects falling through water or viscous media, such as oil, the falling object's weight enters. But this is all just arithmetic that could not have been carried out without Galileo's having first *abstracted* to the case where objects fall through a vacuum.

CONCEPTUAL FRAMEWORKS AND COMMON SENSE

As we noted in our analysis of Figure 2, *seeing that* is seeing or understanding the deep structure in a phenomenon. *Seeing that* is not simply perceptual sight; it includes cognitive organization. Understanding even the simplest everyday occurrence entails a complex interaction between our cognitive and perceptual apparatuses. After all, there is no such thing as direct observation. What we know about physical occurrences in nature tempers the way we see phenomena and so, too, does our intuition, or what we *expect* to notice or to occur. If both a child and a trained physicist see an x-ray tube, they do not "see" the same object. The child sees a glass tube with some metal stuck in it, while the scientist sees a glass tube containing a cathode, an anode, and a heavy metal target that is the source of x-rays. All data, regardless of their source, can be discussed and evaluated only within a framework of knowledge.

Philosophers refer to this observational situation as the "theory-ladenness" of data.

Frameworks of knowledge, sometimes called "conceptual frameworks," die hard. I never cease to be amazed how natural and comforting Aristotelian reasoning is, even to students who ought to know better. Whenever I teach a history or philosophy of science course in which there are physics students, I ask them to imagine a situation in which they are a passenger in a moving train. If they dropped a stone, where would it land? No one immediately volunteers the correct reply (the stone would land directly under where it was dropped). Instead, there are murmured queries regarding wind resistance and other superfluous things. Consider another example: A stone is tied to a string that is then twirled around. If the string is cut, which way will the stone go? Aristotelians reply that it will shoot off along the circle's radius. Those versed in Galilean-Newtonian physics will answer (correctly) that it goes off along the circle's tangent. Yet another example is to ascertain the trajectory of a bomb dropped from a moving airplane (it would be parabolic). Even professional physicists are sometimes fooled by so-called naive physics questions, getting lost in complexities that are completely beside the point. Galileo's genius lay in his seeing the forest for the trees.

Scientists interested in the foundations of their subject, of the ilk of Galileo, Newton, Einstein, Niels Bohr, and Werner Heisenberg, are able to cull away inessentials and cut to the core of a problem. They do not get mired in technicalities. These sorts of minds turn up in other fields, too. Mozart had so mastered the technical aspects of music that he could soar above mere technicalities to produce sublime melodies. Unencumbered by mathematical and physical trivia (from their point of view), Bohr, Galileo, Einstein, and Heisenberg concentrated on if and why questions. They moved with ease about the world of hypotheticals, somehow knowing when to stop asking "why." So, for example, did Galileo realize that the key problem in understanding matter in motion was not how objects are blown about by the wind or fall through viscous media like water or oil. *The essential point is to figure out how objects fall through a vacuum; the rest is arithmetic.*

We now turn to how the world view of Galilean-Newtonian physics replaced the common sense of Aristotelian science.

GALILEO AND NEWTON BECOME COMMON SENSE

Bertolt Brecht's play *The Life of Galileo* concludes with Galileo's manuscript of his book of 1638, *Two New Sciences*, being smuggled across the frontier of totalitarian Italy into the light of the free world. The sight of the jackbooted border guard tells it all. Galileo's book holds a doubly seditious secret: Not only does the Earth move, but it is not the center of the universe. Published by Elsevier in The Netherlands, *Two New Sciences* found fertile ground in northern Europe, in the hands of Christiaan Huygens in The Netherlands, and most importantly Isaac Newton in England.

The new science of Galileo and Newton struck the learned world as nothing short of miraculous. A few statements accepted as valid without experimental or theoretical proof served as the basis for a theory covering terrestrial *and* extraterrestrial phenomena. In Aristotelian science, these two realms were assumed to be separate: the former vulgar, the latter divine. While Aristotelian physics predicted very little, the Newtonian framework explained and predicted comets, produced reliable tidal tables, and predicted the existence and approximate location of new planets, among other marvels. So wonderful was this new scheme of things that it gave humankind the hubris to believe that it covered all matter in motion with rational mathematics. At last we were in control of the universe, and apparently our destiny too. The appearance of Newton's *Principia* signaled the end of the Renaissance and the onset of the Enlightenment, known also as the Age of Reason.

Anyone who has ever seen the *Principia* knows how daunting it is. Whereas Galileo's *Dialog* is written in flowing Italian prose and meant for the educated lay public as well as scientists, Newton's *Principia* is written in academic Latin. In calculating how planets move under the influence of his most far-reaching hypothesis, universal gravitation, Newton chose not to use the sleek new calculus he had invented. Instead, he turned to complex geometrical methods. By the late eighteenth century, in the hands of such scientists as Leonhard Euler, Pierre Simon de Laplace, Joseph Louis Lagrange, and the Bernoulli family— Daniel, James, and John—Newton's mechanics took the form we associate with it today. With the appearance of more accessible texts and popularizations, Newtonian physics has become intuitive to us, a matter of common sense. Once we have taken a course in basic physics, doesn't the world seem very different? Much of what was abstract becomes second nature to us. As we have noted, representing the world about us is a complex phenomenon that is the product of our senses, or perceptual and cognitive apparatuses. Let us look into the Galilean-Newtonian representation by comparing its visual images with those of Aristotelian science.

Representing phenomena means literally *re*-presenting them as either text or visual image, or combination of the two. But what exactly are we *re*-presenting? What sort of visual imagery should we use to represent phenomena in the world? Did not the visual imagery of Aristotelian science turn out to be misleading? A good example of this is in Figure 4, which shows how Aristotelian physics pictures a cannonball's trajectory. On the other hand, Galileo realized that specific motions should not be imposed on nature. Rather, they should emerge from the theory's mathematics. Figure 5 is Galileo's own drawing of the parabolic fall of an object pushed horizontally off a table, as is the case with stone 2 in Figure 3.

Nature gives us data. Both Aristotle and Galileo saw a stone falling down. But how we interpret this phenomenon, that is, how we read nature, depends on one's conceptual framework: For Aristotle, the data were that objects eventually fall to the ground in order that they seek their natural place. For Galileo, the data were measurements of times of fall through

FIGURE 4

An Aristotelian representation of a cannon ball's trajectory. It illustrates the Aristotelian concept that the trajectory consists essentially of two separate motions, unnatural (away from the ground) and natural (toward the ground). In this figure the transition between unnatural and natural motions is a circular arc. From G. Rivius, *Architechtur, Mathematischen, Kunst,* 1547.

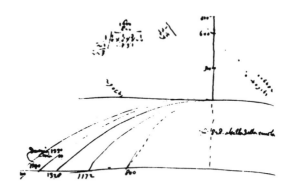

FIGURE 5

Galileo's drawing made in 1608 which can be interpreted as his experimentally confirming conservation of the horizontal component of velocity, and of the decomposition of the vertical and horizontal components to give the parabolic trajectory of a body projected, in this case, horizontally. Galileo was beginning to think along the lines of free fall through a vacuum.

certain distances. Whereas Figure 4 is what Aristotle *imagined*, Figure 5 is what we imagine as a *result* of knowing Galilean-Newtonian science.

As another example of the difference between Newtonian and Aristotelian science, consider a situation that has become "intuitive" in the new meaning of this term in Newtonian science. This apparently straightforward situation hid a Pandora's box of problems for physicists in the late nineteenth and early twentieth centuries. Consider two cars riding along a highway. You are in car A (Figure 6). Your speedometer reads 40 mph. What does this number mean? I ask this because, in Newtonian science, velocity is a relative quantity. Your speedometer indicates 40 mph, which means that you are traveling at 40 mph *relative* to the road. But what about the velocity of the other car, car B, *relative* to yours? Car B is moving at some other velocity relative to the road—say 50 mph. What is the velocity of car B *relative* to A? Its velocity is the difference between car B's velocity relative to the road and yours, which is 10 mph. Should you catch up with the other car and ride alongside it, then the velocity of car B *relative* to yours is zero, while both cars travel along at 50 mph *relative* to the road. We have just hit upon a key part of Newton's mechanics: his law for adding velocities, which plays a central role in this chapter.

The example we have just considered illustrates a new intuition of Galilean-Newtonian motion: You can catch up with whatever you want,

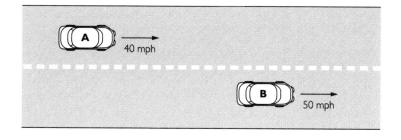

FIGURE 6.

Car B moves with a velocity of 50 mph relative to the road, while car A moves along at 40 mph.

and when you do, the relative velocity becomes zero. Just step on the accelerator. There is no ultimate speed limit. This example is basic and the results apparently straightforward. Yet its simplicity is deceptive. The reason is that behind Newton's addition law for adding velocities is a concept of time that was the accepted common-sense notion until Einstein's special theory of relativity replaced it. In Newtonian science, time does not depend on a clock's motion, whereas Einstein's relativity theory predicts otherwise. How this occurred revolves around Einstein's realizing a connection exists between Newton's velocity addition law, time, and the nature of light. This may seem surprising, and indeed it was even to scientists at the beginning of the twentieth century.

Let's stop for a moment and review the cycles of common-sense beliefs/new knowledge we've examined so far. Through thought experiments, as well as certain actual ones, Galileo convinced himself that Aristotle's physics contained internal inconsistencies. Galileo's results were far-reaching beyond science per se, because they indicated limitations to our common-sense view of nature. He decided that the only way to proceed was to jettison Aristotle's physics and to formulate a new theory of motion based on such terribly counterintuitive hypotheses as the weight independence of objects falling through a vacuum. A little over two hundred years later, it would turn out that there are limitations to the new common sense of Galilean-Newtonian science, which would be revealed by a thought experiment conceived by Einstein.

Einstein's thought experiment concerns the ramifications of taking seriously the fact that although the velocity of light is enormously fast, it is not infinitely fast as common-sense intuition would lead us to believe. Consider, for example, the result of switching on a light bulb. The room seems to light up at once, as if the light came on instantaneously. As we show in the next section, this turns out not to be the case. If so, then what supports light in transit? Moreover, what exactly do we mean when we say that "light is a wave phenomenon"? Where are the "waves" in light? Addressing these questions about the nature of light will lead us to reexamine the foundations of Newton's physics, including Newton's law for adding velocities.

THE WAVINESS OF LIGHT WAVES

The need for a medium to transport light had been apparent ever since Christiaan Huygens had investigated a wave representation for light in the seventeenth century. If light is a wave, there has to be some physical substance to wave. The milieu that waved was called the ether. By the mid-nineteenth century, experimental evidence was overwhelming that light is a wave phenomenon with properties like those of water waves (see Chapter 2). The problem of detecting the ether—the medium through which the waves presumably traveled—became paramount.

The ether served another purpose, too. A point of contention regarding light had always been whether it traveled at an infinite or finite velocity. Common sense indicated an infinite velocity. After all, while the muzzle flash from a cannon reaches the eye instantaneously, the sound comes noticeably later. Although early on in his career Galileo agreed with this consensus, by the 1630s he changed his mind: The velocity of light was very fast, but not infinite. Galileo's counterargument, in *Two New Sciences*, was that: "no more can be deduced than that the sound is conducted to our hearing in a time less brief than that in which the light is conducted to us." So much for common-sense reasoning.

Determining whether light traveled at a finite or infinite velocity was of great scientific importance. A contemporary of Newton, the Danish astronomer Olaf Roemer, made the first accurate measurement of the velocity of light in 1676 through clever use of the eclipses of Jupiter's satellites. Roemer noticed that the intervals of time between eclipses of one of Jupiter's satellites increased as the Earth receded from Jupiter and decreased as it approached Jupiter. Taking a large number of observations extending over one year, Roemer found a mean difference of 996 seconds between eclipses, which he attributed to the finite velocity of light. Consequently, Roemer could argue that it requires 996 seconds for light to traverse the diameter of the Earth's orbit, from which he calculated the velocity of light to be 2.2×10^8 m/sec (meters/second), not bad compared with today's measured value of 2.9979×10^8 m/sec. In 1727, the British astronomer James Bradley used observations on the star γ Draconis to measure the velocity of light. Bradley's measurement took into account the Earth's known orbital velocity about the sun (30 km/sec). His result substantiated Roemer's. Then, in the mid-nineteenth century, it became possible to carry out accurate measurements of the velocity of light in terrestrial laboratories. The results were close to those of Roemer and Bradley, thereby confirming that light travels at a finite velocity. From now on we will denote with the letter c the velocity of light obtained from measurements by the astronomers Roemer and Bradley and then substantiated in laboratory experiments in the nineteenth century.

Through the beginning of the nineteenth century, the prevailing representation of light was the particle one proposed with great vigor by Newton. Many optical experiments could be explained with a particle theory of light, including those of Roemer and Bradley. But due primarily to experiments by the British scientist Thomas Young in 1803, which we will take up in Chapter 2, the nineteenth century witnessed the rise of wave theories of light. As Young and other wave enthusiasts emphasized, there has to be a milieu in which light travels. After all, you cannot have water waves without water. Underlying these scientific reasons for an ether was the anthropocentric need for an ether, based on our experiences of the world in which we live. All physical processes seem to occur by means of bodies acting directly on one another, rather than by any "action at a distance." Most scientists believe that this is universally the case even for phenomena beyond our daily experiences. Consequently, scientific theories should not contain any actions at a distance. As Newton wrote on January 17, 1693, to his sometime amanuensis Richard Bentley:

[Action at a distance] is to me so great an absurdity, that I believe no man, who has in philosophical matters a competent faculty of thinking, can ever fall into it.

Yet, the lack of proper mathematical concepts for analyzing how disturbances are transmitted through a medium forced Newton to formulate his mechanics in empty space. However, in empty space, disturbances are transmitted instantaneously. As Newton's scientific descendent, Einstein, respectfully put it: "Newton, forgive me; you found the only way which, in your age, was just about possible for a man of highest thought and creative power." Einstein echoed Newton's sentiment that

non-physical thought knows nothing of forces acting at a distance [because] such a concept is unsuited to the ideas which one forms on the basis of the raw experience of daily life.

Einstein believed that scientific theories are a means for extending our intuition into realms beyond our sense experiences. In this way so, too, is our notion of what is common sense transformed. This theme is one I return to throughout the book.

We can anticipate certain scientific aspects of the ether with a particularly interesting result of the great Swiss psychologist Jean Piaget on how children construct knowledge. Piaget asked why is it that on a sunny day it becomes cooler when there is a breeze? Most children around seven years of age believe this occurs because the wind blows about sunlight. With the addition of the concept of light rays, this is essentially the explanation nineteenth-century scientists gave for how the ether affects the velocity of light. A more detailed version goes as follows. For ease in exposition, scientists imagined that instead of the Earth moving around the sun, the Earth is at rest and the ether sweeps by it, like a wind. Then light from either stellar or earthly sources will be affected by the ether wind just like swimmers in a fast-running river. If you swim in the direction of the current, you will be perceived by an observer on the shore to be moving faster than if you were swimming upstream.

Since the ether was a basic ingredient of wave theories of light, it became essential to experimentally demonstrate its effects. The enormous problem experimenters faced was that any effects of the ether were extremely small for at least two reasons: (1) Optical equipment and most optical experiments were designed without any regard for the Earth's motion through an ether, and the results were perfectly acceptable; (2) the results of widely varying measurements of the velocity of light concurred to sufficient accuracy. Yet to fully substantiate a wave representation for light required experimental evidence for the ether. Accuracy in these experiments had to be better than four decimal places.*

Figure 7 is a simplified drawing of one such experiment which is meant to illustrate the effect of the ether wind. An observer O, standing on an imagined immobile Earth, measures the velocity of light emanating from a star. If there were no effects from an ether wind, then the velocity of starlight would, by assumption be c, which was taken to be the mean of measurements by Roemer, Bradley, and other experiments done in terrestrial laboratories. Assuming that there is an ether wind, then the measured velocity of light—call it c' (dotted line in Figure 7)—ought to differ from c by the common sense or intuitive addition law for velocities. The observer in the figure must point his telescope along the dotted line in order to observe the star. A measurement of the deviated velocity of light gives the velocity of the ether wind and so direct evidence for the ether's existence. This measurement was the goal of the so-called ether-drift experiments performed in the late 1800s. We will

---

*This was ascertained as follows. A rough estimate for the velocity of the ether wind is the Earth's orbital velocity v about the sun, which is 30,000 m/sec. As a round number, take c to be 3 × 10^8 m/sec. Wave theories of light predicted that any effect of the ether wind would be in the ratio $(v/c) = 10^{-4}$.

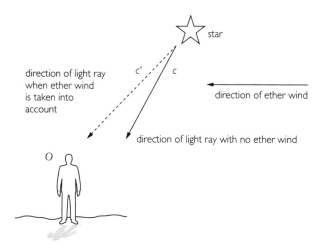

FIGURE 7.

The ether wind was assumed to blow light away from the direction in which it would have been observed had there been no ether wind. The observer O is on the Earth. The velocity of light would be c if there were no ether wind. The quantity c' is the velocity of light affected by the ether wind, where c' differs from c.

discuss those experiments in a moment. However, the measured value for the velocity of light always turned out to be c, and so in the direction undeviated by the ether wind. How can this be? Is Newton's law for the addition of velocities incorrect? Might there be no ether at all?

Any reply to these questions necessitates delving into how theories of light were set up in the nineteenth century. This line of exploration takes us to the very foundations of mechanics, where, for the first time, we encounter one of the most important principles in the physical sciences: the principle of relativity.

LIFE ON A MOVING PLATFORM

The first step in setting up a theory of light was to write the theory's equations from the viewpoint of someone on a platform which is at rest in the ether. *By assumption*, if observers on this platform measured the velocity of light emitted from a source at rest on their platform, they would obtain a definite numerical value that did not depend on any relative motion between them and the source of light. This velocity of light is designated by c in Figure 7. Difficulties appeared when the problem was shifted to observers who were in motion relative to the ether, such as the observer O in Figure 7. We recall that in the experiment in Figure 7, the measured velocity of light is c', which is related to the velocity of light c relative to the ether and to the velocity of the ether wind, through Newton's law for adding velocities. Consequently, the measured result is predicted to differ from c, thereby providing evidence for the existence of the ether.

Such a result would have been welcomed because a concept as important as the ether ought to have direct experimental consequences. But a severe problem arises if c' were measured to be different from c. Why this is the case brings us to the principle of relativity.

Like physicists at the turn of the century, let us restrict our discussion to measurements made on platforms that move with constant velocity in a straight line forever. These platforms are called "inertial reference systems" and are the very basis of Newtonian physics. In order to support this statement, Galileo had proposed ingenious thought experiments. Newton provided the mathematics. Inertial reference systems in Newton's theory of motion have an interesting property: No experiment based only on mechanics, such as falling bodies or pendula, can demonstrate the motion of such a system.

For example, suppose I am on an inertial reference system and can communicate by the phone with someone at rest on the ground (Figure 8). We agree to perform the same experiment, say, to measure how long it takes a stone to hit the ground when dropped from a height of six feet. First we calculate the time of fall using Newton's laws of motion, which we compare with the measured ones. What is so amazing here is that our own predicted and measured times agree on our own platforms *and* with each other's, too. The upshot is that I might as well be at rest as

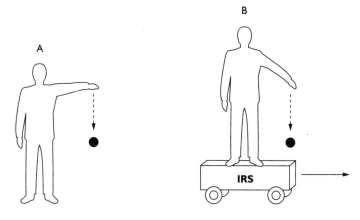

FIGURE 8.

Two experimenters A and B perform identical experiments. Experimenter B is on an inertial reference system (IRS), which moves to the right relative to A. They both drop a stone from the same height and record the times it takes for the stones to fall through identical distances.

in an inertial reference system. This result is called the "principle of relativity," which is counterintuitive to Aristotelian physics in which every movement ought to be experienced somehow.

An important part of the analysis to follow is to consider the Earth as an inertial reference system, which it is to a good approximation. This is the case even though the Earth hurtles around the sun at the rate of one revolution per year at a mean speed of 67,000 miles/hour, while revolving on its axis at the rate of once every 24 hours at a mean rate of 1,040 miles/hour at the equator. Despite these accelerated motions, engineers and scientists can still do the majority of their calculations using Newton's laws as if the Earth were moving in a straight line at a constant velocity. For example, any effects of the Earth's orbital acceleration on laboratory experiments performed on the Earth depend on terms in Newton's gravitational law containing ratios of the Earth's radius to that of the Earth's distance from the sun. Detecting such effects requires accuracy to at least six decimal places. It is apropos to defer to Chapter 3 why, for many situations in mechanics, we can neglect effects caused by the Earth's spinning about its axis. We discuss the origins of the principle of relativity in Chapter 3. Now we turn to the role the principle of relativity played in nineteenth-century theories of light.

A PROBLEM WITH RELATIVE MOTION

Something very fundamental is awry in nineteenth-century theories of light based on an ether, because they allow for measurements that would indicate motion of an inertial reference system. Measure the velocity of light in one of them and calculate by how much it differs from the value c, which, by definition, became the velocity of light relative to a platform at rest in the ether. Any difference indicates that you are in an inertial reference system and so is a violation of the principle of relativity from Newtonian physics. Consequently, every nineteenth-century theory of light had the potential to violate the principle of relativity from mechanics.

But measurements of relative velocities can always be made of one inertial reference system relative to another without any regard for this constant relative velocity affecting measuring instruments in either system. If we could not measure relative velocities, then Newton's science would be worthless. What is so special about measurement of a relative light velocity? To answer this question, we return to Figure 8. The fact that both observers made the same predictions from Newton's laws, which were borne out experimentally, means that Newton's laws have the same mathematical form in every inertial reference system. This is another statement of the principle of relativity. But what about the situation in experiments on the nature of light?

To reply to this question let us recall how the equations of optics were formulated. The first step was to write the equations with regard only to reference systems fixed in the ether. By definition, in these absolutely resting systems, the velocity of light is always c, the mean of the value measured by astronomers and substantiated in laboratories on the Earth. Consequently, right from the beginning, in wave theories of light the velocity of light played a very special role among all other velocities: Despite the way it is measured by observers at rest in the ether—that is, no matter what the relative motion is between the source of light and any of these observers—its value is always measured to be the same, c. But unlike the situation in mechanics, the equations of optics in inertial reference systems do not have the same form as they do when set up in a reference system fixed in the ether. Reference systems fixed in the ether play a preferred role. In mechanics there is no ether and so no preferred reference systems like in theories of light. Since the form of the equations of optics differs in inertial reference systems from ones fixed in the ether, then there is no principle of relativity in optics: Inertial motion can be detected. The key terms that distinguish the equations of optics in inertial reference systems from the ones written relative to the ether contain the relative velocity of light c' through Newton's addition law for velocities, an example of which is in Figure 7.

Measurements of the relative light velocity c' were attempted in the ether-drift experiments, some of the most beautiful pieces of experimental science ever performed. They carried the imprimaturs of the great theoretical and experimental physicists of the nineteenth century, men like James Clerk Maxwell, who formulated the first consistent electromagnetic theory of light; Hermann von Helmholtz, the titan of German physics; the greatest physicist in The Netherlands since Christiaan Huygens, H. A. Lorentz, who honed Maxwellian electromagnetic theory into the form we use today; the American physicist Albert A. Michelson, who was the paragon of high-precision measurements; and Henri Poincaré, one of the greatest mathematicians in history, working as well in the front ranks of physics and philosophy.

The cast of characters was international, the measurements highly accurate. But the experiments were magnificent failures. The most accurate ether-drift experiments were performed in the laboratory. They were carried out on state-of-the-art equipment in which light beams are split in two and then recombined after having been deflected several times by mirrors and perhaps even passing through media other than air. The effect of recombination—optical interference—was examined for any discrepancies from what would be expected if there were no ether drift. Every experiment yielded a value for the velocity of light as if the platform—the Earth—were at rest in the ether, in direct contradiction with the predicted result. It seemed, therefore, as if some sort of principle of relativity ought to apply in optics, yet the ether ran counter to it. What were these great scientists to do? The most straightforward alternative was to suppose that there is no ether wind. But does the Earth not move around the sun? Another reply is to assume that the Earth drags all of the ether in its vicinity and so there is no relative velocity between the Earth and the ether. Theories incorporating this suggestion were so unwieldy that they were not taken seriously. So why not drop the ether altogether?

This strategy was as abhorrent to nineteenth-century physicists as dropping the notion of God from science was to their seventeenth-century brethren. Earlier we discussed the scientific and psychological reasons for maintaining an ether. The reasons can be summarized as the need for some medium to support light in transit. Wild and desperate hypotheses were offered to bring theory into line with experiment, rescuing as well Newton's law for adding velocities while constructing a principle of relativity for optics. In 1892, almost apologetically, Lorentz tried to explain one of the failed ether-drift experiments by proposing that bodies contract in the direction of their motion through the ether. Lorentz's hypothesis of contraction, as it was referred to, turned out to be on the right track when clarified by Einstein's special theory of relativity.

Some two decades after formulating special relativity, Einstein recalled his rather audacious mindset in 1905 when he was a Patent Clerk Third Class in Bern, Switzerland. Einstein concluded that all of the great physicists of the day were "theorizing out of their depth." In order to begin to fathom the Patent Clerk's bold conclusion, we turn to Einstein's own view of relative motion.

ANOTHER WAY TO THINK
ABOUT RELATIVE MOTION

The physicist and philosopher Philipp Frank, who succeeded Einstein in 1911 at the German University in Prague, recalled an incident in which he was late for an appointment with Einstein. Frank apologized profusely for wasting Einstein's time. But Einstein shrugged off Frank's apology with the comment that the wait gave him time to think, a process that did not require sitting at a desk. Like several of the great scientists who wrote introspections on their creativity at the turn of the century, Einstein pointed out that he could think anywhere. What is so striking about Einstein's pre-general relativity papers are piercing physical insights that do not require complex mathematics. For Einstein, thought experiments were of the essence, and these can be done anywhere. So it was in 1895 when Einstein was at a preparatory school in Switzerland that emphasized the role of visual thinking. The 16-year-old boy framed the key problem in late nineteenth-century physics in a bold new way.

Einstein's 1895 thought experiment can be developed from Figure 6, where car A is replaced by a thought experimenter's laboratory, and car B by a point on a light wave. The thought experimenter tries to catch up with a point on the light wave labeled *c*. According to Einstein's own description, we can draw his 1895 thought experiment as in Figure 9. His thought experiment beautifully encapsulates every possible ether-drift experiment.

Intuitively, according to Newtonian science, there ought to be a measurable relative velocity between the light wave and the thought experimenter's laboratory. According to Newton's addition law for adding velocities, the thought experimenter ought to measure a decreasing relative velocity. But this relative velocity would violate any widened principle of relativity pertaining to phenomena that include mechanics *and* optics. Physicists responded to this situation with numerous hypotheses especially formulated to rescue Newton's law for adding velocities.

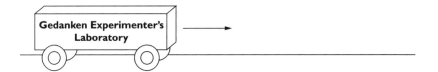

FIGURE 9.

Einstein's thought [*Gedanken*] experimenter tries to catch up with a point on a light wave labeled *c*, which is the velocity of light relative to a source at rest on the ground.

Although, at first, Einstein did not know quite what to make of the problem situation, by 1905 he analyzed it in terms of two possibilities:

1. The theoretical prediction of a varying measured relative velocity of light permits the observer on the moving platform to catch up with a point on the light wave. The result would be a wave that vibrates up and down but does not progress linearly. Such a phenomenon had not been observed, and Einstein believed it never would be.

2. On the other hand, according to the thought experimenter's "intuition," there ought to be no measurable effects, and so the moving observer ought to measure the same velocity of light as someone at rest relative to the light's source.

Einstein recalled that he rephrased options (1) and (2) as a "paradox":

On the one hand—
(1) The velocity of light can be compounded with other velocities according to Newton's addition law for velocities.
While on the other hand—
(2) The velocity of light is always *c*, regardless of any relative motion between source and observer.

Clearly, this shows that the two options are incompatible. Einstein's strategy for paradox busting was to declare option (1) to be simply wrong, with no further ado. His reason for choosing (2) was the thought experimenter's intuition. By this, Einstein meant that the thought experimenter's *common-sense* belief in the principle of relativity for mechanics supported its extension to phenomena involving light as well. Any violation of the principle of relativity in a mechanics experiment had become *counterintuitive*. Einstein raised the ante by claiming it is counterintuitive to all of physics. We may say that Einstein considered the problem situation on which everyone else was working in 1905 to be insoluble—they were "theorizing out of their depth." So, by rejecting (1) outright, he redefined the problem situation into one that was soluble. While other physicists were attempting to construct a principle of relativity for optics by proposing a large number of mutually supporting hypotheses, Einstein declared outright that an exact principle of relativity covered optics as well as mechanics.

This was a daring move because it proclaimed straightway that the ether-drift experiments were preordained to failure. They flew in the face of nature, according to which the principle of relativity, simply and with great finality, slammed the door in the face of any attempts to measure violations from what is expected from inertial reference systems.

Einstein meditated on this thought experiment for 10 years before realizing in 1905 that it contained the "germ of special relativity," as he recollected some years later. This is a fine example of a situation in which a scientist with great boldness extends and then elevates a hypothesis to the exalted status of an *axiom*, putting it beyond experimental or theoretical proof. In this case, the hypothesis is a principle of relativity covering all of physics axiomatically.

In making this claim, Einstein redefined our concept of what is intuitive regarding motion and time. How can it be, for example, that something could pass you on the road and when you attempt to catch up with it, you find that no matter how fast you move, the other object's velocity relative to you remains the same? Is this not counter to Newtonian intuition? Indeed, it is! The culprit here turned out to be the intuitive or common-sense notion of time in Newtonian physics.

TIME AND LIGHT

The hidden crack in the superstructure of Newtonian physics lay in an assumption implicit in the mathematical foundation of Newton's law for adding velocities: the absoluteness of time. This assumption, imbued in our Newtonian intuition, says that properly functioning clocks are not altered by their motion. Suppose you are at a friend's house and you and your friend set your watches to read the same time: Your watches are synchronized. Then you go off on a car journey, while your friend stays home. Later, you return to your friend's home, compare watches, and find that they remain synchronized.

As Newton wrote in the *Principia*, time is "absolute, true . . . and from its own nature flows equably without relation to anything external." By the nineteenth century, the Newtonian notion of time had been stripped of its poetics and theology and was defined simply as something unchanging. This conclusion made sense. In an Aristotelian vein, physicists declared that nothing had ever been measured otherwise. There is no reason to assume that a clock's motion affects its working (excluding, of course, smashing it). Thought experiments, such as the one we just described in Figure 9, led Einstein to conclude that the "absolute character of time, viz., of simultaneity, unrecognizably was anchored in the unconscious." Instilled into our "unconscious" is the Newtonian common-sense notion of time based on measurement with clocks traveling at velocities characteristic of our daily experiences. These velocities are insignificant compared to that of light, and so we tend to disregard ramifications of a theory of motion based on taking the velocity of light as finite but huge. In Einstein's special relativity theory, the velocity of light turns out to be the ultimate speed limit.

With the aid of certain "imagined experiments," as Einstein wrote in his 1905 relativity paper entitled "On the electrodynamics of moving bodies," he concluded that time is a relative quantity. These imagined, or thought, experiments concerned using light signals to synchronize clocks at rest in an inertial reference system. Einstein then studied whether these clocks are in synchrony with those in another inertial reference system in motion relative to the first one. This turned out not to be the case. An immediate consequence is that two events that are simultaneous in one inertial reference system need not be simultaneous to observers in another inertial reference system, to whom one event can precede the other. This relativity of simultaneity is counterintuitive to Newtonian common sense. In this way, Einstein's 1905 special relativity theory serves to extend our intuition into phenomena indigenous to the realm of velocities comparable to that of light. Incidentally, Einstein's 1905 relativity theory is called the "special theory of relativity" because it refers only to measurements made in inertial reference systems. Einstein was extremely sensitive to this restriction, which he referred to as a "logical weakness." Ten years later he removed it in his general relativity theory, which permits measurements to be made in accelerated reference systems as well.

Another ramification of not neglecting the high speed of light that Einstein deduced in 1905 is a new addition law for velocities, according to which the velocity of light is always the same no matter in which inertial reference system it is measured. This result agrees with a widened principle of relativity. Newton's law for adding velocities emerges in the limiting case when the velocity of light becomes infinite. This is the case that parallels our daily experiences and so our Newtonian intuition.

In achieving these results, Einstein was led to consider the time-honored ether as "superfluous," as he wrote in his 1905 relativity paper. Yet Einstein knew full well that the view of space and time he formulated was provisional because, among other points, which include its emphasis on inertial reference systems, it contained no medium for the propagation of light. This, too, was remedied in the 1915 general theory of relativity, in which light rides on the structure of space-time. The key to this magnificent theory would be a thought experiment that he conceived in 1907 and to which I defer analysis until Chapter 8. General relativity once again redefined the notion of intuition and consequently common sense, as well.

In summary, much as Galileo had done, through thought experiments Einstein explored a possible world in which space and time are *relative* quantities. Like Galileo's, Einstein's possible world turns out to be an actual one. As with Galileo, changes in the concept of intuition are accompanied by increased abstraction (see Figure 10).

The relativity of time was particularly hard for physicists to accept. Take Henri Poincaré, whose opinions are somewhat indicative of the era, although more sophisticated than most. Poincaré objected that on physical grounds there were no experimental data to back up the relativity of time. He demurred on the psychological ground that this phenomenon is beyond our sense perceptions and so need play no role in

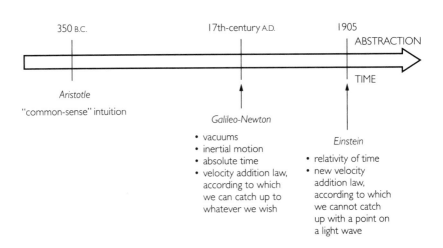

FIGURE 10.

The horizontal axis indicates increasing time and, coincidentally, increasing abstraction of intuition.

physical theory. Then Poincaré marshaled what became known in philosophy as the "underdetermination thesis": Every set of experimental data is open to explanation by an in-principle infinite number of hypotheses. This is a fact of life that we will discuss further in Chapter 3. Lorentz's already finely honed theory of the electron seemed to cover the same set of data as Einstein's. Then why not just elaborate on Lorentz's, whose basis squared with sense perceptions? As we shall see, the situation between Poincaré and Einstein was highly interesting.

The counterintuitive nature of the relativity of time also struck the influential French philosopher Henri Bergson. Starting in 1921, in scientific popularizations written with his usual flair, Bergson contended that psychological time is real while that of the physicist is not. He tried to support this reasoning with physics based on fundamental misunderstandings of relativity. Essentially Bergson put himself in the physically uncomfortable situation of having each foot in a different inertial reference system. Einstein tried setting him straight in 1922 in Paris, emphasizing that it was the very high velocity of light that tricked us to conclude that, for example, occurrences simultaneous in one inertial reference system remain so for other inertial reference systems. Despite their differences on relativity, Einstein and Bergson became friends. But with regard to Bergson's philosophical view of relativity, Einstein once commented "*Gott verzieh ihm* [God forgive him]."*

CONCLUDING COMMENTS

In little more than three hundred years, developments in the physical sciences have been brought about by, and in turn have led to, transformations in the concept of intuition and what we call common sense. The catalysts for these changes have been thought experiments in

---

*Bergson's disagreement with the relativity of time was taken seriously at first and became a factor in Einstein's not being awarded the Nobel prize in 1921 for relativity.

which scientists abstract to situations beyond their sense perceptions. Despite the startling changes in intuition and common sense, all theories discussed thus far are based on visual imagery abstracted from the world of sense perceptions. Through the first decade of the twentieth century, scientists assumed that this would always be the case, even though they were studying phenomena way beyond sense perceptions, such as the newly discovered electron. This wishful thinking would soon come to an end.

At stake would be such time-honored essentials of physical theory as certainty and visual imagery. If science is another means for understanding nature, then it must represent her as well with visual imagery. Yet developments in atomic physics would not bode well for representing what turned out to be a bizarre and unexpected world of atoms and light, one that defies visualization. In atomic physics, the concept of intuition underwent its most profound transformation. This greatly disturbed the scientists most closely involved in developing atomic physics because they worked in a cultural milieu in which intuition is linked with visual imagery.

How can visual imagery of a world beyond imagination be obtained? The history of scientific thought incontestably bears witness to the desire of scientists for visual imagery. This is the principal theme running throughout this book. Pursuing it will entail our exploring how scientific research is done, the notion of metaphor, of visual imagery and the problem of vision itself, of scientific creativity, and the common struggles of artists and scientists to represent the world about them, both visible and invisible.

2
THE
INTUITION
OF
ATOMS

In this chapter our story gets more complex in addition to taking a strange turn. The physical theories we've examined so far have all been at least amenable to visualization. In your mind's eye you can *see* planets orbiting the sun, and you can draw them on paper. Even four-dimensional space-time can be visualized by analogy with two-dimensional representations. The representation of unseen phenomena through analogy with what we actually witness began to come into question with Einstein's hypothesis in 1905 of a particle nature for light. But then there was the pleasing visual imagery in Bohr's atomic theory of

1913 in which atoms emerge as minuscule solar systems. Soon after, the situation in atomic physics began to deteriorate, and visualization was abandoned, to be replaced by nonvisualizable mathematical formalisms. Heisenberg's new quantum mechanics of 1925 is the nadir of visualization and intuition as the basis for a scientific theory. The passionate struggles that ensued between Bohr, Heisenberg, and Schrödinger culminated in Heisenberg's uncertainty principle and Bohr's complementarity, neither of which offers any visualization of atoms and electrons. Both call for bold new transformations in the concept of intuition and consequently in the meanings of terms such as "particle" and "wave" in the atomic domain.

Whereas the world of special relativity extended our intuition, its visual imagery is drawn from the world of sense perceptions, as in Newton's mechanics. Thus, the imagery of tables, chairs, and electrons drawn like charged billiard balls still seemed the most appropriate visual imagery to *impose* on physical theories. This method for handling science is a good example of how theories were constructed by imposing visual imagery abstracted from the daily world onto scientific considerations.

But Is Light Really a Wave?

The experimental setup in Figure 1 is similar to the one used in 1803 by the British polymath Thomas Young. A source of light falls on a slit S_0 in screen A and then passes through two narrow slits, S_1 and S_2, bored through screen B. Figure 2 contains the data from registering light intensities over screen C.

Nature gives us a source, screens, and the data in Figure 2. The data can be explained through the analogy with how water waves can produce interference (see Figure 3) and with mathematics based on the source of light emitting light waves. Other optical data are also explainable in this manner.

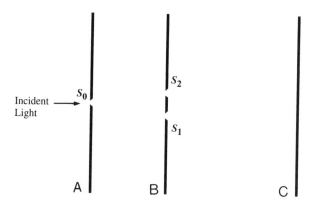

FIGURE 1

This shows the experimental arrangement in which light is incident on the slit S_0 on screen A. Another screen, B, with slits S_1 and S_2 is in front of screen C.

FIGURE 2

The data contained in the photograph of the effect on screen C from Figure 1 caused by light incident first on screen A and then on screen B.

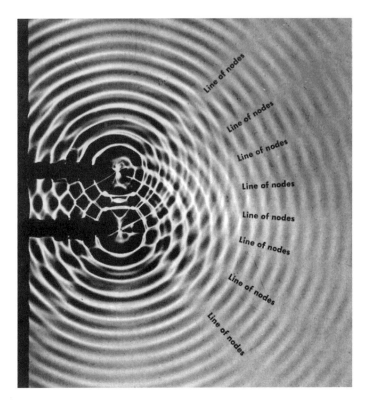

FIGURE 3

The interference of water waves in a ripple tank. The waves are generated by two vibrators that strike the water's surface synchronously, producing two expanding spherical waves. There is destructive interference along the lines marked "lines of nodes" and constructive interference between these lines.

Consequently, the scientist "infers to the best explanation" that the data (*effect*) are *caused* by light waves emitted by the source: Light is *really* a wave phenomenon. This sort of reasoning is neither induction nor deduction. "Induction" is reasoning from experimental data to hypotheses. "Deduction" works the other way: The scientist derives predictions from an accepted axiom system and then tests them by experiment. "Inference to the best explanation," as the philosopher Gilbert Harman coined the phrase, is what the nineteenth-century American philosopher Charles Pierce called "abduction," making a hypothesis suggested by perhaps only a single experiment.

But hypotheses must be backed up by theory. To this end, Young applied the powerful tool of *reasoning by analogy*, which makes use of more familiar and well-understood phenomena for explaining the process in question. Reasoning by analogy brings into play a less abstract intuitive phenomenon, in this case water waves. By analogy with water waves, scientists "see" and understand light as a wave phenomenon. Yet our eyes do not see any light waves emitted from a source. For example, when we switch on an electric light, it seems to illuminate the room instantly. We do not see spherical waves emitted from the bulb, like the waves receding from a stone thrown into a still pond.

Almost any good science museum has a demonstration of Young's reasoning by analogy. The apparatus is composed of a tank filled with water in which there is a screen with two holes drilled in it, just like in Figure 3. A needle is set vibrating in the tank, and spherical water waves concentric with the needle spread out. When these waves strike the two pinholes in the screen, two new sets of waves appear, which spread out and cross over each other. Corks in the tank begin to bob up and down. Some of them bob up and down with maximum amplitude (this is constructive interference of the two sets of waves); some remain at rest (this is destructive interference); while some bob up and down with amplitudes intermediate between the maximum and minimum. These phenomena are analogous to the data in Figure 2 and is actually demonstrated with water waves in Figure 3.

This being the case, Figure 1 can be redrawn as in Figure 4, which is standard physics textbook fare. Figure 5 is Young's rather amazing 1803 rendition of double-slit diffraction. Once again, in analogy with water waves, Young drew the diagram in Figure 5. Holding your eye close to the edge of the figure's left-hand side, and observing from a grazing angle along the figure, reveals that along the lines tending toward C, D, E, and F, there is cancellation of the waves, while between them is reinforcement. Abstracting to light waves, Young hypothesized

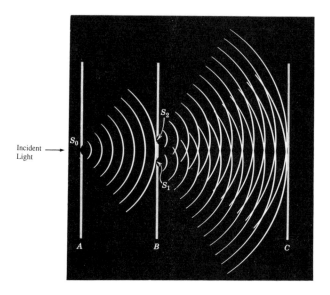

FIGURE 4

Figure 4 is Figure 1 redrawn after inferring to the best explanation that the cause of the data in Figure 2 is light waves. The pattern in this figure is called a *Huygen's construction.*

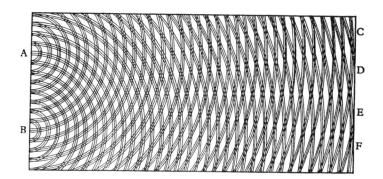

FIGURE 5

Thomas Young's original drawing from his Bakerian Lectures of 1803 for demonstrating interference effects in overlapping waves. At A and B are sources of waves that spread out and interfere like those in Figure 3. This can be seen by placing your eye near the left edge and sight along the figure. Lines of nodes are along C, D, E, and F.

42

that if a screen is placed anywhere across the superimposed waves, we can expect to see alternate bright and dark spots as in Figure 2.

Experiments such as these, in conjunction with certain mathematical developments for a wave theory of light, led to the wave theory's eventual acceptance by the late 1820s. This theory's rival was a particle theory of light, favored by Newton and his followers. Newton envisioned light particles to have mass and so to behave along the lines of his mechanics, suffering attractions and repulsions inside matter. But Newton knew this was insufficient to account for all known optical data.

The coup de grâce for particle theories of light was supposedly given in 1849 by the French experimentalist Léon Foucault. He established that light travels more slowly in denser media than in air, whereas Newton's particle theory predicted otherwise. But as the French scientist and philosopher Pierre Duhem explained in 1906, there is no such thing as a crucial or critical experiment, because the "physicist can never subject an isolated hypothesis to experimental test. . . . Physics is not a machine which lets itself be taken apart." This point is wonderfully illustrated by Einstein's 1905 hypothesis that light can be represented both as a hail of shot or particles, also known as light quanta, and as waves. It turned out that Foucault had eliminated only particle theories of light in which the particles behave like Newtonian things. Einstein's light quanta are literally worlds away, having no mass. It then remained to show how light quanta explain optical interference, which is one of the defining characteristics of light. As we see in Figures 1 through 5, wave theory offers an explanation by analogy with how water waves interfere.

Since no visual image could be constructed for how light quanta produced optical interference and the double-slit diffraction pattern, light particles were deemed to be counterintuitive and thus of no use. As the American physicist O. W. Richardson asked in 1916, how can radiation be described as "though it possessed at the same time the opposite properties of extension and localization"? Light quanta were resisted on purely conceptual grounds until 1927. Physicists thought it more than fortuitous that in the atomic realm visual imagery abstracted from macroscopic phenomena continued to prosper.

Let us leave the physicists in this state of short-lived innocence while we explore the relations between intuition and visual imagery.

THE VISUAL IMAGERY OF INTUITION

Modern physics has linked intuition with visual imagery partly through the rich philosophical lexicon of the German language. Relativity and atomic physics in the early 1920s were formulated almost exclusively by scientists educated in the German scientific-cultural milieu. As the historian J. T. Merz wrote at the time, the "German man of science is a philosopher." Philosophy was an integral part of learning in the German school system, particularly the ideas of Immanuel Kant. No matter what the proclivity of the German adult scientist might be, the youngster had been thoroughly grounded in Kantian philosophy. Even such an arch anti-Kantian as the Viennese philosopher-scientist Ernst Mach recalled that Kant's *Prolegomena to Any Future Metaphysics* "made at the time [when Mach was 15] a powerful and ineffaceable impression upon me, the like of which I never afterward experienced in any of my philosophical reading."

Immanuel Kant was one of the great intellectual forces at the turn of the nineteenth century. It was said that he never traveled more than 30 miles outside his home town of Königsberg, because everyone came to see him. A dinner with Kant was considered the height of intellectual society. Kant spun an intricate philosophical system whose goal was to place Newtonian physics on a firm cognitive foundation. He struggled with such deep problems as "How is pure mathematics possible?" and "How is pure natural science possible?" The Kant scholar Michael Friedman writes that "Kantian thought stands as a model of fruitful philosophical engagement with the sciences." Few philosophers of science then or now can claim such distinction.

In setting the basis for his philosophical system in his monumental book of 1781, *The Critique of Pure Reason*, Kant carefully separated intuition from sensation.* His goal was to separate higher cognition from the processing of mere sensory perceptions. To get the flavor of physics in the early twentieth century, we should discuss a bit of German philosophical vocabulary. In German, the word for intuition is *Anschauung*, which can be translated equally well as "visualization." To Kant, intuitions or visualizations can be abstractions of phenomena we have actually witnessed.

On the other hand, we can try to investigate physical phenomena in a more concrete way. For example, magnetic lines of force can be demonstrated by placing iron filings on a sheet of paper held over a bar magnet. Kant refers to this sort of visual imagery as "visualizability." In German, "visualizability" is rendered as *Anschaulichkeit*. In Kantian terminology, we say that *Anschaulichkeit* is what is immediately given to the perceptions or what is readily graspable in the *Anschauung*. So *Anschaulichkeit* refers to properties of an object, which exist whether or not we look at it or make measurements on it. The step of abstracting lines of force to entities that fill all of space and that are mathematically described by certain symbols in the equations of electromagnetism raises them from a visualizability to a visualization, from an *Anschaulichkeit* to an *Anschauung*. In Kant's philosophy, the visual imagery of visualizability (*Anschaulichkeit*) is inferior to the images of visualization (*Anschauung*).

In Newtonian physics, visualization and visualizability are synonymous because there is no reason to believe that experimenting on a system in any way alters the system's properties. The object's properties, in other words, are unaltered by the act of measurement. For example, when we ascertain how long it takes a dropped stone to reach the ground, we watch the stone and assume that the act of observation—

---

* The first edition of *The Critique of Pure Reason* was so difficult to understand that in 1783 Kant wrote the much smaller *Prolegomena* as a key to it, and more as well.

comprised of observing the light reflected off the falling stone—does not affect the stone's descent.

So far so good, but this must also apply to objects that right from the start were never visible, like electrons. Prior to research into electrons in 1897, physicists worked with systems that with some justification were assumed to be amenable to perceptions and for which the usual space and time diagrams from Newtonian physics were applicable. For example, the visual representation of light as a wave phenomenon was referred to as visualization because it is a bold abstraction from the usual concept of water waves. The equations of electromagnetism turned out to be workable with the visual imagery imposed on them of electrons behaving like charged billiard balls. Consequently, through the first decade of the twentieth century, the visual imagery of visualization and of visualizability were one in the same. As was the case for objects treated in Newtonian physics, there was no reason to believe that experimenting on the electron in any way alters its characteristics.

Surprises were in store in the atomic realm, in which the order of importance of visualization and visualizability would turn out to be reversed. The reversal came about in an interesting manner. Since the "actual" visual image of elementary particles is so counterintuitive to anything we can imagine, only the mathematics of theories pertaining to them can offer us a glimpse into properties of the objects themselves. To everyone's surprise, it turned out that no imagery can be *imposed* on modern atomic theories. That these developments first occurred in the German scientific-philosophical milieu is not surprising, due to the heavy premium it placed on visual imagery, and more generally, on visual thinking, by scientists and engineers in this culture.

It is not surprising, therefore, that Einstein conceived of his 1895 thought experiment in highly visual terms while a student at the Canton School in Aarau, Switzerland. This school was set up by followers of the Swiss education pioneer Johann Heinrich Pestalozzi, who wrote in 1801 of his greatest achievement as having "fixed the highest, supreme principle of instruction in recognition of visualization [intuition] as the absolute foundation of all knowledge." To Pestalozzi, visual thinking is a fundamental and powerful feature of the mind. As the psychologist of art Rudolf Arnheim has observed, "The scientist, like the artist, interprets the world around him and within him by making images." At Aarau, Einstein realized that his mode for creative thinking was visual imagery. The visual imagery in his 1895 thought experiment is one of visualization rather than visualizability (see Figure 9 in Chapter 1).

Another concrete illustration of the terminology introduced here involves Einstein's first published use of the Kantian term "intuition." In his 1909 paper "On the development of our intuition [*Anschauung*] of the existence and constitution of radiation," Einstein explores the *counterintuitive* dual nature of light as wave and particle. He writes that the

> relativity theory altered our intuition of the nature of light, because it interpreted light not as a consequence of the state of a hypothetical medium [or ether] but as existing independently of matter.

Light in its particle mode does not need a milieu in which to propagate, as is true of particles generally. This alters our intuition of light behaving only as a wave.

Our discussion here has shown that the concept of visual imagery takes on a particularly philosophical meaning in the German-language cultural milieu, where the engineers and scientists receive a thorough grounding in the subject. It serves as a guide through the maze of research into the terribly counterintuitive atomic realm. The scientists' interpretation of Kant's philosophy did not always match that of Kant scholars, or even that of Kant himself. Nevertheless, having been a scientist, Kant would have understood how desperate the situation can be on the frontiers of research. As Einstein recalled, "Every scientist has his own convenient Kant."

Atomic physics provides more examples of the power of visual thinking and its link with intuition.

ATOMS AND SOLAR SYSTEMS

In October 1911 the 26-year-old new Danish Ph.D. Niels Bohr looked forward to spending time as a postdoctoral student in the famous Cavandish Laboratory at Cambridge University. He came to study with the Nobel laureate physicist J. J. Thomson, discoverer in 1897 of the electron. Eager to make a good first impression, upon entering Thomson's office Bohr pointed out in halting English several errors in a recent book by the great man. This was not an auspicious beginning. Bohr persisted in such behavior, and so eventually Thomson went out of his way to avoid Bohr. A short time later, Bohr had the occasion at Cambridge to attend a lecture by Ernest Rutherford, visiting from Manchester, in which he discussed his new atomic model. Bohr decided that Rutherford's laboratory was the place to be. In contrast to the rather reticent Thomson, Rutherford was volcanic in energy, enthusiasm, and self-expression. As a colleague recalled, Rutherford was a man who "had no cleverness—just greatness."

Paradoxically, having just been awarded the 1908 Nobel prize in chemistry for work on radioactivity, in 1909 Rutherford embarked on what turned out to be the greatest research of his career.* It came about owing to a hunch he had when observing two of his researchers, Hans Geiger and Ernest Marsden, experiment with the scattering of α-particles from thin foils of aluminum and gold. For no apparent reason, Rutherford suggested that the pair of researchers change their setup to test whether any α-particles were scattered backward off the foils, instead of just passing through. To Geiger's and Marsden's amazement, this occurred. Rutherford was stupefied. He recalled that

> It was quite the most incredible event that has ever happened to me in my life. It was almost as incredible as if you

---

* An excellent example of Rutherford's wit is his acceptance speech at a banquet during the Nobel ceremonies. With reference to his surprise that a professor of physics should be awarded a Nobel prize in chemistry, he said that of all the transformations he dealt with, the quickest one was his own from a physicist to a chemist.

fired a 15-inch shell at a piece of tissue paper and it came back and hit you.

This was so incredible because the most widely accepted atomic model of the time was J. J. Thomson's, in which positive and negative charges are distributed like plums in a pudding in such a way that the atom is stable against small changes in the positions of its constituent positively and negatively charged particles. However, the problem common to any atomic model with moving charges is radiative instability. The reason is that when charged particles accelerate, they emit light, or radiation. Since radiation has energy, the atom's constituent electrons are constantly losing energy and the atom will eventually disintegrate. Yet most atoms are stable. How do we know this? Well, if they weren't, you would not have time to read this passage before disintegrating. Thomson assumed that atomic stability had to be explained. This turned out to be a mistake.

Geiger and Marsden's scattering experiments sounded the death knell for Thomson's atomic model because, as Rutherford carefully reasoned, back scattering of the positively charged α-particle could occur only if it encountered a concentration of positive electricity. By 1911 emerged the visual image of the atom as a miniscule solar system with a positively charged center (which Rutherford referred to in 1912 as the nucleus) and electrons in planetary orbits around the nucleus. The problem of radiative instability persists even here, because the atom's electrons rotate about the nucleus. Most scientists felt sure that further tinkering with Newton's mechanics and Lorentz's electrodynamics would lead to a detailed atomic theory in which a reason for atomic stability would also emerge. Disappointment was just around the corner.

ATOMIC INTUITION AND VISUALIZATION

At the beginning of the twentieth century, physicists believed that the language of our daily world, with its accompanying visual representations, could be extended into the world of the atom. After all, this approach to formulating theories had worked well thus far. There was no reason for the newly discovered electron not to be like a charged billiard ball. And the visualization of Rutherford's atom, with its billiard-ball electrons moving continuously around the nucleus, was striking.

But the nagging problem of radiative stability persisted. By early 1913 Bohr resolved it much as Einstein had dealt with similar problems in 1905: He declared them to be incorrectly put, and he redefined the problem into a soluble one. Whereas Thomson took stability as something to be explained, Bohr took it as a fact of life. In Bohr's atomic theory of 1913, stability is axiomatic, meaning that it is a given, not to be questioned. Bohr's theory is a perfect example of how scientists build physical theories by counterintuitive axioms.

For the purpose of ensuring the stability of the solar system atom, Bohr had somehow to introduce a basic length. In this way he could prevent the atom from shrinking due to its orbiting electrons' falling into the nucleus. He accomplished this with the help of a constant of nature that had already played a key role in physics, Planck's constant, named after the doyen of German physics, Max Planck. Planck's constant is an extremely small quantity, 6.63×10^{-34} joule-seconds (the joule is a unit of energy). The fact that it is not zero would turn out to wreak havoc with attempts to understand the properties of atoms and light. It played a key role in Planck's 1900 theory of the characteristics of light emitted from the interior of hot substances ("cavity radiation"). Planck's theory was disagreeable to most physicists, including its originator, and only began to achieve acceptance through Einstein's work begun in 1906 on the capacity of substances to retain heat (specific heats). Einstein's inclusion of Planck's constant into the defining characteristics of light quanta was ignored.

What turned most everyone, except Einstein, against Planck's theory was its violations of the principles of mechanics and electromagnetism. Electromagnetic theory assumes that all processes involving exchanges of radiation are continuous. To his horror, Planck found that the only way he could give a fundamental underpinning to his theory was to assume that electrons in matter can absorb and emit energy as radiation discontinuously in certain discrete or quantized amounts. Planck's constant is the quantity that connects the radiation packet's energy with its frequency. Coincidentally, something equally bad had occurred. Because if there are restrictions on energy exchanges, then the electron's positions are similarly restricted. This violates mechanics, according to which electrons in matter can occupy any position permissible by Newton's laws. All of these intrusions can be encapsulated by the realization that if Planck's radiation law were valid, then discontinuity had entered physical theory.

Bohr's atomic theory, on the other hand, was openly predicated on discontinuities. Bohr realized that a suitable combination of the electron's charge and mass with Planck's constant resulted in a length that was of the order of magnitude expected for an atom, about 10^{-8} centimeters. With this result in hand, as a first step, Bohr set up a means to insert Planck's constant into Newtonian mechanics in order to calculate what he referred to as "allowed orbits" or "waiting places" for the atom's electrons. Like Planck's radiation theory, this move violates Newtonian mechanics, which permits the electron to occupy any orbit (Figure 6(a)).

An atom's electrons can be elevated into higher allowed orbits by heating the atom or hitting it with projectiles of proper energy. A transition downward is accompanied by a burst of radiation with a frequency measurable as one of the atom's spectral lines (see Figure 6(b)). This hypothesis violates Maxwell's electromagnetic theory, which states that all processes involving radiation are continuous. Incidentally, Bohr went out of his way to exclude light quanta by stressing that the burst of radiation emitted by an atom is "homogeneous." Instead of trying to *explain* why atoms are stable, he *postulated* a lowest allowed orbit below which atomic electrons cannot drop—*ergo* stability.

Bohr's theory achieved astounding results, such as equations for the measured spectral series in the hydrogen atom as well as stunning and unexpected characteristics of atomic structure. Einstein lauded

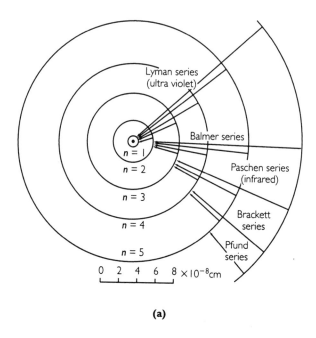

(a)

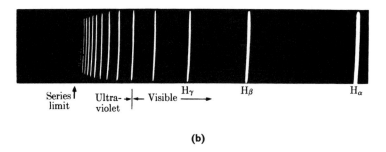

(b)

FIGURE 6

In (a) is the representation of the hydrogen atom with its single orbital electron from Bohr's 1913 atomic theory. The number *n* is the
principal quantum number and serves to tag the atomic electron's
permitted orbits. A downward transition of the atom's electron from
a higher to lower orbit is accompanied by a burst of radiation. The
frequency of this emitted radiation is measurable as a spectral line.
Lyman, Balmer, etc., are names for the series of spectral lines, an example of which is given in (b), which contains the Balmer series.

Bohr's theory as "an enormous achievement." But success was bought dearly. Orbit jumping, or quantum leaping, is a process in violation of mechanics and electromagnetism. Even stranger, since the electron's trajectory in space and time between allowed orbits cannot be traced using mechanics, the atom's electron seems to disappear and then reappear like the Cheshire cat.

We recall that a wave representation for light emerged as a result of Young's working from effect (interference pattern) to the effect's source. Something similar happens in atomic theory. From an atom's spectral lines (effect), scientists try to work back to explore the structure of the line's source, the atom. In the case of Bohr's 1913 theory, this is not historically correct. Bohr considered atomic stability first, rather than spectral lines. (Rather extraordinarily, he was unaware of issues concerning spectral lines until *after* he had begun to formulate the theory. Then, he recalled, everything fell into place.) After this initial phase, scientists puzzled over what surprises spectral series offered regarding the atom's structure and properties. The loss of visual imagery, which occurred in 1923, was disturbing, because visualization is linked to intuition.

VISUALIZATION LOST, INTUITION REDEFINED

Between 1913 and 1923, Bohr's theory, with its exciting visual imagery, had many triumphs (Figure 7). Physicists waxed ecstatic. One of the pioneers of atomic theory, Max Born, wrote in 1923 that

> A remarkable and alluring result of Bohr's atomic theory is the demonstration that the atom is a small planetary system. . . . The thought that the laws of the macrocosmos in the small reflect the terrestrial world obviously exercises a great magic on mankind's mind; indeed its form is rooted in the superstition (which is as old as the history of thought)

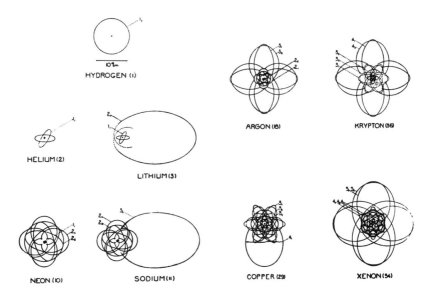

FIGURE 7

Representations of the atom according to Niels Bohr's 1913 atomic theory.

that the destiny of men could be read from the stars. The astrological mysticism has disappeared from science, but what remains is the endeavor toward the knowledge of the unity of the laws of the world.

Such exaltation is reminiscent of ancient neo-Platonic texts extolling the virtues of a sun-centered universe. So long as the intuitive visual imagery of visualization could be maintained, the rupture between the time-honored mechanics and electromagnetism seemed tolerable.

But augurs of doom for the wonderful visual imagery of Bohr's theory began to appear in 1921. Whereas the properties of the hydrogen atom appeared to be under control, nothing useful was obtainable from the solar system model of the next lightest atom, helium. Then, in 1923, data accrued that thoroughly discredited the solar system atom, stripping atomic physics of its visual imagery. At this point physicists applied the term "intuition" to the mathematical apparatus of the variants of Bohr's theory that appeared. The visual image of the solar system hydrogen atom was replaced by one in which the atom's electron was represented by as many electrons on springs—harmonic oscillators—as there are possible atomic transitions. Since there are an infinite number of stationary states, then there are an infinite number of possible transitions. Mathematics became the guide for physicists, and so, not unsurprisingly, the term "image" became associated with the atomic theory's mathematical framework.

By mid 1925, the Bohr theory had collapsed entirely and atomic physics lay in ruins. Some physicists thrive in such periods; others give up. The 25-year-old Wolfgang Pauli had already made several key and lasting contributions to atomic physics, but vented his opinion that perhaps he ought to go into movie making at this point or, at least, intensify his escapades in Hamburg café society. On the other hand, his 24-year-old colleague Werner Heisenberg thrived and went on to formulate the modern atomic physics called quantum mechanics. Heisenberg's quantum mechanics was based on nonvisualizable particles and so avoided any space and time description. Heisenberg meant it to be based only on measurable quantities, a programmatic intent that physicists in Bohr's circle had taken since 1923. Thus, the atomic electron was "described" by the radiation it emitted during transitions, which is measured by its spectroscopic lines.

Problem after problem that had resisted solution within the old Bohr theory was solved. Yet what bothered physicists of the ilk of Bohr and Heisenberg was that not only were the intermediate steps in calculations not well understood, but, even more fundamental, the atomic entities themselves were of unfathomable counterintuitivity. In addition to the wave/particle duality of light—by now (1926) accepted—in 1923 the French physicist Louis de Broglie had suggested that electrons also have a dual nature. What this means in terms of intuition needs to be explored.

INTUITIVITY: THE CENTRAL ISSUE

In 1926, 43-year-old Erwin Schrödinger, professor of physics at Zurich and of Viennese origin, added to the confusion over concepts by offering another version of atomic physics based on the wave nature of light and electrons. He was an outsider geographically, by his age, and in temperament and thought to the relatively younger band of quantum physicists who clustered about Bohr in Copenhagen, Arnold Sommerfeld in Munich, and Max Born in Göttingen. Besides being a serious classicist, Schrödinger's bent in physics and philosophy was closer to that of Einstein and other physicists at the University of Berlin than to Bohr and company. Not surprisingly, in 1927 Schrödinger succeeded Planck in Berlin.

Schrödinger made it abundantly clear in 1926 why he decided to formulate a wave mechanics:

> My theory was inspired by L. de Broglie . . . and by short but incomplete remarks by A. Einstein. . . . No genetic relation whatever with Heisenberg is known to me. I knew of his theory, of course, but felt discouraged not to say repelled, by the methods of the transcendental algebra, which appeared very difficult to me and by the lack of visualizability [*Anschaulichkeit*].

Imagine publishing such highly critical remarks in a scientific journal! Clearly, the stakes were high in 1926. At issue was nothing less than the intuitive understanding of physical reality itself, replete with visual imagery. The visual imagery Schrödinger offered was atomic electrons surrounding the atom's nucleus like vibrating strings charged with electricity (Figure 8).

In addition, Schrödinger replaced the discontinuities in Heisenberg's quantum mechanics with continuous transitions between allowed states. A useful analogy is to think of the vibrational modes in a drum-

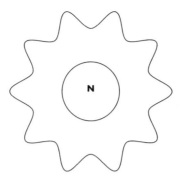

FIGURE 8.

According to Schrödinger's original interpretation of wave mechanics, an atomic electron is represented as a charged vibrating
string surrounding the atom's nucleus (N).

head. Striking the center of a drumhead with sand granules on it will
cause concentric rings of sand to form. The rings are analogous to the
stationary states available to electrons. As the vibrations ebb, the rings
coalesce continuously, one into the other. This is Schrödinger's visualization of how electrons make continuous transitions between allowed
states.

In a paper of March 1926, Schrödinger expressed his disapproval
of a physical theory based on a "theory of knowledge" in which we "sup-
press intuition [*Anschauung*]." Although objects that have no space-time
description may exist, Schrödinger was adamant that "from the philo-
sophic point of view," atomic processes are not in this class.
Schrödinger formulated an atomic theory that offered the possibility of
recovering the concepts of classical physics suitably reinterpreted.
Physicists such as Einstein and Lorentz were jubilant. On May 27, 1926,
Lorentz wrote to Schrödinger that

> if I had to choose between your wave mechanics and the
> [quantum] mechanics, I would give preference to the former,
> owing to its greater visualizability [*Anschaulichkeit*].

The battle lines were drawn. The prize was intuition. Heisenberg lamented over its loss. He writes in a paper of late 1925 that the present theory labors

> under the disadvantage that there can be no directly intuitive [*anschauliche*] geometrical interpretation because the motion of electrons cannot be described in terms of the familiar concepts of space and time.

Heisenberg was privately furious over Schrödinger's work and its rave reviews from the physics community. The reason for this, in large part, was that the mathematics underlying wave mechanics was much easier to deal with than the mathematical formalism of quantum mechanics. To make matters worse for Heisenberg, Schrödinger had proven the equivalence of the two theories, which he drove to a conclusion suitable to his own viewpoint—when discussing atomic theories, one "could properly use the singular." Heisenberg and others, including Bohr, did not fully accept Schrödinger's equivalence proof because of his physical interpretation of the wave mechanics as containing no discontinuities. Heisenberg vented his frustration in a letter dated June 8, 1926, to Pauli:

> The more I reflect on the physical portion of Schrödinger's theory the more disgusting I find it. What Schrödinger writes on the visualizability [*Anschaulichkeit*] of his theory . . . I consider trash.

Highly personal assumptions on what is an acceptable scientific theory emerge from these exchanges. In print, Heisenberg objected to Schrödinger's imposing onto quantum theory "intuitive pictures" of the sort that previously had led to confusion. Elsewhere in 1926, Heisenberg suggested setting limitations upon any discussion of the "intuition problem [*Anschauungsfrage*]." Heisenberg recalled this period as one of great psychological stress. The atmosphere among the quantum physicists was highly charged. Elsewhere in print, he struck out in an almost ad hominem manner against not only Schrödinger, but others of his colleagues who, in Heisenberg's opinion, had gone over to Schrödinger's camp.

Heisenberg expressed antipathy toward Schrödinger's view of a return to classical concepts—somewhat reinterpreted, of course—in a remarkable paper of September 1926. It is Heisenberg's first fundamental analysis of the problems that beset atomic physics. Heisenberg writes that the basic problems of quantum mechanics are rooted in the fact that our "ordinary 'intuitive view'" cannot be extended into the atomic domain. This is so, he writes, because

> the electron and atom possess not any degree of physical reality as the objects of daily experience. . . . Investigation of the type of physical reality which is proper to electrons and atoms is precisely the subject of atomic physics and thus also of 'quantum mechanics'.

Thus, the basic problem confronting atomic physics concerns the concept of physical reality itself on the atomic level. Compounding the problem is that physicists must use everyday language, with its perceptual baggage, to describe atomic phenomena. They must explain how something can be simultaneously wave and particle, in addition to having other strange attributes that emerge.

One year made a huge difference. In early 1925, atomic physics was in a state of shambles. By mid 1926, there were two apparently dissimilar theories: Heisenberg's was based on nonvisualizable particles and couched in difficult mathematics; Schrödinger's claimed a visualization and was set on familiar mathematical grounds that resulted in calculational breakthroughs. And yet a gnawing problem emerged: No one really understood what either formalism meant. The only thing on which Heisenberg and Schrödinger agreed was that basic issues in physics verged on the philosophical and centered on the concept of intuition.

ANOTHER REDEFINITION OF INTUITION

During late 1926 into early 1927, Bohr and Heisenberg struggled to come to grips with physical reality on the atomic level. As Heisenberg wrote to Pauli on November 23, 1926, "What the words 'wave' or 'corpuscle' mean we know not any more." Heisenberg recalled that these discussions left us in a "state of almost complete despair." Heisenberg and colleagues needed to determine how to use ordinary language to describe entities that defy imagination.

Linguistic difficulties had been present earlier in quantum theory and noted chiefly by Bohr, among others. For example, the energy and frequency of a light quantum are related through Planck's constant.* But equating energy, which connotes localization, with frequency, which connotes nonlocalization, is not unlike equating apples with oranges. De Broglie's hypothesis of the wave/particle duality of electrons poses a similar paradox. According to de Broglie, the electron's momentum—a localized quantity—is related to its wavelength—a nonlocalized quantity—through Planck's constant.

Deciding to take a rest from their discussions, Bohr took a skiing holiday in February 1927. Away from the pounding of Bohr's relentless dialectics, Heisenberg realized a way out of the morass, which he published in a hallmark paper in the history of ideas, "On the intuitive [*anschauliche*] content of the quantum-theoretical kinematics and mechanics." (Note the use of the term "intuitive" in the title.) Heisenberg immediately launches into a linguistic analysis: "The present paper sets up exact definitions of the words velocity, energy, etc. (of the electron)." Heisenberg turns to relativity for lessons about distrusting our intuition. After all, relativity has shown that "intuitively based concepts" cannot be extended to high-speed processes or to phenomena in large regions of space-time. Consequently, proper concepts for quantum

---

* I shall also refer to the light quantum as a "photon," a term introduced in 1926 by the American physicist G. N. Lewis.

physics are "derivable neither from our laws of thought nor from experiment." The initially peculiar mathematics of the quantum mechanics is a clue that "we have good reason to be suspicious about the uncritical application of the words 'position' and 'momentum'."

Here Heisenberg states that the "numbers" used in quantum mechanics are of a nonstandard sort. The arithmetic we deal with in our normal life possesses the property of commutativity, which means that $3 \times 2 = 2 \times 3$. We say that numbers like 2 and 3 commute. But this is generally not so for quantum theory, where quantities like position and momentum do not commute. Just as in the wave/particle duality, once again Planck's constant enters. If Planck's constant were zero, then position and momentum would commute. In this case, atomic phenomena would be like those in daily life where "particle" means a localized entity and "wave" means something extended and periodic, the two being antithetical. But we have no other lexicon than that of our macroscopic world. Herein lies the real problem.

Heisenberg's resolution of the paradoxes involved in extrapolating language from the world of sense perceptions into the atomic domain is to let mathematics be the guide. He reasons that the "revision of fundamental concepts" of position and momentum "follow directly from the equations of the quantum mechanics," which produce the uncertainty relation. Since mathematics must be the guide, then in the atomic realm it is essential to make a clear distinction between what it is "to be understood intuitively" and what the visual imagery of atomic processes is used to portray.

Heisenberg explains that the mathematics of quantum mechanics offers the proper restrictions on the words "position" and "momentum." The restriction is Heisenberg's uncertainty principle, which today even decorates T-shirts with wording such as "Heisenberg may have slept here." Heisenberg's uncertainty principle for momentum and position relates to experiments in which one tries to measure simultaneously both of these quantities. It says that the product of the errors in measuring momentum and position is not zero. Rather, the product of these two errors is Planck's constant. Consequently, the better we know an electron's momentum, the less well we know its position. Incidentally, in the uncertainty principle paper, Heisenberg proved that Schrödinger's interpretation was incorrect—there are discontinuities in wave mechanics, too.

On the other hand, in Newtonian physics, it is permissible to say that we can abstract to a world in which all measurement errors become infinitesimally small, and essentially zero. Also, all errors are systematic. For example, in measuring how long it takes for a stone to drop to the ground, I may have to take sightings through the crosshairs of a telescope. Consequently, I can only know that, for instance, the stone is 3 meters ± 1 mm off the ground at some instant of time. The error ± 1 mm is a systematic error due to the small, but nevertheless nonzero, width of the telescope's crosshairs. But we can imagine a limit in which systematic errors vanish. This is similar when I measure the stone's momentum, which requires sighting the change of position and recording times of fall. In nonquantum physics, the product of the errors in any experiment where position and momentum are measured can be *imagined* to be zero.

Upon reading Heisenberg's paper, Bohr became furious, demanding that it not be submitted for publication. His main contention was that Heisenberg had reached all results by considering only one-half of the picture, namely, the particle and not the wave nature of light and electrons. Heisenberg's rationale lay in his insistence on the key role played by discontinuities in quantum theory. For Heisenberg, discontinuity meant particles, not waves. After studying Heisenberg's paper, Bohr concluded that both modes had to be taken into account. A tense atmosphere prevailed, which we can glean from a letter Heisenberg wrote to Pauli on May 16, 1927, where he states that there are "presently between Bohr and myself differences of opinion on the word 'intuitive'." And well there were, because throughout Bohr's writings one finds either emphasis on the visual imagery of atomic physics or lamentations of its lack thereof. In contrast, Heisenberg seemed willing to renounce visual imagery altogether and forever.

As history has shown, the interpretative phase was of critical importance to quantum theory. After Heisenberg's initial breakthrough in June of 1925, followed by Schrödinger's wave mechanics less than a year later, key problems were solved quickly, but a deep understanding of the formalism eluded physicists. For this purpose, Heisenberg produced the uncertainty principle and Bohr offered the notion of complementarity, to which we now turn.

VISUALIZATION REGAINED, IN PART

In what can be described only as a tour de force of tightly knit reasoning, in September of 1927 Bohr revealed his viewpoint of complementarity, refined in heated conversations with Heisenberg. It focuses on the wave/particle duality of light and matter. In Bohr's view, this paradox cannot and should not be avoided. Bohr's approach to paradox busting is as unique as Einstein's was in 1905: Instead of choosing one "side" and trying to quell the other, he took on *both* as acceptable. In the end, he found that there was no paradox at all, but a conceptual error rooted in extending words like "wave" and "particle" into the atomic world while retaining their meanings from the realm of sense perceptions. The wave and particle aspects of light and matter complement each other in that the totality of these attributes is needed for a description of atomic entities.

Bohr reasoned that the seemingly paradoxical situation arises from the equations that relate an atomic entity's energy and momentum (localized properties) with its frequency and wavelength (nonlocalized properties). Bohr's great realization was that this state of affairs is paradoxical only if we understand "particle" and "wave" to refer to objects and phenomena from the world of sense perceptions. In the atomic realm, however, these words refer to totally different classes of entities, which can be wave and particle simultaneously. Try to imagine something like this. You cannot.

Bohr found that the clue to an understanding resides in Planck's constant, which relates particle and wave concepts. Taking Planck's constant to be zero decouples the wave and particle modes of existence. The extremely small but nonzero, value of Planck's constant signals to us that we cannot rely on sense perceptions to understand atomic phenomena. Doing so led physicists into erroneously concluding that the wave/particle duality was paradoxical. Similarly, the large value of the velocity of light had tricked scientists into neglecting the relative nature of time. Bohr also emphasized that the wave/particle duality of light and matter is at the basis of the uncertainty relation as well.

Bohr concludes his first exposition of complementarity with a tribute to Einstein and with a guide to the term "intuition," which must again be redefined as referring to pre-quantum mechanics visual imagery:

> Indeed, we find ourselves here on the very path taken by Einstein of adapting our modes of perception [*Anschauungsformen*] borrowed from the sensations to the gradually deepening knowledge of the laws of Nature. The hindrances met with on this path originate above all in the fact that, so to say, every word in the language refers to our ordinary perception.

To reach this conclusion, Bohr was led into ever-deeper levels of analysis of atomic phenomena, from an examination of physics per se into an analysis of perceptions and then into an analysis of thinking itself. In this way, Bohr recognized the fullness contained in the wave/particle duality of light and matter, thereby raising atomic physics out of the abyss. Bohr followed his favorite aphorism from Schiller: "Only fullness leads to clarity/And truth lies in the abyss."

Thus, the summer of 1927 ended a heroic age of physics. The young men of Bohr's circle went off to make their own lives in physics. To Bohr's great disappointment, Heisenberg went to Leipzig instead of remaining in Copenhagen. On arrival at his new post, Heisenberg wrote an emotional letter to Bohr, recalling fondly his many stays in Copenhagen. Bohr replied on August 24, 1927, "The kind words you wrote about your stay in Copenhagen were a great pleasure to read. Also for me this has been an unforgettable time."

Complementarity provided an underpinning for the results of Heisenberg's linguistic analysis in his 1927 uncertainty principle paper. The need remained, however, for a guide from quantum theory as to how terms whose meanings are tinged necessarily with sensation-based language can be taken into the atomic world. Bohr's correspondence principle offers a way.

EXTENDING INTUITION: BOHR'S CORRESPONDENCE PRINCIPLE

Planck's constant turned out to be ubiquitous. It signaled a break with sense perceptions at the atomic level. Initially, everyone, including Bohr, overlooked its importance for the wave/particle duality. The reason for this oversight is that, like other physicists, Bohr assiduously avoided light quanta and failed to grasp the depth of de Broglie's proposal.

Right from the beginning, in 1913, Bohr obsessed over using the language of Newtonian mechanics in a domain where this theory failed. But Newtonian mechanics was the only theory available, and Bohr had to begin somewhere. To provide a means for moving Newtonian physics into the atom, Bohr demonstrated that in his theory there is a domain of atomic physics in which classical and quantum concepts merge. (Nature is kind sometimes.) This merging occurs for electron transitions between orbits that are very far from the nucleus in the solar system atom. The spectral lines resulting from such transitions are bunched together more and more until they are almost indistinguishable. This is the case for the spectral lines on the far left in Figure 6(b). In this region, the frequency of the emitted light is very close to that from the light emitted by an electron actually orbiting the nucleus—emitted light as predicted from classical electromagnetic theory. As discussed earlier, in stark contrast to this classical theory of electrons radiating, in Bohr's atomic theory electrons emit radiation *only* when they make a downward transition between stationary states.

In 1920 Bohr referred to this merging of classical and quantum concepts as the "correspondence principle." The power of this principle is that it enabled Bohr and his co-workers to explore the atomic world by moving time-honored formulas from classical electromagnetic theory into electron orbits close to the nucleus. The modifications required often had no physical foundation except that they gave correct answers. Bohr's uncanny use of the correspondence principle led one eminent physicist to refer to it somewhat derisively as a "magic wand." Its various formulations guided physicists through the myriad of multiplying difficulties and turned out to be the key to Heisenberg's formulation of quantum mechanics in June 1925.

Until complementarity in 1927, none of Bohr's deliberations made any reference to Planck's constant, denoted by h, vanishing in order to

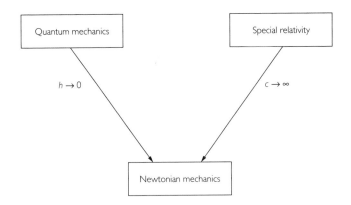

FIGURE 9.

This diagram shows the importance for correspondence principles in both quantum physics and relativity, a point made most eloquently by Bohr in the 1927 complementarity paper. Newtonian physics is based too closely on what our sense perceptions register, which is no relativity of time ($c = \infty$) and no wave/particle duality ($h = 0$).

retrieve results from classical physics (Figure 9). The correspondence principle for Einstein's special relativity theory to revert back to Newtonian mechanics is for the velocity of light to become infinite, which agrees with results expected on the basis of our sense perceptions. One must be wary of taking the limits in Figure 9 in too cavalier a manner. Bohr emphasized this graphically in Figure 10.

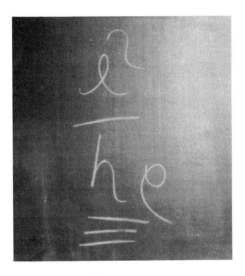

FIGURE 10.

This equation was written by Niels Bohr in 1961 on the blackboard of Professor Edward M. Purcell of Harvard University. The occasion was Bohr taking a few minutes' rest during a visit to Harvard's Physics Department. The equation in the figure is the fine structure constant e^2/hc' where e is the electron's charge, h is Planck's constant, c is the velocity of light, and the factor of 2π, which is missing, is inessential for what follows. The fine structure constant is $1/137.04$, almost the exact ratio of two integers. This quantity sets the scale of atoms and so is essential to the sort of life forms in the universe. Bohr, Purcell, and Professor Norman Ramsey were in the midst of a general discussion about the correspondence principle when, Purcell told me, Bohr commented: "People say that classical physics is the limit of quantum mechanics when h goes to zero." Then, Bohr shook his finger, walked to blackboard, and wrote the equation in the figure. As he made three strokes under h, Bohr turned and said, "You see, h is in the denominator." Later in the day, Purcell and Ramsey arranged to have the board photographed.

Correspondence principles play the role of a conduit to take terms defined in the world of sense perceptions, words like "position" and "velocity," into the atomic realm where Planck's constant can no longer be regarded as zero. For atomic theory, Heisenberg's uncertainty principle provides restrictions on such terms. In special relativity, the relativity of time plays such a restrictive role for the concept of time.

Seeking "correspondence-limit procedures" continued to be a guideline for Heisenberg and Bohr in combining quantum mechanics with special relativity and then formulating a theory for the interaction between light and matter—a quantum electrodynamics. We will return to the correspondence principle in subsequent chapters as an essential tool for formulating new theories and for demonstrating the continuity of scientific change.

CONCLUDING COMMENTS

Heisenberg's subsequent research in atomic and nuclear physics led to further unexpected redefinitions of the concept of intuition in addition to a new visual representation of the atomic world. Astonishingly, it turned out that the atoms of Leucippus and Democritus from the fifth century B.C., and subsequently those of Plato, are closer to the atoms of the twentieth century than the billiard-ball atoms of James Clerk Maxwell and Ludwig Boltzmann. The atoms of Leucippus and Democritus are not Platonic ideal forms incapable of being glimpsed in the mind's eye because they exist only in a mathematical world of perfect shapes. According to Leucippus and Democritus, the drawings of hooked shapes that enable atoms to link up with other atoms are but crude approximations to the way things really are. Rather, in their view, we cannot imagine atoms because of limitations on our perceptual apparatus. Plato's imagery is specifically mathematical rather than mechanical. Consequently, it is philosophically deeper than the view of Leucippus and Democritus, and more modern, too, because one of its consequences is a mathematically structured universe.

We have found that the atoms of the modern atomic physics, or quantum mechanics, are entities that can be both wave and particle simultaneously, in addition to possessing other attributes defying visualization. Heisenberg's uncertainty principle paper teaches us that the only means we have of learning about atoms and elementary particles is through the mathematics of quantum mechanics, while Bohr's complementarity reveals the extreme subtlety of the wave/particle duality. It is apparent that problems of atomic physics have come full circle back to those of the founders of science in the fifth century B.C.

The principal problem with the visual imagery or intuitions, or *Anschauungen*, of Bohr's atomic theory is that the images were abstractions from the world of sense perceptions and so were encumbered with meanings inappropriate for the atomic realm. A clue to the new meaning of "visual image" is that beginning in 1926, Heisenberg used almost exclusively the term "visualizability" or *Anschaulichkeit*, reserving "intuition" or *Anschauung* to denote the prior state of affairs in atomic physics.

Before proceeding headlong into subsequent developments in physical theory, we must pause and examine the complex transformations in common sense and intuition that have emerged. The common-sense Aristotelian intuition has been transformed several times. It is important to determine what this means. Science may be informing us of worlds beyond our sense perceptions. Or maybe we have just been lucky in formulating scientific theories that are increasingly better descriptions of empirical data. Any attempt to resolve these issues requires us to widen our net of inquiry to include the methods scientists use to formulate theories.

3

SCIENTIFIC
METHODS

Our discussion so far has focused on how developments in science have transformed our concept of what is intuitive and thus our notion of common sense. Exploring how these developments came about will provide the necessary groundwork for picking up the strands of our investigation and extending them into the role of image-making in science and art, as well as the nature of scientific creativity.

Much has been written about a scientific method in which the scientist collects data, inspects the data, spins a theory that describes the data, and makes predictions, which then are tested to decide the theory's ultimate fate. The new scientific theory is then accepted by consensus. This is essentially the way the process of science is presented in professional journals. In this chapter, we will examine if this is the way it is actually done.

THE VERDICT OF EXPERIMENT OR NOT?

An exciting confrontation occurred at the September 1906 Meeting of German Scientists and Physicians in Stuttgart, which shows how complicated the experimental situation gets when it comes to the acceptance of empirical data. The chief experimental protagonist was Walter Kaufmann. In 1906, Kaufmann was a 35-year-old associate professor of physics at the University of Bonn. He had an illustrious record of research on electron physics, marred only by his missing out on the accolades for the electron's discovery in 1897, all of which went to J. J. Thomson. (We will return to this episode a little later in this chapter.) Kaufmann's supporter was his close friend Max Abraham, a 31-year-old theoretical physicist at the University of Göttingen. Kaufmann and Abraham had been colleagues at Göttingen from 1899 to 1902. Although almost forgotten today, Abraham was a force to be reckoned with in the first decade of the twentieth century. Abraham's acid criticisms and satirical wit were well known throughout the German physics community. Only physicists who were secure in their understanding of the subject appreciated him and could withstand his critiques. Einstein often came to the defense of "fearsome Abraham." The third main figure in this episode was Max Planck. In 1906 the 48-year-old Planck was a professor of physics at the University of Berlin. The three men knew each other quite well. Planck had been Abraham's Ph.D. supervisor at Berlin. Besides periods of time studying in Berlin, Kaufmann had spent the years from 1896 to 1900 there as a researcher.

After missing out on the electron, Kaufmann was hell-bent on making an outstanding discovery. So, in 1901, he embarked on a series of experiments that established him as the first high-energy experimental physicist. Using radium salts as a source of electrons traveling faster than 90 percent of the speed of light, Kaufmann set out to measure how the electron's mass varies with its velocity. Explanation of Kaufmann's data with theories of the electron was at the time the very essence of deep physics. The "electromagnetic world-picture" sought a unified description of the then-known forces, electromagnetism and gravity, and the electron was its basic entity. Lorentz and Poincaré were among the physics elite involved in the quest. In his crowning work of 1906, Kaufmann published data that disconfirmed Lorentz's theory of the electron while favoring a theory proposed by Kaufmann's close friend and former colleague, Max Abraham.*

According to the British philosopher Karl Popper, in this sort of situation Lorentz ought to have renounced his theory as falsified in the face of Kaufmann's high-precision data.† Lorentz did just that. He wrote to Poincaré on March 8, 1906, that "unfortunately my [theory of electrons] is in contradiction with Kaufmann's results, and I must abandon it. I am, therefore, at wits' end." But scenarios proposed by philosophers for how science proceeds often fail the test of historical veracity, not to mention their absence of contact with actual science. Theory, not experiment, is the adjudicator in scientific disputes. With a major theory and world view at stake, Poincaré suggested that his discouraged colleague wait and ask for further experiments. Other major physicists, such as Planck, concurred with Poincaré.

This is one of several stratagems that scientists use to save a theory. Another is to change the theory. But Poincaré had sharpened and polished Lorentz's theory to perfection—in his opinion.

---

* According to Lorentz's theory, the electron can be imagined to be a balloon smeared with charge, which contracts when moving. On the other hand, in Abraham's theory the electron is a charged rigid sphere.

† Karl Popper advocated that robust theories are capable of making novel predictions such as Lorentz's on the mass of high-velocity electrons. But should the prediction be disconfirmed, then the theory is falsified and ought to be discarded. This is Popper's view of scientific progress. We will see that scientists usually don't work in this manner.

In the German-speaking scientific community, the major annual event was the Meeting of German Scientists and Physicians, convened every September. The proceedings of the science section were published in the journal *Physikalische Zeitschrift*, complete with discussion sessions. They make fascinating reading because they contain the polished lectures as well as the often spirited exchanges that followed. In his exchanges with Kaufmann and Abraham, the soft-spoken and diminutive Planck sided with Lorentz's theory for reasons that went beyond physics per se.

As a physicist, Planck sought the absolute, the grand principles overarching all of physical theory. Lorentz's theory contained one of them: the principle of relativity. The previous year Poincaré had proven with great generality that only Lorentz's electron theory, not Abraham's or anyone else's, included this principle. An unknown Patent Clerk Third Class in Bern, Switzerland, Albert Einstein, offered another approach to theories of motion also based on a principle of relativity. Everyone interpreted Einstein's viewpoint as a generalization of Lorentz's, thereby further confirming Lorentz's theory of the electron. Kaufmann and others began referring to a Lorentz–Einstein theory of the electron. The patent clerk did not demur at the reverse alphabetical order, even though he knew that his viewpoint differed as drastically from Lorentz's as night and day. He could wait.

After scrutinizing Kaufmann's data, at the September 1906 Meeting of German Scientists and Physicians at Stuttgart, Planck forthrightly concluded that he could find nothing wrong. However, taking into account the delicateness of Kaufmann's measurements and analysis, Planck cautioned that "in my opinion [Kaufmann's data are] not a definite verification of [Abraham's theory] and a refutation [of Lorentz's]." After Planck concluded, Kaufmann circulated his impressive data obtained from his exquisitely designed equipment, machined and assembled with the care of a watchmaker (Figures 1 and 2). Kaufmann described his data as "two branches symmetrical as if by a draftsman." The symmetry of the two wings of the gull-shaped curve in Figure 2 indi-

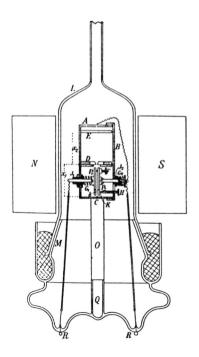

FIGURE 1.

Kaufmann's schematic of his "measurement chest," which is the rectangular box situated in an evacuated enclosure between the pole faces of an electromagnet (labeled N and S). The measurement chest is of dimensions $2 \times 3 \times 4.5$ cm. The high-speed electrons emerge from a granule of radium chloride (1 mm long by 0.3 mm thick) placed at C. The electrons are acted on by an electric field generated by the capacitor plates K and magnetic field. Both fields are parallel. The dimensions $x_1 = 2$ cm and $x_2 = 2.07$ cm. The electrons then strike a photographic plate at E.

FIGURE 2.

The actual size of the data plates that Kaufmann took to be his best measurement. We can assume that this was one of the plates he passed around at the 1906 meeting at Stuttgart. How the electron's mass depends on its velocity was ascertained through measurements made on the coordinates of points on the fuzzy gull-winged curve in the figure.

cates the purity of the vacuum in Kaufmann's apparatus, in Figure 1. This is essential because it took 24 hours for electrons to form each wing of the curve in Figure 2. On the basis of Kaufmann's reputation, these data were deemed first-rate.

Abraham's criticisms of Lorentz were framed with a satirical wit that served to entertain and then to sharply divide the audience. Consider the first sentence of his contribution to the discussion session. The point of contention was the factor-of-2 disagreement between the predictions of Abraham and Lorentz–Einstein according to Kaufmann's measurements of the moving electron's mass. Abraham's coda was a facetious statement about the two theories:

> When you look at the numbers you conclude from them that the deviations of the Lorentz theory are at least twice as big as those of mine, so you may say that the sphere theory represents the deflection of β-rays twice as well as the *Relativtheorie* [relative theory].*

A great deal of laughter was reported to have followed this remark. Tempers flared in the course of this dispute. Planck was accused of being a pessimist and not appreciating the extraordinary difficulties Kaufmann faced.

The session chairman, Arnold Sommerfeld, tried to defuse the situation with a bit of levity, but it was too late. Planck was overwhelmed and, in the end, could only whisper that he was "attracted" to the Lorentz–Einstein theory. His correspondence from this period reveals that he was strongly affected by this exchange. On July 7, 1907, he wrote to Einstein that "so long as the supporters of the principle of relativity constitute as tiny a band as they do at present, it is doubly important that they agree with each other."

---

* At this session Planck coined the term *Relativtheorie*, a variant of relativity theory, to distinguish the Lorentz–Einstein theory of the electron, with its firm basis on a principle of relativity from Abraham's "*Kügeltheorie* [sphere theory]," or theory of a rigid spherical electron. Einstein later noted that he would have preferred the name *Invarienten Theorie*, or invariant theory, rather than relativity theory, with its too easily made ideological connections.

Astonishingly, while Lorentz, Planck, and Poincaré were concerned over the possible falsification of Lorentz's theory, Einstein set about generalizing the view he presented in 1905 to include gravity. Einstein never considered the principle of relativity to be disconfirmed for electron physics. As he explained in a paper of 1907,

> In my opinion [other electron theories] have a rather small probability, because their fundamental assumptions on the mass of moving electrons are not explainable in terms of theoretical systems which embrace a greater complex of phenomena.

Einstein's intuition served him well.

In 1908 a colleague of Kaufmann's at Bonn, Alfred Bucherer, produced data in favor of Lorentz's theory and went on to note errors in Kaufmann's experiment and data analysis. It turned out that Kaufmann's gull-winged curves were not as symmetrical as he claimed, which meant serious procedural errors had occurred. Bucherer's experimental apparatus, far simpler in design and operation than Kaufmann's, was the one Kaufmann should have used. It is possible Kaufmann did not because it was the sort J. J. Thomson had used in his discovery of the electron, for which Thomson had received honors Kaufmann may have thought belonged to him as well. Generally, Kaufmann had a fatal attraction for complex apparatus and methods, and in this instance they were far too complex for the measurements he sought. After Bucherer announced his results, neither Lorentz nor Poincaré called for further data.

A few years later, Lorentz suggested that Kaufmann might have felt driven to produce data in favor of his highly opinionated and dominant friend Max Abraham. Sometimes it is dangerous for an experimentalist to know beforehand what results to expect.*

---

* Kaufmann faded out of the limelight in 1908 and died in 1947 after holding academic positions in Königsberg and Freiberg. Einstein considered Abraham's criticisms of his early efforts toward a general theory of relativity as enlightening. In 1922 Abraham was finally offered a professorship. On the train journey to take up his post in Aachen, he collapsed from what was diagnosed as a brain tumor and died in great pain shortly after.

But the story is not over yet, because further experiments were performed for the purpose of improving Bucherer's accuracy. Meanwhile, the electromagnetic world-picture fell by the wayside, pushed there by the emergence of Einstein's special theory of relativity, which predicted an electron mass identical to Lorentz's.* Bucherer's data were eventually taken as support for special relativity. But a surprise was in store. In 1938 two American experimentalists, C. T. Zahn and A. H. Spees, were involved in a series of experiments exploring the nature of electrons emerging in the β-decay of nuclei. Knowing some history of physics, Zahn and Spees realized that they were essentially repeating with better equipment the older experiments of Bucherer, as well as the ones that purported to test Bucherer's accuracy. Amazingly, in all of these experiments they found design errors that nullified their data as a means for distinguishing between competing electron theories. As Einstein wrote of experiment being the arbiter of competing theories, "The theory must not contradict empirical facts. However evident this demand may in the first place appear, its application turns out to be quite delicate." So, even after data are gathered, assessing their quality is not clear cut.

DATA, DATA EVERYWHERE . . .

Nature provides data. We have noticed this in the case of double-slit interference in Chapter 2 (see Figures 1–5). The scientist works backward from effect (data) to cause (source) in order to formulate theories about the source. All scientists see the same processes, yet some scientists interpret them very differently. What is simpler than a falling stone? Yet how very different were the ways in which Aristotle and Galileo saw it.

* Basically, the electromagnetic world-picture did not succeed because of the wave/particle duality of light, which had no place in Lorentz's theory of electromagnetism. Lorentz's theory could accommodate only the wave nature of light. Einstein was already aware of this defect in 1905.

Aristotle saw an object falling "in order that" it reach its natural place, which is on the ground. The millennium-long journey from Aristotle to Galileo, containing a complex combination of thought experiments, real experiments with pendula, the needs of artillery batteries, and astronomical observations, led Galileo to formulate the basis for a new theory of motion. Galileo "saw" the process in terms of distances covered by the stone during lapsed time intervals. Aristotle "saw as" while Galileo "saw that."

Examples abound of seeing data differently. In 1905 Poincaré and Einstein were presented with the same set of data, for which they produced essentially identical mathematical formulations. Whereas Poincaré interpreted the mathematics as improving on Lorentz's electron theory, Einstein interpreted it as the special theory of relativity. Examples as dramatically drawn as this are hard to find, and the issues go beyond mere priority.

Every set of experimental data is open, in principle, to an infinite number of hypotheses. This is a fact of scientific life; it is called the underdetermination thesis. In sciences with little or no mathematical basis, such as biology and psychology, the underdetermination thesis poses severe difficulties, for one must choose among a very large number of possibilities. Physics contains rich enough mathematical guidelines to let us cull out all but a few choices. The problem is whether the correct hypothesis has been chosen; determining what we mean by the "correct" one is also a problem. But even before the grist of empirical data is fed into the mathematical mill, we need to know whether the proper data are being collected. The choice of data appears to be rational. After all, experimentalists read meters and record numbers.

Clearly, data selection is a key point in theory construction. As Henri Poincaré emphasized, "Experiment is the sole source of truth." Is science as straightforward as this? Poincaré knew better. He wrote often on the not strictly rational manner in which creative science is accomplished and on aesthetic criteria that ought to hold in formulating theories. Even for the data-gathering stage, questions of aesthetics are of the essence, writes Poincaré:

> It is, therefore, the quest of this especial beauty, the sense of the harmony of the cosmos, which makes us choose the facts most fitting to contribute to this harmony, just as the artist chooses from among the features of his model those which perfect the picture and give it character and life.

Our experimental considerations are not free of bias, not even for everyday observations. The more knowledge we have, the less straightforward is our representation of the world. Science should not be any different, especially for systems that are not directly measurable. (A moment's reflection will reveal that nothing is "directly" observable.)

For example, making any sense at all of the characteristics (data) of rays in a cathode ray tube (a simple version of a television tube) requires a theory of electromagnetism. The fact that there is no such thing as "raw" data, that is, data free of any interpretation, is expected from the theory-ladenness of data. Even the data of vertical fall were interpreted by Aristotle within a conceptual framework comprised of such assumptions as that there is no vacuum and there are four elements, each seeking its natural place, and so on. There is no theory-neutral language with which to describe observations.

UNWANTED PRECISION

To complicate matters further, sometimes less accurate data are more informative. Examples from Galileo's research are particularly relevant. A thought experiment on which Galileo spent a great deal of time in the *Dialog on the Two Chief World Systems* involves someone who, standing on the mast of a moving ship, drops a stone. Galileo pondered where it will land (Figure 3).

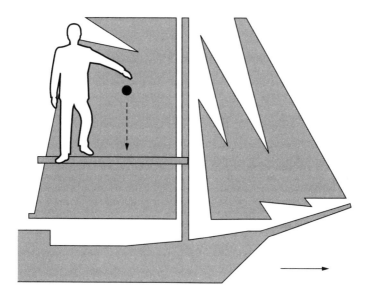

FIGURE 3.

A person standing on the mast of a ship, which is moving to the right, drops a stone. Where will it land?

Aristotelians argue that the ship moves out from under the stone and so the stone does not fall directly under the person who dropped it (Figure 4). Galileo, on the other hand, argues that the stone will land directly below the person who dropped it (Figure 5). Similarly, someone standing on the shore will observe the stone to fall directly below the person who dropped it. But whereas the person on the ship sees the stone fall in a straight line downward, the observer on the shore sees the stone fall in a parabolic trajectory [Figure 5(b)].

Yet, if Galileo had extremely precise measuring instruments, he would have found that the stone does not fall directly under the person who dropped it. Its deviant course is due to a force whose origin is in the Earth's rotation about its axis: the Coriolis force. The effects of this force depend on where you are on the rotating Earth and how fast you are moving. The Coriolis force is of key importance in the motion of large air masses such as cyclones, tornados, and trade winds, causing

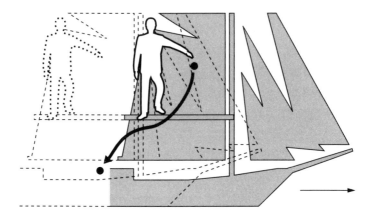

FIGURE 4.

Aristotelians argue that according to common sense the ship moves out from under the stone. Consequently, the stone falls to the left of the person who dropped it because the ship moves to the right.

them to move in different directions in the northern and southern hemispheres. But for falling bodies on Earth, its effects are extremely small. In Galileo's experiment from Figure 5, the deviation from vertical fall due to the Coriolis force for an object dropped from a mast of height 30 feet at a latitude of 45° is only .02 inches. Luckily, Galileo did not have highly sensitive measurement devices. Any measurements revealing this deviation* would only have confused matters, because a first step in understanding how objects fall is to formulate a theory of motion in which the Coriolis force is *not* present. This is what Galileo did by restricting all considerations to inertial reference systems. As we discussed in Chapter 1, the fact that effects due to the Earth's rotation are small permits us to consider to a very good approximation that the Earth is, indeed, an inertial reference system.

---

* There is another force arising from the Earth's rotation called the "centrifugal force," which depends only on where you are on the rotating Earth. One of its effects is to flatten the Earth at its poles. This flattening produces a variation in the acceleration of free fall depending on your latitude. Any corrections are at most in the third decimal place. So, for most calculations, the Earth can be considered to be a sphere.

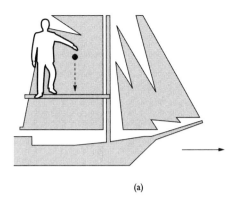

(a)

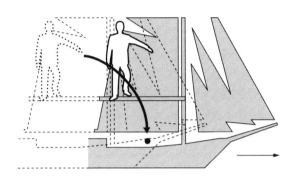

Observer on shore

(b)

FIGURE 5.

According to Galileo, the stone will land directly below the person who dropped it as in (a). The far-reaching additional ingredient in Galileo's solution is in (b), in which this result is also observed by someone at rest on the shore. Instead of seeing a vertical descent, the shore observer sees the stone fall in a parabolic trajectory.

Similarly, Galileo's experiments with pendula produced flawed data because he pulled the pendulum bob far enough from the vertical so that the motion was not simple harmonic. Better clocks for timing purposes would have been deleterious to his goal of using pendula to convince himself that all bodies fall with the same acceleration in a vacuum, regardless of their weight.

Then we have Galileo's telescopic observations of Saturn, which he found to be an oblate spheroid. Generalizing from his experiments on the shape of spinning viscous matter, Galileo concluded that Saturn's rotation accounts for the flattening at its poles. This conclusion was congenial to Galileo's goal of providing evidence for a Copernican universe in which all planets rotate like the Earth and so are not perfect spheres. In a Copernican universe, there is a democracy among the heavenly bodies: They are all blemished. But if Galileo's telescopes had had greater resolving power, he would have observed Saturn's rings, which would have caused problems for his explanation. With poor telescopic resolution, Saturn's rings merge into the planet itself, giving the illusion of an oblate spheroid.*

Another example of the abstractness of scientific theories that nevertheless bear directly on how we understand phenomena in the world about us concerns a concept of key importance in the theories of Einstein, Galileo, and Newton—the inertial reference system. To a good approximation, the Earth can be considered an inertial reference system, despite the fact that the Earth spins on its axis while moving in an elliptical orbit about the sun. This is fortuitous, because if it were not the case, then it would have been very difficult, indeed, for Galileo to have formulated a theory of motion alternative to Aristotle's.

Since Newtonian science is based on inertial reference systems, we would expect to see examples of them. Surprisingly, such systems exist solely in principle, that is, in the mind of the thought experimenter. The

---

* Examples of this sort abound. Had Geiger and Marsden used higher-energy α-particles in their work, then Rutherford's route to a solar system atomic model would have been circuitous to say the least, because there would have been negligible back scattering.

reason is that in order for a platform to move along in a straight line at a constant velocity forever, all of the forces acting on it must add up to zero. Such a system is a fiction. Imagine a situation in which you hear about an inertial reference system somewhere in your vicinity, and you decide to verify this claim. But, as soon as you approach the moving platform, you unbalance the forces acting on it—then it is no longer an inertial reference system.

In essence, we are faced with a situation in which the very basis of Newtonian mechanics does not exist in the real world. We say that this theory is approximate, and so we do not blame the collapse of bridges on there being no real inertial reference systems. Einstein's special theory of relativity is also based on the inertial reference system in the sense that it deals only with measurements made in such systems, which Einstein himself took to be its "logical weakness." And so special relativity is also approximate. Einstein repaired this situation in 1915 with his general theory of relativity, in which measurements can be made in accelerating reference systems.

In addition, philosophical assumptions can alter interpretations of data. While J. J. Thomson was experimenting on cathode rays in early 1897, Walter Kaufmann, whom we met earlier in this chapter, was independently performing exactly the same experiments in Berlin and with better equipment. But Kaufmann was a follower of the Viennese philosopher-scientist Ernst Mach, who advocated that atoms cannot be physically real because they cannot be directly *observed*. Consequently, Kaufmann interpreted his experimental data as the effect of charged ethereal rays. Thomson, with only scientific expediency in mind, interpreted cathode rays as composed of particles of electricity, or electrons. Thomson found this interpretation the easiest to deal with because it fit with Lorentz's well-developed electromagnetic theory. In the end, Kaufmann's philosophical bias cost him a Nobel prize.

We have seen thus far that data are theory-laden and sometimes open to philosophical bias, that imprecise data are sometimes beneficial because more exact data would have confused issues, and that theories are underdetermined. What can we say about scientific rationalism? Do experimentalists "massage" data, as the saying goes, picking and choosing certain data rather than others for publication? As we will see, sometimes this is true.

PICKING AND CHOOSING DATA

The American physicist Robert Millikan attempted during 1909 and 1910 to measure the electron's charge. A detailed study of Millikan's notebooks by the historian of science Gerald Holton reveals a less than objective choice of published data points. Next to some points in a laboratory notebook, Millikan writes "Beauty. *Publish* this surely, *beautiful*" (italics in original).

Millikan discarded other points. From calculations based on previous measurements of the electron's properties, performed principally by J. J. Thomson, Millikan had a good hunch as to what he was looking for. But Millikan's goal was more than merely to measure the electron's charge. He sought evidence that it was the basic unit of electricity. I am not accusing Millikan of maliciously massaging bad data to make them look good, because Millikan's data are reproducible. His intent, which had firm scientific underpinnings, was not the publication of every data point.

Just as in the Thomson-Kaufmann episode, simultaneously and independently of Millikan, and with better equipment, Felix Ehrenhaft was performing almost identical experiments in Vienna. But being a follower of the anti-atomist Ernst Mach, Ehrenhaft did not believe in a basic unit of electricity. So Ehrenhaft published all of his points, in each of which he had great confidence. Ehrenhaft's equipment, as Holton shows, "was rather more sophisticated than necessary."*

* Incidentally, Millikan was wrong because today we know that certain basic building blocks of matter—quarks—can have electric charges on the order of $\frac{1}{3}$ and $\frac{2}{3}$ of the electron's charge.

I will indulge in one more example. Galileo's telescopic observation that the planet Venus goes through a complete set of phases like the moon was too late for inclusion in *Starry Messenger*. Such evidence could not be fitted into the geocentric system in which Venus was so tightly locked into an orbit between the Earth and the sun that it could not show a one-half or three-quarter phase. Any changes in its orbit would reverberate through the Earth-centered system like an earthquake. Galileo knew this and pounced on the new data as definitive evidence in favor of a sun-centered universe. Here is his argument: If the universe is sun centered, then Venus shows phases; Venus shows phases; therefore, the universe is sun centered. Students of logic know that this is the error of confirming the consequent.* The point is whether there is any other planetary system that can account for Venus's phases. And there was one, namely, that of the Danish astronomer Tycho Brahe, in which all of the planets except the Earth revolved about the sun. In turn, everything revolved about the Earth. Although Brahe never provided any mathematics to back up his view, he nevertheless offered a way to include the new observations on Venus in a geocentric universe. The Jesuits in turn pounced on Galileo for his outrage against logic.

Surely, if science is a rational process, then Galileo would never have committed this silly mistake in logic. We can suppose that he chose to do it because he was desperate for any evidence in favor of a sun-centered universe. Scientists do this sometimes, and not for malicious reasons.

SOCIOLOGY INFLUENCES SCIENCE

Calling for more data to save a theory brings into play some sociological aspects of science. If the Lorentz electron theory, for example, did not have the backing of scientists like Poincaré, further experiments might not have been performed to rescue it.

---

* In symbolic form, the error of confirming the consequent is like this: If p, then q; q, therefore p. Roughly speaking, this is akin to arguing backward.

In 1957, the American physicists Richard Feynman and Murray Gell-Mann produced a theory of weak interactions which clashed with certain measurements that were accepted as state of the art for their precision. With chutzpah, Feynman and Gell-Mann stated flatly that these "simple theoretical arguments [of their theory of weak interactions] seem to the authors to be strong enough to suggest that the disagreement with the He^6 recoil experiment and some other less accurate experiments are wrong." This turned out to be true. But had scientists of lesser stature than Feynman and Gell-Mann made the same assertion, the offending experiments most likely would not have been repeated, at least not so soon.

We have an earlier example of this very point. In 1906, as a patent clerk in Bern, Switzerland, Einstein published a paper in which he suggests an experiment with low-velocity cathode rays to distinguish between competing electron theories. At the paper's conclusion he wrote that it would please him if a physicist became interested in actually doing the experiment. No one did. Contrast this with Einstein's suggestion only nine years later that experimenters test his gravitational theory by measuring the sun's deflection of starlight. As soon as World War I ended, two expensive expeditions were outfitted. One went to the village of Sobral in Brazil and the other to Principe, a Portuguese island off the coast of West Africa. The outcome of these experiments proved to have a spectacular sociological dimension: Einstein achieved worldwide fame as no other scientist had before him.

Einstein's 1915 general theory of relativity made the astounding prediction that light bends out of its hitherto assumed linear path when passing near massive objects (Figure 6). The best way of verifying this is during a total eclipse of the sun. Frank Dyson, the Astronomer Royal, noted that on May 29, 1919, a total solar eclipse would occur and it could be observed particularly well in Sobral and Principe. The expedition to Sobral was headed by C. Davidson and the one to Principe by the well-known theoretical physicist Arthur Stanley Eddington. Although the weather was far better at Sobral, Eddington insisted on em-

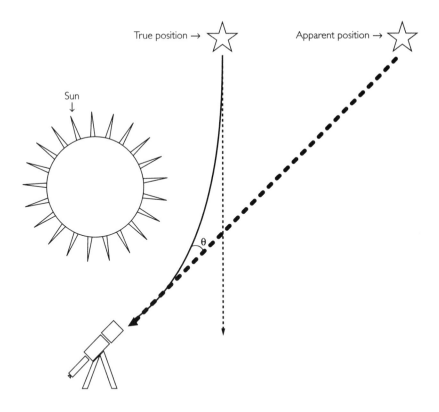

FIGURE 6.

A star's true position is taken as its position when the sun is not in the line of sight. According to Einstein's 1915 general theory of relativity, when the sun is in the line of sight, then light from the star in question bends so that there is a discrepancy of θ. The observer's telescope would then have to be pointed along the direction of the bent rays, and so the star would appear in its apparent position. Comparison of apparent and true positions gives the angle of deviation, θ. (Deflection exaggerated.)

phasizing data from Principe. Clearly, a key part of this experiment is to use the same instruments at the same site for comparing data for the apparent and true star positions, but this was not done. The comparison photographs were instead taken several months later at Oxford.

To be sure, this is an extremely complicated experiment in which accurate comparison photographs are essential. Even if the comparison photographs were taken at Sobral and Principe six months later, the equipment would have undergone changes in attitude owing to weather conditions. Indeed, the atmosphere at these two sites would also have changed. All in all, the errors in the data are of the order of 10 percent, or 1.5 standard deviations.*

The results of the Joint Permanent Eclipse Committee were presented at a historic session of the Royal Astronomical Society on November 6, 1919. Newton's portrait hung in its place of honor as Dyson reported that Einstein's theory was favored over Newton's. J. J. Thomson, discoverer of the electron, Nobel laureate, and at the time President of the Royal Society, hailed Einstein's theory as "one of the greatest—perhaps the greatest—of achievements in the history of human thought."

The results had an aura of inevitability. The next day all errors disappeared and the *Times of London* reported a "Revolution in Science . . . Newtonian Ideas Overthrown." Einstein was deified. I think it is easy to see what happened. To the war-weary world appeared a man who, sitting in his study, had apparently divined new laws of the universe. The stars were not where they were supposed to be. But, he locates them for you in a four-dimensional warped space-time. What could be more esoteric? To add intrigue and authority, the man looks like he glimpsed the Creation itself. A sociological phenomenon has occurred.

At least one other canonical experiment is surprisingly flimsy. In 1887 Albert A. Michelson and Edward W. Morley performed an etherdrift experiment that promised to measure effects of the ether on light rays whose source is on the moving Earth. Already Michelson had the reputation of a paragon of accuracy. In 1907 he was awarded a Nobel prize for his high-precision measurements. The world of physics was surprised when he measured no change in the direction of light rays whose source is on the Earth, which is supposedly moving through the

---

* Two other experiments proposed by Einstein to test his theory have been verified to high accuracy. They are the advance of the planet Mercury's perihelion (point of its orbit furthest from the sun) and a displacement toward the red of spectral lines from the sun (the so-called red shift).

ether; there seemed to be no ether drift. But not only was Michelson and Morley's accuracy not as good as they claimed, they essentially never completed the experiment. They should have conducted it over at least six months to achieve maximal effect of the ether on light rays. Instead, they took data over only five days, for a total of just six hours! Nevertheless, owing to Michelson's reputation as an experimenter par excellence, all errors were soon forgotten and the Michelson-Morley experiment became a true null experiment, which is that rare jewel of a high-precision experiment that measures absolutely nothing.

Thus far we have found that while the data-gathering stage of science is far from straightforward, the interpretation phase by the scientific community and the lay public can be even more complicated.

Assuming we have acceptable data, proceeding requires generalization, for which mathematics is the proper tool. But we cannot write down equations in any random fashion. There are guidelines for proceeding. However, taking certain statements as guidelines means that we agree not to question their veracity. We agree not to question their origin, not to wonder on whose authority they are elevated to axiomatic status, not to ponder how scientists uncovered these guidelines. The route to some of these guidelines was circuitous, almost magical. Arthur Koestler was not far from the mark when he described creative scientists as sometimes behaving like sleepwalkers, stumbling onto momentous discoveries. But the times must be right and the mind prepared, and a little bit of luck helps, too. Two such axioms are the principle of relativity and the conservation of energy, to which we now turn.

THE PRINCIPLE OF RELATIVITY

An appreciation of the importance of the idea of the relative nature of motion is old. We can imagine its message brought home to Galileo in Pisa. Many times Galileo crossed and recrossed bridges over the Arno River. No doubt he took the opportunity to rest for a moment in the glorious Italian sunshine and contemplate a sailor on a boat passing under the bridge. Galileo might have wondered what the sailor's view was like. The only difference between their platforms was their relative velocity. The sailor might easily imagine himself to be at rest, with Galileo moving away in the opposite direction. This being the case, then all phenomena must be seen as the same for both of them.

Ultimately, however, Galileo's moving platforms were ships moving on a sea "as smooth as glass," as he wrote in the *Dialog*. Although Galileo advocated overturning a centuries-old view of the universe, his idea of a continuous motion was one in a circle and not a straight line. *Philosophically*, the reason is Galileo's insistence on continuity with Greek thought, in which circular motion is the preferred one. In science, continuity is of the essence. Theories emerge one from the other, continuously. But Galileo was a man of the sixteenth century, and so there was another equally important reason: *Theologically*, it is the horror of a straight-line motion going on forever. To paraphrase the historian of science Alexandre Koyré, to incorporate a straight line, Galileo would have had to move from a finite to an infinite universe. To remove the Earth from the center of the universe and set it moving about the sun is one thing; to be lost in an infinite void is quite another. Although some 2,400 years had passed since the days of Hesiod and Homer, superstition and fear of the unknown still gripped the Renaissance mind. Stories of the Creation and of our place in the scheme of things had not yet been stripped completely of myth. As Johannes Kepler wrote in his *De stella nova* of 1607:

> The very cogitation [of an infinite universe] carries with it I don't know what hidden horror; indeed one finds oneself wandering in this immensity, to which are denied limits and center and therefore also all determinate places.

This lack of "all determinate places" as a result of straight-line motion at constant velocity forever—inertial motion—is a concise statement of the principle of relativity, and it evoked horror in the Renaissance mind.

Although Kepler was prepared to defend a sun-centered universe, an infinitely large one was quite another matter. In an infinite universe, any point is the center, and so there is no center. We wander through the void like lost souls. This was a serious matter. In 1600, the Dominican priest Giordano Bruno was burned at the stake in Rome for speculating in print on an infinite universe, a rotating sun, and the possibility that the fixed stars are actually suns. All of this has been found to be true. Totalitarian regimes, be they political or religious, demand fearsome retribution for differing views.

Newton properly set down the concept of inertial motion, the central idea of the principle of relativity, but it caused him great angst. Early on in the *Principia*, which he wrote ultimately for theological reasons, Newton states that his goal is to distinguish between "true" or "absolute motions" that are relative to God's sensorium, from the ones relative to us: "For to this end it was that I composed it." Newton wrote in a letter of December 10, 1692, to Richard Bentley that he "had an eye upon such principles as might work for considering men for the belief of a Deity, and nothing can rejoice me more than to find it useful for that purpose." Among these principles are ones concerning relative and absolute motions. Although for the purposes of scientific calculation, Newton's principle of relativity (Corollary V in the *Principia*) demands an infinite universe, Newton thought of the universe as God's sensorium, or God's playground, which cannot be infinite. If it were, he reasoned, how would souls reach the new city of Jerusalem on Judgment Day? Newton agonized over the tension between physics and theology.

As one of Newton's biographers, Frank Manuel, writes somewhat lyrically:

> Throughout his life Newton arrived at his conclusions whether in physical science or world chronology, by intuition—by the intrusion of the Holy Spirit—after long concentration upon a single idea; he was in direct relationship with his [dead] father, and things were revealed to him as

they had once been to the Hebrew prophets and the apostles and the legendary scientists of the ancient world whom he identified with one another. Though Newton could speak freely in praise of Moses and Toth, Thales, Pythagoras, Prometheus and Chiron the Centaur, he only rarely had good words to say about either living scientists or his immediate predecessors—and then not without equivocation. History had begun anew with him.

So much for the image of Newton as originator of the Age of Rationalism.

Whereas Newton's theological musings were essential to his creative processes, Newtonian mechanics was cleansed of them mainly through Kant's raising Newton's concepts of space and time to properties of the intellect, and Pierre Simon de Laplace's setting Napoleon Bonaparte straight on the role of God in physics: "Sir, I have no need of that hypothesis."

The story of the principle of relativity is one of scientists at odds with their scientific beliefs. Nevertheless, these scientists fervently believed that any mathematical generalization of empirical data must conform with the principle of relativity, and they proposed mathematical recipes for this purpose.

CONSERVATION OF ENERGY

Another powerful guideline for creating scientific theories is the law of conservation of energy, which has origins at least as strange as those of the principle of relativity. It is rooted in *Nature Philosophy*, a view of phenomena that emerged from central Europe, where science lay dormant for 100 years after Kepler's death in 1630. The dominant rationalism of the French and English versions of Newtonian science repelled the Germans, who were attracted instead by mysterious forces of vitalism with roots in alchemy as well as in primitive forms of Christianity practiced in central Europe. From these two pursuits of knowledge emerged Nature Philosophy, based on two tenets:

1. All processes in nature are the result of struggles between opposites, such as light and darkness, good and evil, death and regeneration, or, more generally, thesis and antithesis.

2. There is a unifying vital force in nature to which all forces can be reduced. This world spirit permeates the universe.

Despite their vagueness and mystery, these principles offered themes that are still pursued in modern-day physics, such as the quest for unity of forces. They are a restatement of goals set down thousands of years before.

The second tenet led the Danish scientist Hans Christian Oersted, in 1820, to discover the generation of magnetism by electricity. While preparing a demonstration for his physics class concerning the effects of electricity running in a wire, Oersted noted that the orientation of magnetic compasses in the wire's vicinity suddenly shifted from due magnetic north when the current was turned on. Others must have noticed this phenomenon, but Nature Philosophy prepared Oersted's mind to "see" its deep structure.

Since all processes in nature have amounts of energy associated with them, it is logical for these energies to be related. Research by Huygens and the German polymath Johann Gottfried Leibnitz had early on indicated the fertility of the concepts of potential and kinetic energies in mechanics, even though Newton never fully grasped the point. The concept of energy is extremely valuable for problem solving, and more importantly for understanding phenomena. For instance, if a block of cement is raised off the ground and then held motionless, it has the potential of doing something, namely falling. Here the term "potential" is not meant in the Aristotelian sense of "potentia," in which the object will do something "in order that." The cement block has a potential energy that depends on its height off the floor. We can give it a numerical value of, say, one energy unit. If we drop the object, it will develop a velocity and so an amount of kinetic energy. But whatever its height from the floor, it will have also some potential energy. The object's *total* energy is the sum of its potential and kinetic energies. The key point here is that if we calculate and then add up the kinetic and potential energies, the result will be one energy unit, which was the object's energy before it began falling.

Two things have happened:

1. Initially, the object possessed only potential energy in the amount of one unit. When it was dropped, some of this potential energy was *transformed* into kinetic energy.

2. This transformation occurred in such a way that the total is always one energy unit. *Energy is conserved.*

The proviso here is that we do not disturb the object in any way. If we hit it on the way down, then the problem situation changes and we have to begin all over again with a new initial energy, which, this time, is kinetic and potential.

The realization that something very basic has just occurred comes when we ask ourselves whether we have actually *derived* this conservation law. This question was asked by Christiaan Huygens, who first discovered a form of this law and then drew back from its far-reaching consequences. As the historian of science Robert S. Westfall writes,

> When [he] encountered Leibnitz's conservation of [kinetic energy] for the first time, however, Huygens' response had all the enthusiasm of a dry cough. "But he cannot pretend that this principle of the conservation of motive forces will be granted as though it had no need of proof."

In fact, it cannot be proven. Studying this situation, the historian of science Charles Gillispie perceptively wrote that "conservation laws are not discovered in the laboratory." Scientists either notice or assume the existence of regularities in nature, and then they see what follows. This is perhaps best illustrated in the generalization of the energy conservation law in the nineteenth century to include heat.

HEAT IS ENERGY

Huygens and Leibnitz restricted their conservation laws to mechanical processes, where teasing out a conserved quantity was more straightforward than dealing with heat. By the 1820s, though, steam engine technology had achieved enormous successes. The ancestors of steam engines appeared with artillery in the early fourteenth century. In a fascinating set of essays on the history of technology, Lynn White rightly points out that "the cannon is a one-cylinder internal-combustion engine, and all modern motors of this type are technically descended from it." White also shows that throughout early developments in heat engines, theoretical science was essentially of no use. Steam engines were the province of iron mongers such as the ingenious Thomas Newcomen in the eighteenth century. Science entered when it came time to build bigger and more efficient engines, which required understanding the nature of heat.

With the second tenet of Nature Philosophy in mind, as well as other homespun philosophical notions such as conservation of cause and effect, in the 1840s the German physician Julius Robert Mayer proposed that heat is just another form of energy. Mayer backed up this claim with a very basic calculation, the intent of which was to force a relation between the work done by a piston while compressing a gas, and the change in the gas's temperature. In physics, "work" is equivalent to a change in energy. Mayer's hypothesis led to three findings:

1. With notions from way beyond science per se, Mayer came up with the far-reaching idea that heat is a form of energy.

2. This being the case, there can be conversions between mechanical and heat energy.

3. Consequently, the energy conservation law must be expanded to include heat energy.

Mayer's results were not received with great acclaim. Their wildly speculative aura made scientists shy away. Only much later would

Mayer be awarded the fame due him. Instead, the honors went to scientists like James Prescott Joule. Independently and simultaneously with Mayer, but with more solid experimental techniques and no philosophical content, Joule achieved early recognition for assuming the equivalence between heat and energy. In fact, the unit of energy in the meter–kilogram–second system is the joule. This is a good example of the occasional resistance of scientists to unorthodox trains of thought. The priority disputes that ensued between supporters of Joule and Mayer exemplify the seamier side of science. Like composers of music and literature, scientists take great pride in seeing their names in print and being considered the first to make a discovery.

Equivalence is the key word. Mayer and Joule *forced* a relation between the units of mechanical work and heat. They *defined* a way to convert the unit of heat to that of energy. This was acceptable because the conversion factor between inches and centimeters contains nothing deep at all, but equating the units of heat and energy means more than merely equating two numbers. It means these two processes are one and the same, and heat is a form of energy.

But one cannot go around forcing relationships between any sort of quantities. Rather, one must see what follows in order to ascertain whether they are just shotgun marriages. What followed from Mayer's work is a version of the energy conservation law that could be widened immediately to include all other forms of the then-known energies: mechanical, electromagnetic, gravitational, and solar. The energy conservation law includes every known energy, which it reduces to a single entity—energy. This is another instance in modern science where the programmatic intent of Nature Philosophy succeeded, although Nature Philosophy itself disappeared from the German scientific literature after about 1854, when Hermann von Helmholtz set down the energy conservation law in the form in which it is used today.

Von Helmholtz formulated and used the energy conservation law in a strictly rational manner, that is, couched in mathematics. This conservation law is a guideline for theorizing, by which I mean a guideline for dealing with data. Any mathematical generalization of data must satisfy conservation of energy. There are mathematical techniques to ensure this.

Sometimes the various pedigrees of energy are not so easily identifiable and the most we can say, to quote Henri Poincaré, is that "there is something that remains constant." Poincaré meant that the total energy in a system had better remain constant. If it does not, either a mistake was made in computing the energies involved, or a new and heretofore unnoticed form of energy may exist and needs to be explored. Poincaré's colleague Henri Becquerel faced this choice while experimenting with uranium in 1896. It seemed as though the small samples of materials with which he and subsequently the Curies, Pierre and Marie, were investigating had virtually inexhaustible amounts of energy. Perhaps energy is not conserved in radioactive materials. Further research by Rutherford and colleagues at McGill University in Montreal revealed radioactivity to be a new sort of energy. This result led to the understanding of radioactive decay as a means for certain atoms to become more stable as they deexcite by moving down the periodic table.

The energy conservation law, in summary, enables scientists to group the known energies of apparently disparate processes under one law and to uncover new energies. Attempts to formulate theories that violate energy conservation have been disastrous.

Regarding the rationality of the scientific enterprise, we have found extra-scientific considerations entering at every level. In the data-gathering stage we saw Poincaré's lyrical assessment of how data are chosen and the influence of both theories and philosophies on Kaufmann and Thomson, Millikan and Ehrenhaft. In the formation of a theory, we saw how the mystical belief system of Nature Philosophy and the homespun philosophy of Mayer led to his realization that heat is a form of energy. We saw how personal loyalties have led scientists to defend or oppose theories: Heisenberg versus Schrödinger, Abraham and Kaufmann versus Planck, and Einstein's pretty much ignoring everyone. And we saw how inexact and even erroneous data can lead to the proper theory: Galileo's flawed pendulum experiments, his erroneous reasoning based on observing Venus's phases, and his telescopic observations of Saturn; and the hasty acceptance of Michelson and Morley's ether-drift results as well as Eddington's confirmation of general relativity. We noted how Bucherer's experimental data turned out to be incorrect for the purpose he intended, even though everyone assumed they had been substantiated in subsequent high-definition experiments. But by then—1938—it was a moot point whether special relativity was consistent with the electron's mass variation. The sociological component in defending a theory is nicely shown in the prestigious physicists Feynman and Gell-Mann in defense of their theory against supposedly high-quality experimental data, which they declared to be wrong.

What emerges from this chapter is that there is no strict rationality in scientific discourse. Scientists sometimes argue from incorrect premises to correct results. Even in the course of debates, the bases of their arguments are sometimes dubious or completely intuitive. And sometimes the winner turns out to be the loser.

We have discussed two guidelines for theorizing, the principle of relativity and conservation of energy. Are these cut and dried (= rationalistic) constraints also constraints on creativity? Do the rules by which music is written limit creativity?

UNIFICATION OF THE SCIENCES

Scientists develop theories that claim to unify processes that seem entirely separate. The very first attempt at unification was made by the man who invented science, Thales of Miletus.

The Greeks lived in a world populated by gods, with human frailties, who every now and then dropped down to Earth to influence the course of human events, particularly wars. We know about this from Homer's poems *The Odyssey* and *The Iliad*. It must have been enormously unsettling to believe oneself at the mercy of capricious forces of nature beyond one's wildest imagination. Perhaps it was the desire to strip away myth from creation stories that led Thales to invent science. As the scholar of ancient Greek philosophy, F. M. Crawford, writes:

> The Milesian system pushed back to the very beginning of things the operation of processes as familiar and ordinary as a shower of rain. It made the formation of the world no longer a supernatural, but a natural event. Thanks to the Ionians, and to no one else, this has become the universal theme of all modern science.

Thales was quite specific as to how science ought to be conducted: A maximum amount of phenomena should be explained by a minimum number of hypotheses. This statement, sometimes called Ockham's razor after the medieval philosopher William of Ockham, remains a crucial guideline in scientific research. Of course, what constitutes the minimum number of hypotheses has never been agreed upon.

Thales and his followers, the Milesian philosophers, went even further. They proposed that there ought to be a single fundamental substance. The oneness or unity in nature was a notion not unique to Thales. It was preached in a variety of ways by such contemporaries of Thales as Siddhartha in India, who became known as the Buddha, and by the Hebrews. And we must not forget that Thales was familiar with the *Enuma Elish* (the Babylonian *Genesis*), written in the second millennium B.C., which describes how it is that superhuman gods emerged from a primordial sea to create our world and us. Although Thales was off the mark regarding the nature of the primordial substance, it was the goal that counted.

Newton offered the first great synthesis of modern science in his *Principia*, where one set of laws accepted as axiomatic referred to phenomena both on the Earth and in the heavens. We have discussed the work of Oersted and Mayer. In England, Michael Faraday, without the Nature Philosophical basis of the Germans, also believed in the unity of electricity and magnetism. In 1831, he established the direction that would pay off handsomely by generating electricity from magnetism. When asked by the then Prime Minister Robert Peel what use could be made of his tabletop electrical dynamos, with great prescience Faraday replied, "I know not, but I wager that one day your government will tax it."

But Faraday was not wholly without philosophical intent. As a young man, he became enamored of a version of Kant's philosophy extolled with great force by the poet Samuel Taylor Coleridge. It was based on the universe being filled with a web of attractive and repulsive forces, convertible one into the other, with their total being conserved. At the bottom, this cosmic web was woven by God. Everything is here in nascent form, including forces to fill the vacuum, as well as conservation of energy. Faraday was particularly taken by Coleridge's statement that "Things identical must be convertible." He believed a relation between electrical and magnetic forces had to exist. As Faraday's biographer L. Pierce Williams writes, "It was the conviction that forces were inherently identical and convertible that inspired Michael Faraday during the major portion of his scientific career."

Sometimes preconceived philosophical notions are not necessary to spot unities. A good hunch will do, or sometimes a good eye for data, as the following example shows. By the 1830s the velocity of light had been measured to be on the order of 3.15×10^8 m/sec. By comparison of electromagnetic and electrostatic units (that is, by indirect means), the velocity of electrical disturbances in wires was calculated to be 3.11×10^8 m/sec. Almost 30 years after these measurements had been in the open scientific literature, only one person, the Scottish scientist James Clerk Maxwell, was struck by the similarity of the two huge numbers. Maxwell concluded in 1861

> that we can scarcely avoid the inference that *light consists in the transverse undulations of the same medium [ether] which is the cause of electric and magnetic phenomena* (Maxwell's emphasis).

What understatement! With this connection, Maxwell formulated a magnificent electromagnetic theory of light in which he postulated unification of light and electromagnetism. Maxwell's electromagnetic theory distinguished itself from all competing theories by predicting electromagnetic waves, verified in 1889 by Heinrich Hertz. Wireless telegraphy came into being overnight.

But not every postulated unification works. Take, for example, the electromagnetic world-picture, which failed because it could not incorporate the wave and particle nature of light. Another failure was Einstein's long-term efforts, beginning in the 1920s, to formulate a unified theory of electromagnetism and gravity. By the 1930s, the discovery of other forces such as the nuclear force had turned this into a quixotic quest. In the late 1960s, Abdus Salam and Steven Weinberg succeeded in unifying the weak and electromagnetic forces. Some of their reasoning will be discussed later. For here it suffices to note that their electroweak theory predicted certain processes hitherto undetected and a new particle, the Z^0, which was subsequently discovered at CERN. For their assumption of unification, Salam and Weinberg were awarded the Nobel prize in 1979.

In the first three chapters of this book, we have examined physics from the most ancient of times almost to the present. We have studied the importance of visual representations in classical and early quantum physics, and then the methods of science itself. These results bear heavily on scientific representations in the twentieth century. To see how this occurred and what its ramifications were for visual representations in science and art, in the next chapter we explore the assumption of an ordered universe.

4

FAITH

IN

AN

ORDERED

UNIVERSE

B ehind all our discussions about science is perhaps the biggest assumption of them all: that the universe is ordered. An ordered universe is one in which causes precede effects, so causality or predictability holds. In 1911, for example, in the face of emerging discontinuities such as those in Planck's theory of cavity radiation, Poincaré asked anxiously, "Will discontinuity reign over the physical universe?" He hoped not.

Happily, we inhabit a benign part of the universe where irregularities from apparently circular motion occur smoothly. For example, Mars sometimes loops backward in a smooth manner instead of zig-zagging across the sky. This continuity in motion permitted Claudius Ptolemy in 150 A.D. to reproduce the observed planetary motions by compounding periodic motions. In the 1820s, the French mathematician Jean Baptiste Joseph Fourier demonstrated how any sort of motion can be generated from many periodic ones. Continuity is one of the basic assumptions of classical physics. Leibnitz emphasized an underlying continuity in nature with the statement *natura non facit saltum* [nature makes no jumps]. Subsequent advances in the physical sciences began to gnaw away at Leibnitz's desideratum. Underlying continuity was threatened by atoms and finally ruptured by quantum physics.

In order to probe these topics, we move to another promise of Newton and his followers, indeed of what became of classical physics and certainty, with the emergence of Bohr's atomic theory.

CERTAINTY AND/OR KNOWLEDGE?

One of Galileo's great achievements was to realize that the Aristotelian universe precluded any systematic mathematical formulation of a theory of motion. As a step toward alleviating this situation, Galileo suggested separating "primary" from "secondary" qualities. Primary qualities can be described mathematically and so are desirable in a scientific theory. Some examples are velocity and acceleration. Secondary qualities cannot be translated into mathematics. Examples of such subjective entities are tendencies, essences, and odors. With this distinction in hand, Galileo could declare a separation between matter dead and alive, with the former being grist for mathematical theories of motion in which predictions over time intervals prevail. The cannoneer is well aware that the cannon ball will return to Earth, which is predicted by Aristotelian teleology. The main problem is where it will land.

Because it contains no proper concept of time, Aristotelian physics was incapable of predicting the projectile's trajectory. It considers time to be inseparable from its method of measurement, be it the motion of stars or of clocks. Galileo took the giant step of abstracting time as a quantity to be used in mathematical equations that discuss phenomena that are not necessarily cyclical, such as in problems of falling bodies. He assumed time to be on a mathematical par with space, and thus useful as a coordinate. Only in this way could a physics be formulated which is capable of making predictions in time and space.

Newtonian science offers the spectacular prospect of absolute certainty. The fundamental assumption here is that the initial conditions can be ascertained to any accuracy subject only to systematic errors such as the width of crosshairs in a telescope. In principle, inserting a system's initial conditions (initial position and velocity) into Newton's laws of motion produces an equation capable of predicting, with perfect accuracy, the system's subsequent motion in space and time. This is Newtonian causality. Since the system will occupy any point on its path with certainty, Newton's science is said to be deterministic, too. In Newton's mechanics, causality and determinism are one and the same. All experimental errors are assumed to arise from imperfections in the measuring apparatus and are referred to as systematic errors, as we discussed in Chapter 1. Consequently, any physical entity is assumed, in principle, to be open to our knowledge with perfect accuracy.

Laplace considers the possibility of a completely determined universe in his book *The Analytic Theory of Probabilities*, published in 1814. He describes a superintelligent being with the ability to know at a glance the velocity and position of every object in the universe. Since this intelligence was also expert in Newton's physics, the "past and future alike would be before its eyes." Clearly, this is only an ideal, because the complete state of the entire universe at any given moment cannot be known. Yet scientists clung to the notion of a completely determined universe, perhaps because it gives the same comfort as the geocentric universe. The story of Laplace's omniscient being fits a recurrent definition of science as the control of nature. The Laplacean view remained an illustration of the extremes of Newtonian science until twentieth-century atomic physics revealed it to be merely a good story.

THE SIGN OF THE TIME

Cracks in the edifice of Newtonian causality were noticed in the latter part of the nineteenth century, when such figures as William Thomson, raised to the peerage as Lord Kelvin, looked afresh at its foundations. In the 1840s, Thomson found something rather peculiar and astounding, and its realization needed no mathematical pyrotechnics. Newton's second law of motion contains an interesting and disturbing symmetry: When the sign of the time is made negative, Newton's second law retains its form as force equals mass times acceleration. Scientists find this sort of symmetry interesting because, in this case, it means that Newton's second law is unchanged under time reversal. If the second law completely describes the physical universe, we ought to see phenomena running in reverse everywhere. We ought to see waterfalls running backward, bubbles deflating, and divers returning to their diving boards from the water. In short, we ought to be able to see life as if a film were running backward. Instead, it is as if there is an arrow of time pointing relentlessly in one direction. Consequently, concluded Thomson, something is missing in Newton's mechanics. The identity of this something had already been glimpsed in 1824 by the young French engineer Sadi Carnot.

CARNOT'S IMAGINARY ENGINE

The truism that science owes more to the steam engine than the reverse is nowhere more evident than in Carnot's research. In his 1824 book *The Motive Power of Fire*, Carnot explores the problem of a steam engine's efficiency. At the time, the popular concept of heat was metaphoric: Heat flowed *as if* it were a fluid called caloric. But this was more than mere analogy. It was taken as pretty close to the real thing. After all, it is commonsensical to say that heat "flows." We do it all the time. Fluids like caloric were extremely useful in the infancy of such subjects as heat and electricity. These phenomena are less amenable to the sort of reasoning in Newton's mechanics that deals with localized entities we can touch and see, and envisioning them as fluids offered a way around this problem.*

In the second of the *Principia's* three books, Newton systematized the study of fluids—hydrodynamics. Consequently, treating heat and electricity as due to the flow of some sort of fluid made them accessible to the mathematical methods of hydrodynamics and allowed these sciences to get off the ground mathematically. Remnants from eighteenth-century theories of electricity persist in our vocabulary: "Current" and "resistance" both evoke images of fluids flowing. Voltage was referred to as pressure, and batteries were envisioned as pushing current through a circuit. Although these fluids would, in time, be discarded, they played an essential role in quantifying electricity and heat. Even when scientific theories turn out to be untenable, something in them often survives. For Carnot, it would be the second law of thermodynamics.

Carnot conceived of a thought experiment in which gas in a cylinder is expanded and compressed with a piston while transferring heat from a hot reservoir to a cooler one. In Carnot's view, heat flow is the same as caloric flow, which must be conserved. This means that all the heat taken from the hot reservoir must be transferred to the cooler one (Figure 1).

Although this assumption turned out to be incorrect, the faulty reasoning nevertheless permitted Carnot to relate heat flow to work the piston does in expanding and compressing the gas. Carnot found that an engine can do work only by transferring heat between hotter and cooler reservoirs. Once equality of temperatures occurs, no more work can be done, and so the engine stops. Carnot's result is universal in nature because it is independent of what sort of gas is in the cylinder. He

---

* For this reason, Newton's mechanics held a privileged position in the sciences and there were attempts to reduce all explanations of matter in motion to it. The philosopher-scientist Ernst Mach pungently pointed out that the heretofore assumed primacy of mechanics was a historical accident. But Mach notwithstanding, it was a good accident because it offered to such developing sciences as heat and electricity a means to become established mathematically.

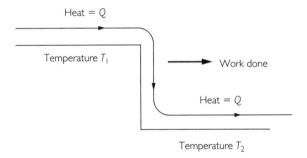

FIGURE 1.

Carnot's original concept of a universal heat engine. The analogy is with water flowing over a waterfall and turning a wheel, thereby performing work. For Carnot, an amount of heat Q obtained from a heat reservoir at temperature T_1 flows to a reservoir at a lower temperature T_2. Work is performed while the quantity of heat remains unchanged. Although conservation of heat turned out to be incorrect, Carnot's basic assumption—that when both reservoirs reach the same temperature, no further work is accomplished—is correct and profound.

inferred that only the ideal engine of the thought experimenter can be 100 percent efficient. The engines of our world must lose heat; for example, through friction.

Idealizations, or extreme abstractions, are the essence of theoretical physics. Galileo abstracted to the world of vacuums, which turned out to be an actual world. Newton set the magnificent framework of his mechanics on the concept of an inertial reference system, which does not exist. The Carnot engine is to thermodynamics what the inertial reference system is to Newtonian mechanics and special relativity. (Another abstract engine that underlines the importance of abstraction is the Turing machine, which is, in principle, every modern computer.)

Carnot soon realized his error concerning conservation of caloric, and he abandoned the concept. He rephrased his result to say that due to the work done by an engine, a lesser amount of heat is given to the cooler reservoir. As before, the two reservoirs will eventually reach the same temperature, and no further work can be accomplished. All of this rests on the empirical fact that heat flows and coolness does not. Carnot's research made it clear that certain processes in nature are irreversible. The story does not end here, because Carnot's results contain the one-way arrow of time missing from Newtonian mechanics.

The first step toward this stunning result is due to work done in the late 1840s and early 1850s by William Thomson and the German physicist Rudolf Clausius. They realized that there are two independent laws of thermodynamics:

1. Energy is indestructible, that is, energy is conserved. Joule's investigations had been instrumental in this work by showing the mutual convertibility of different forms of energy and also in demonstrating that heat is not a substance but a form of motion.

2. There is a universal tendency in nature to the dissipation of mechanical energy. Stated otherwise and equivalently: The *tendency* of heat to flow from hot to cold makes it impossible to obtain the maximum amount of mechanical work from a given amount of heat.

The dissipation of energy seems to impress a directional character on the universe, an arrow of time. The ultimate result of the dissipation principle is that all mechanical energy will be turned into heat and, as von Helmholtz and others prophetically put it, "The universe from that time forward would be condemned to a state of eternal rest." Clausius referred to this condition as "heat death." According to Carnot, when the entire universe reaches the same very low temperature, no more work can be done, and the universe will literally come to a standstill. Along with this pessimistic long-range prediction, the concept of probability, with its incipient uncertainty, entered physics immediately.

THE BIRTH OF UNCERTAINTY

Max Planck's exploration of the microscopic roots for his law of cavity radiation laid bare the shaky foundations of what would soon be called classical physics, in comparison with the modern physics of relativity and quantum theory. For a while physicists politely ignored Planck's theory, whose inherent discontinuities were anathema to the more commonsensical continuity of Newton's mechanics and Maxwell's electromagnetic theory. A similar situation occurred some 50 years earlier in the 1850s, when Clausius investigated, as he phrased it, in 1857, "the kind of motion we call heat." By this he describes heat as a measure of the kinetic energy of atoms. How did Clausius proceed, given that there was no atomic theory available for him?

Clausius focused on the second law of thermodynamics, which he set out to formulate specifically in terms of directionality of heat flow. Being an amateur linguist, he liked to invent scientific terms that could be transliterated from German into English. Sometimes he succeeded. In 1854, to denote a measure of time directionality, he coined the term *Entropische*, or "entropy" (meaning transformation): The entropy of an isolated system always tends to increase. For example, left to themselves, gas molecules in a corner of a box will eventually spread out throughout the box rather than remaining clumped together. There are more possible configurations of the gas molecules spread out throughout the box, and so this configuration is one of highest entropy, or disorder.

Clausius restated the two laws of thermodynamics to say that, first, the energy of the universe is constant; and second, the entropy of the universe *tends* toward a maximum. In 1862, Clausius attempted to set up a molecular basis for entropy but collided with the reversibility paradox. We discussed this earlier in the work of Thomson in which he realized the time-reversal symmetry of Newton's laws of motion yet noted that we do not observe such phenomena.

The problem that beset Clausius was to deduce certain well-established results of thermodynamics from the assumption that gases consist of molecules. One such result is the relation between a gas's pressure, volume, and temperature, known as the ideal gas law. This law states that the product of a gas's volume and pressure varies directly with changes in temperature, and so volume and pressure vary inversely with each other. The quantities—pressure, volume, and temperature—do not at all depend on the gas's constitution being atomic. This is a fundamental characteristic of the laws of thermodynamics: They are independent of the gas's constitution. An attempt to provide a microscopic basis for the ideal gas law immediately encounters the problem that any given amount of gas contains a huge number of molecules. Some model for the gas molecule is required, and the usual one had it behaving like a billiard ball. The goal was to connect the three quantities in the ideal gas law (pressure, volume, and temperature) through considerations of how billiard-ball molecules strike the container walls and each other. This involves calculating how their kinetic energy changes. Because of the huge number of molecules involved, Clausius used methods of Newtonian mechanics and certain averaging techniques. But the reversibility paradox rears its head as soon as Newton's mechanics is used.

James Clerk Maxwell, and subsequently the Viennese physicist Ludwig Boltzmann, also pursued the problems in gas theory that preoccupied Clausius, but from a very different viewpoint. As Josiah Williard Gibbs, the late nineteenth-century American expert in thermodynamics, wrote while laboring in splendid isolation at a quaint New England college called Yale: "In reading Clausius, we seem to be reading mechanics; in reading Maxwell, and in much of Boltzmann's most valuable work, we seem rather to be reading in the theory of probabilities." Probability was something truly new in physics. It entered as follows, according to Maxwell. Since there are a large number of molecules, then it is difficult to calculate a definite value for the velocity of each one. We can, however, say that the molecules in a gas can have any one of a large number of available velocities. For each temperature, there is a distribution of possible velocities in which certain ones are more probable than others. Maxwell set up methods for calculating the probability of an individual molecule in a gas at some definite temperature to have a certain velocity. With other assumptions specifically designed to obtain the ideal gas law, the science of *statistical* mechanics was born. Among these assumptions is that, in principle, individual gas molecules can be followed throughout their motions.

But if probability is part of the second law of thermodynamics, then is there a probability that the law can be violated? Maxwell offered a thought experiment to show how it can be done. He considered a box whose volume is separated into two parts by a wall. On each side of the divider, the gas is at a different temperature. The molecules on each side can have a distribution of velocities, from zero up to some maximum value that increases with the gas's temperature and then tails off to zero again. Consequently, some gas molecules on the lower-temperature side can have a greater velocity than some on the higher-temperature side. Maxwell imagined that there is a sliding partition with a weightless door so that no work or change in energy of the system occurs when the door is opened and shut (Figure 2).

He imagined further than the sliding partition is operated by a being with incredibly quick reflexes. This "Maxwell demon" performs the following feats. When it sees a molecule in the higher-temperature gas approaching the partition at a very low velocity, it opens the door and lets the molecule through. Conversely, the demon permits higher-velocity molecules from the low-temperature side through. At the end, the same number of molecules remain on each side: The demon has arranged for the hot side to grow hotter and the cool side cooler *without doing any mechanical work*. This would be akin to a cup of hot coffee on your desk becoming hotter, which of course does not happen. Yet Maxwell's thought experiment shows that there is a *probability* for it to occur. This violates the second law of thermodynamics, making it a universal law of nature that can be violated.

The core of this baffling situation is evident by noting something else that has occurred during this process: By arranging for all the faster molecules to be on one side and the slower ones on the other, the demon has transformed a disordered collection of molecules into an ordered one. Since an ingenious contrivance is required to move a system from a disordered to an ordered state, we would expect that the dissipa-

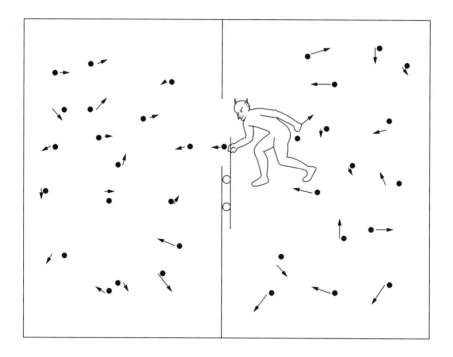

FIGURE 2.

A box containing gas molecules is divided into parts. The gases in the right and left portions are at different temperatures. A being with incredibly sharp reflexes—a Maxwell demon—operates a sliding partition in such a way that the cooler gas becomes cooler and the hotter one hotter.

tion in energy we *usually* see in nature entails increasing a system's disorder. Further exploration into what the concept of entropy means will clarify why I stressed the word "usually."

Clausius introduced the term "entropy" to be the measure of a system's tendency toward disorder. The more disordered a system, the higher its entropy. Shuffling a new deck of cards offers a good illustration of what entropy means. A newly purchased deck of 52 cards is in its lowest entropy state because it is ordered as to suit and number value. One shuffle throws the deck into some disorder, increasing its entropy. But the second shuffle can return the deck to its ordered state: Such a feat is improbable but *not* impossible. It is more probable that each successive shuffle will further disorder the deck, which is the assumption underlying card games. Yet each shuffle beyond the first can potentially restore the deck to its original order. The point here is the overwhelming tendency for the deck to reach a disordered state because there are many, many more possible configurations of disorder than order. In fact, the deck has only one ordered configuration.

Another example involves injecting a drop of black ink into a glass of water. The ink drop spreads out, and very soon the water turns black (Figure 3). The system has been transformed from one of order (ink drop with a well-defined boundary) to one of disorder (ink diffused throughout the water). The state of disorder is more probable because it contains more possible configurations for the molecules constituting the ink drop. Yet a very small probability exists for the ink drop to coalesce into its original shape, giving the appearance of time running backward—locally, that is.

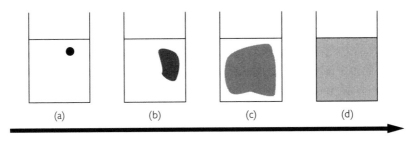

Arrow of time
(Most probable direction for time)

FIGURE 3.

An ink drop is injected into a container of water in (a). The sequence (b), (c), (d), in which the ink drop spreads out into the water, is the most highly probable one because it offers more possible configurations than, say, for the ink drop to remain localized. Consequently, the sequence (a), (b), (c), (d) provides an arrow for time. It is highly improbable that at some later time the system will run through the sequence (d), (c), (b), (a), thereby reversing time locally.

One further example utilizes a box with a partition, on one side of which is a gas. If a hole was punched in the partition, we would expect the gas molecules to distribute themselves throughout the entire volume available to them because increasing the volume increases the configurations available.

In these three examples, the one direction that is overwhelmingly probable coincides with the arrow of time—the one given by increasing entropy. According to the second law of thermodynamics, whatever we can conceive of happening in nature can happen. But certain occurrences are much less probable than others. Thus, entropy can suddenly decrease locally, but this requires an exchange of heat with the surroundings, which, when taken into account, shows that entropy is increasing overall. There are local decreases in entropy everywhere about us. Every ordered system, from a snowflake to a human, is a local decrease in entropy. But when the chemical processes inside humans are considered, entropy is seen to increase globally, which it will continue to do until the end of time.

A Mal du Siècle over the Fin de Siècle

Dissipation of all forms of energy, the heat death, tendency toward chaos: All these scientific terms became catchwords for the *fin de siècle* mood that closed the nineteenth century. Entropy and evolution were the scientific topics that most caught the public's imagination. As the historian of science Charles Gillispie put it:

> If the robber baron and the strong man armed took license to ruthlessness for the theory of natural selection, the intellectual who shrank from such successes found in entropy an excuse to indulge his *mal du siècle.*

The second law of thermodynamics gained notoriety in late nineteenth-century Europe and was horribly misinterpreted to apply to politics and social conditions that were deemed to be disordered and disintegrating. The most widely known misinterpreter was the American historian Henry Adams, who regarded himself as the perhaps last and most decadent member of an illustrious family that began with President John Adams. For posterity, Henry Adams wrote his autobiography in the third person.

Adams thought himself more as a philosopher of history than a historian: "I don't give a damn what happened. What I want to know is why it happened." He believed that the laws of thermodynamics offer a clue to why events occur as they do and advised that historians ought to take account of the latest developments in physics, which flew in the face of evolutionists, who preached nothing but eternal progress. Instead of progress toward the perfection of the human race, Adams saw a movement toward a degenerate society without class structure, gradations, and nobility, or culture—social randomness, and disillusionment resulting from the soured dream of progress. Adams was clearly an elitist and pessimist, having lost faith in liberalism and educational values. In his view, things were going from bad to worse and the human race would soon be extinct. The music of the new age was the pessimism of Gustav Mahler and Anton Bruckner.

While today's cosmologists have dreamed up such fantastic ways to beat the second law as building designer universes and hopping into them ad infinitum, the philosopher Friedrich Nietzsche had another idea, which was a bit more far-reaching, at least in intent. Nietzsche wanted to triumph over the fundamental philosophy that he believed went hand-in-hand with atomism: mechanistic materialism. By "mechanistic materialism," he meant that physical systems move according to the laws of Newton's mechanics (mechanistic) and they are comprised of atoms (materialism).

Nietzsche sought to revive the old idea that history repeats itself. Throughout time, this view has been intimately connected with man's place in the universe. It implies that human actions and even historical events by themselves have no meaning but can be understood as the working out of timeless archetypal patterns of behavior in the mythology of society. It has philosophical roots in classical Greek and Roman art and architecture, where there is no consciousness of past or future (that is, no progress) but only of principles and values. In the face of the dark mood in *fin de siècle* Europe, Nietzsche promoted the idea of eternal recurrence in the hope of stimulating a revival of the heroic qualities needed for mankind to return to a Golden Age. History repeating itself is sometimes referred to as the "paradox of return." Within his own philosophical framework, Nietzsche had come across it, as had Poincaré by a very different route. Nietzsche's antiscientific agenda is self-defeating. Eternal or cyclical recurrences can be achieved only if, in fact, the masses in the universe move according to some sort of mechanics.

Notwithstanding the general malaise that permeated turn-of-the-century society, which tried to force laws of physics onto human nature, the second law of thermodynamics is an example of a universal law of nature that can be violated by the occurrence of highly improbable random events. Its introduction of probability concepts into physics was interpreted as reflecting our ignorance of the dynamics in phenomena involving a large number of particles. But two pressing questions remained: Since atoms are invisible, might they really not exist? And if they do exist, what about our assumption of an orderly world, that is, a causal world? The next chapter addresses the first problem. Let us take up the second one here.

CAUSALITY IN QUANTUM PHYSICS

Whereas probability in the classical gas theory of Maxwell and Boltzmann was deemed to reflect our ignorance of individual processes, quantum probabilities are intrinsic to atomic systems. An auger of this result is the role played by Planck's constant in the uncertainty relations and the equations linking the wave and particle modes of light and matter. If Planck's constant were zero, then the wave/particle duality would not exist, nor would quantum probabilities and discontinuities of the atomic realm.

To Bohr and Heisenberg, these discontinuities were essential. But they came at a high price. As Heisenberg wrote to his friend and confidant Pauli on November 23, 1926:

> That the world is continuous I consider more than ever as totally unacceptable. But as soon as it is discontinuous, all our words that we apply to the description of facts are so many c-numbers [classical expressions]. What the words "wave" or "corpuscle" mean we know not any more.

In Chapter 2 we saw how Heisenberg reacted to this linguistic morass by formulating the uncertainty relations. But another problem arose: If the atomic world is discontinuous, then how can cause and effect be related? In addition, Heisenberg was startled by what the uncertainty principle implied about causality.

According to Newtonian causality, exact knowledge of initial conditions is required to predict accurately future events. Yet, the uncertainty principle seems to contradict Newtonian causality. According to the uncertainty principle, the better a system's position is measured, the less accurate its momentum is known. Therefore, a system's initial conditions cannot be ascertained with perfect accuracy. This result drove Heisenberg to conclude, in his 1927 paper setting forth the uncertainty principle, that the "invalidity of the law of causality is definitely established by quantum mechanics." We ought to question the worth of a scientific theory incapable of making predictions. In a nonsequitur, Heisenberg adds that we ought not conclude that quantum mechanics is "an essentially statistical theory in the sense that only statistical conclusions can be drawn from specified data," because the conservation laws of energy and momentum, which remain valid for atomic processes, still permit accurate predictions.

Despite the wave/particle duality, the validity of the conservation laws of energy and momentum permits collisions between elementary particles to be represented as if they were billiard balls hitting one another. For example, if a light quantum strikes a target electron, these particles fly off in different directions, just like a billiard ball at rest struck by the cue ball. Using energy and momentum conservation, the experimentalist can calculate where to set up detectors to catch the light quantum and electron. So, what's wrong with not knowing to any desired accuracy both position and momentum?

There is no problem so long as position and momentum are not measured simultaneously in the same experiment. More basically, however, the fundamental equations of quantum mechanics, such as Schrödinger's equation, are completely different from the equation fundamental to Newtonian physics. Newtonian physics produces analytical expressions for a system's space and time coordinates; Schrödinger's equation produces a so-called wave function that gives the probabilities for the system's state, that is, its position and momentum among other attributes of atomic systems. So, in quantum physics we have the seemingly contradictory statement that an equation (Schrödinger's) informs us how probability varies in space and time but does not inform us directly about position itself. The notion of a space-time description of physical processes in quantum physics is therefore dubious. This had been suspected right from the beginning in Bohr's 1913 atomic theory when physicists could not follow the course of an atomic electron as it dropped from a higher to a lower stationary state in Bohr's planetary model of the atom. In the spirit of classical physics, in 1918 Bohr injected probabilities into his theory essentially as an expression of ignorance. In the new atomic physics, according to Bohr, the concept of probability was totally nonclassical, being a consequence of the wave/particle duality of light and matter.

Complementarity attempted to restore some order to this situation. Perhaps as Hume did for Kant, Heisenberg's renunciation of causality, confused as it was, served to awaken Bohr from his dogmatic slumber. Bohr realized that the causality renounced by Heisenberg was the causality of Newtonian physics with a visual representation that had been prematurely discredited in quantum theory. Complementarity, on the other hand, links causality with the energy-momentum conservation laws and not with space-time pictures. This, according to Bohr, is the result of taking Planck's constant seriously despite its smallness.

What this means is that there are both a classical causality and a quantum causality. Heisenberg confused the two in his uncertainty principle paper. But despite the possibility of exact prediction in certain cases, Bohr tells us that on the atomic level statistics reign. As Heisenberg states in his uncertainty principle paper, any speculations on the existence of a real world "beyond the perceived statistical world . . . are sterile." Einstein thought otherwise and offered his famous retort, "God does not throw dice." What bothered Einstein, as he stated so eloquently in 1931, is that "The belief in an external world independent of the perceiving subject is the basis of all natural science." Two questions perplexed him: Why should we not be able to know a particle's position and momentum to any desired degree of accuracy? Do these attributes not exist, somewhere?

To Bohr these are nonquestions, because measurement is the essence of quantum physics. Without performing an actual experiment, it is nonsense to ask what is the state of a hydrogen atom. According to quantum physics, *before measurement* the atom has the potential to be in any one of a number of states. The measurement process snaps the atom into one state or the other. In classical physics, we can *pretend* to separate the system on which a measurement is being made from the measurement apparatus. In quantum mechanics, we cannot even pretend.* According to Bohr, it is in the inextricable connection between measurement apparatus and system being measured—characterized by Planck's constant—that probabilities enter and the quantum system is irrevoca-

---

* Even in classical physics one can wonder what is "really" being measured. Consider someone asking you what the temperature is of that cup of tea on the table over there. The way to determine it is to insert a thermometer. But are you actually measuring the temperature of the cup of tea in this way? No. You are measuring the temperature of the system comprised of thermometer and tea. You assume (pretend) that any interaction between thermometer and tea is negligible.

bly changed. For Einstein, quantum mechanics was a splendid theory, but it was not the last word in the atomic domain.

The Image of Light

To illustrate a typically counterintuitive element of quantum physics which bothered Einstein a great deal, let us return to the double-slit diffraction apparatus from Chapter 2. We recall that Young used this device to argue for the wave nature of light. But we know that light has a particle nature as well. Instead of assuming that light waves impinge on the double-slit system, we can arrange for light quanta to impinge one at a time. Given sufficient data, that is, over a long enough time, the result in Figure 2 of Chapter 2 appears. This seems to imply that light quanta interfere with themselves. To inquire into this, we can set up a detector at one of the slits to tell us which slit the light quantum went through. In this case we find that the data in Figure 2 disappear and a pattern appears of the form we would expect from shooting bullets from a machine gun at a single slit: Light acts like a particle because we have set up a particle-detection experiment. This is expected from complementarity. When we remove the detector, the double-slit pattern reappears. As the American physicist Richard Feynman said, understanding the double-slit diffraction phenomenon—understanding how the light quanta knew the other slit opened up and where to go on the screen—is the key to quantum physics. No complex mathematics is needed to display the counterintuitivity of this experiment and so, too, the mysteries of the quantum world.

The mathematics of quantum mechanics explains the data in Figure 2 of Chapter 2 as exhibiting a probability distribution. Light quanta will go most probably to the central maximum and less probably to the maxima flanking it. The intensity of the maxima is a measure of the probabilities. This may be all there is to it. Some of us think otherwise and search for yet another definition of intuition.

A double-slit diffraction experiment with light quanta of the sort in Figure 1 of Chapter 2 is truly a thought experiment. An equivalent real laboratory experiment was done by Alain Aspect and coworkers in Orsay, outside of Paris, with a Mach-Zehnder interferometer, starting in 1982. The apparatus is sketched in Figure 4.

According to classical electromagnetic theory, if instead of single light quanta a light wave were incident on the first beam splitter, then the incident light wave would be split into two waves. A beam splitter is a half-silvered mirror. One light wave would move to the right and the other would progress vertically. These rays would reflect off the two mirrors and then be recombined at the second beam splitter. Depending on what distances they travel—that is, on their optical path differences—the combined rays will produce either destructive or constructive interference, and an interference pattern like the one in Figure 2 of Chapter 2 will appear.

To see what happens with light quanta, Aspect used light quanta from the decay of calcium atoms. He was able to ensure that single light

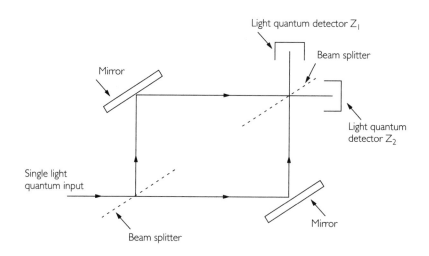

FIGURE 4.

Aspect's Mach-Zehnder interferometer experiment with single light quanta.

quanta are incident at the first beam splitter. Unlike electromagnetic waves, light quanta are indivisible and cannot be split by a beam splitter. A host of experimental data from biology, chemistry, and physics would be virtually unintelligible if this were the case. Consequently, each incident light quantum proceeds along one of two possible optical paths. Upon reaching the second beam splitter, it is deflected into one of the two light quantum detectors. Aspect's incredible result appears in Figure 5.

By varying the optical path length in the Mach-Zehnder interferometer just as with light waves, Aspect built up interference patterns over time as light quanta were detected and recorded. The results are in Figure 5, where Z_1 and Z_2 refer to counts in the light quantum detectors Z_1 and Z_2, and, for example, .01 s/canal means .01 seconds per channel (*canal* is French for "channel"). Aspect emphasizes that the interferograms in Figure 5 from the two detectors are *complementary*.

Understanding this result is not intuitive, for it seems as if light quanta are interfering with themselves. How can this be? Aspect rightly points out that complementarity seems to take care of the situation because "the selected behavior depends on the selected apparatus": If a quantum system is subjected to an experimental apparatus for waves, then it will behave like a wave. So, apparently we should assume a visual image where "something" split at the first beam splitter and then recombined at the second one. This seems impossible, though, if light quanta are indivisible. Perhaps the light quantum "knows" that it should behave like a wave in this experiment and therefore has a "premonition" of the second beam splitter's presence and changes of optical path length. Although the mathematics is consistent for either a wave or particle situation, no visual image abstracted from phenomena we have actually witnessed can give a proper explanation for Aspect's experiment.

Regarding the problem of understanding quantum phenomena with proper visual imagery, Aspect concludes as follows:

> The problem then only arises when we ask the question of what image we must choose to describe the light (wave, or

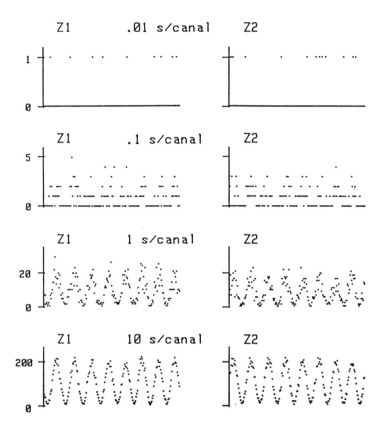

FIGURE 5.

Aspect's data reported in 1989 from the Mach-Zehnder interferometer experiment in Figure 4. The dots indicate the number of detected counts in the light quantum counters Z_1 and Z_2, and .01 s/canal means .01 seconds/channel or detector. The four sets correspond to different counting times at each path difference. Over longer counting times enough light quanta are counted to form a typical interference pattern obtainable from light waves. But we must recall that according to quantum theory, "classical light waves" are, in fact, comprised of huge numbers of light quanta. So, counting single light quanta over a long period of time gives a pattern just like the one obtainable from a classical light wave. But how do individual light quanta "know" precisely where and when to go into each counter in order to produce complementary wave patterns?

particle?). Is it "a foolish question"? We do not think so, because our experience of physics is that images are useful for imagining new situations. But we have to be extremely careful with images in quantum mechanics!

Most physicists consider complementarity in conjunction with the consistent mathematical apparatus of quantum mechanics as providing sufficient reasons for experiments such as Aspect's. So, they conclude, any questions about a proper imagery for light are foolish. Yet experiments such as Aspect's have been the catalyst for a small but influential group of physicists to reexamine the foundations of quantum mechanics, particularly the role of the external observer. This is essential, writes the Nobel laureate physicist Murray Gell-Mann, "especially since it has become increasingly clear that quantum mechanics must apply to the entire universe."

QUANTUM EFFECTS AND THE CORSICAN BROTHERS

The theoretical work to which Aspect's experiment can be traced is an extremely weird result of quantum physics Einstein obtained in 1935. This is the famous Einstein–Podolsky–Rosen (EPR) thought experiment. It goes like this. Imagine that an elementary particle in London decays into two electrons that go off at very large velocities. In New York City a measurement is made on one of the particles. According to quantum mechanics, this measurement changes a certain characteristic of the other particle, which by this time could be in Moscow. Einstein's challenge was to find how a measurement on one particle could instantaneously affect characteristics of the other. What is the nature of these long-range correlations, which we accept as one of nature's regularities? Is this situation not acausal?

In contrast to the eloquence of the EPR paper, Bohr's reply was pure party line and heavy handed. According to Bohr's colleague Léon Rosenfeld, "This onslaught came down upon us as a bolt from the blue. Its effect on Bohr was remarkable. . . . [A]s soon as Bohr heard my report of Einstein's argument, everything else was abandoned." As James Cushing has pointed out recently, what was a suggested resolution in 1927 (complementarity) by 1935 had become catechism and was referred to as the "Copenhagen Interpretation." It was not until 1964 that Einstein's thought experiment was rephrased to a point where actual measurements could be carried out. This was presented in a monumental paper by John S. Bell at CERN, with vital hints from work of David Bohm of Birkbeck College, London. Based on Einstein's notion of cause and effect as it is understood in classical physics, Bell produced a set of inequalities, which ought to be violated by quantum physics. Almost 20 years later, the requisite experiments were again carried out, in Orsay, by Alain Aspect. Bell's inequalities were violated in a way that supported quantum mechanics. Long-range correlations do exist.

The problem is to understand these long-range correlations. One "explanation" is that since the decay particles in the EPR experiment are born from the same particle, their properties are entangled from the start and remain so: They are like the Corsican brothers of Dumas' novel of the same name. But is entanglement not just a slogan, or a metaphor? It is surely not an explanation, even with some mathematics behind it. On the other hand, maybe it is. The recently deceased Bell even speculated on the long-range correlations indicating that special relativity had to be overhauled, and he advocated a return to ether-based versions of light. Most quantum physicists are content with a mathematical explanation along the lines that we can expect entanglement when the measured quantities are represented by noncommuting numbers (or operators), and there is nothing wrong with special relativity. Notwithstanding whether or not terminology such as entanglement strikes a resonant chord, may we not seek a more "intuitive" explanation, in some new definition of the term? Perhaps there is none. I am not alone in thinking otherwise.

Clearly, we have to investigate further what we mean by reality and atoms. After all, we can explain just about anything with invisible entities. Why not just conjure up ghosts? The next chapter examines science from the viewpoint of scientific realism.

5

SPEAKING REALISTICALLY ABOUT SCIENCE

Do atoms really exist? We can't see them. Instruments that claim to detect atoms are enormous in size and complexity, and so are many layers removed from a world in which 10^{-8} cm is huge. So let's begin this chapter by looking at what's real in the daily world. Then we'll move to the debate between realists and antirealists, using the atom as a concrete case. We'll explore the atomic concept from the ancient Greeks to a heuristic device in eighteenth- and nineteenth-century chemistry, to the real physical object in the theories of Boltzmann, Einstein, Maxwell, and Poincaré. Einstein's paper on Brownian motion firmly established for scientists the real existence of atoms. For philosophers the situation was different, however. What emerged is the tension between antirealist and realist philosophies, with antiscience movements always in the background. The chapter closes with a discussion of C. P. Snow's "two cultures" and then present-day postmodernism.

Let's Have a Reality Sandwich

Bumper stickers on automobiles are all the rage. We see serious ones like "Baby on Board" and "Trekkie" ones like "Beam Me Up, Scotty." My privately awarded prize is for "No One on Board." This got me thinking of how I would establish that someone actually is on board (the automobile had all the properties of being operated). Can I accomplish this purely by sensation? The eighteenth-century Irish philosopher George Berkeley tried his hand at this. He succeeded in cogent and far-reaching criticisms of certain philosophical notions in Newton's physics in which Newton seemed to have overreached his grasp. Berkeley took aim at Newton's assumptions of an absolute space and time, which are unmeasurable and so, in Berkeley's opinion, have no place in mechanics.

As mentioned in Chapter 1, Newton's aim in the *Principia* was to set up a formalism that could distinguish between relative motions (i.e., motions of entities relative to other perceptible entities) and absolute motions (i.e., motions relative to absolute space, which is God's sensorium). The problem is that absolute motions seemed to be undetectable. Electromagnetic theory filled God's sensorium with the ether; thus, absolute motion became synonymous with motion relative to the ether. But that too turned out to be undetectable. Absolute time, according to Newton, "flows uniformly on, without regard to anything external." Poetically pretty, to be sure, but only the relative time on clocks is measurable.

Berkeley's philosophical descendent Ernst Mach referred to Newton's absolute space and time as "metaphysical monstrosities." When Berkeley put on his Bishop's miter, however, he argued himself out of existence, asserting that an object's primary qualities (e.g., position and velocity) cannot exist without its secondary qualities (e.g., tendencies and essences), which are ultimately put into one's mind by God. Consequently, the world has no material existence—life is but a dream. The irascible Dr. Samuel Johnson claimed to refute Berkeley by kicking a stone. Amusing as Johnson's antics were, he entirely misunderstood the situation. Berkeley was trying to make the much deeper point that the representation of objects results from our perceptual and cognitive apparatus. And we must be capable of generating higher abstractions, particularly when we do science.

So, in our rudimentary case of bumper stickers, variations on Dr. Johnson's kicking a stone won't do! Maybe what I read on the bumper sticker ("No One on Board"), the other sensations of the automobile onto which it is stuck, in addition to the other surrounding traffic, gas fumes, and so on, are just reducible to states of my mind and so perhaps are mere illusions. This view is called idealism. Bishop Berkeley was an idealist. What is wrong with this view is that other humans in other cars have essentially identical thoughts on what they are observing. Moreover, they can communicate these thoughts, and so we can all agree on the occurrence of certain phenomena. Consequently, these occurrences are not mere figments of my imagination. There must be physical entities out there which, *pace* Dr. Johnson, I can bump my car into.

If I instead choose to focus Descartes' dictum *cogito ergo sum* totally inward on myself, thereby declaring myself as the only existing being, I become a solipsist, for whom, as Karl Popper wrote, "The world is my dream." Kant called solipsism a "scandal to philosophy," but not a trivial one. Kant's demolition of solipsism is one goal of his magisterial cognitive view of science. In Kant's system, we interact with things whose properties generate sense perceptions that are grist for the mill of our intuitions of space and time.

The following quote from Poincaré perhaps best summarizes everyone's view on the existence of *large-scale* objects external to ourselves:

> Nothing is objective that is not transmissible, and consequently . . . the relations between the sensations can alone have an objective value. . . . In sum, the sole objective reality consists in the relations of things.

When he wrote these words in 1902, Poincaré still doubted the existence of atoms. At that time he referred to atoms as a "metaphor," by which he meant that certain measurable phenomena seem *as if* they were caused by submicroscopic entities. Yet, in the end, all traces of atoms ought to disappear from theories and so be regarded as nothing more than auxiliary quantities useful for facilitating calculations. As Poincaré stated at the International Conference of Philosophy in 1900, the "question of the *réalité du monde extérieur* would best be placed in section 1 (Metaphysics)." Most philosophically inclined scientists at the end of the nineteenth century were antiatomists and focused on information reducible to sense perceptions. What are the roots and justifications of antiatomism, which seems so alien to us today?

ATOMS IN ANTIQUITY

The ancient Greek philosopher-scientists were hairsplitters. They wondered about everything. Take, for example, change. How can we explain occurrences in a world where everything is changing? Can we explain change on the basis of a single primeval substance, or ought there to be several basic substances? Is change real? After all, a change from one state to another entails passing from "what is" through "what is not" in order to reach the end state. But, how can we speak of what is not, that is, of nothing? Some concluded that there can be no change.

Around 470 B.C., Zeno of Elea thought deeply about such problems and went on to devise paradoxes that made one wonder how anything could catch up with anything else, or for that matter why anything moved at all. Zeno's most famous example is of a race between Achilles and a tortoise. Since the tortoise moves at some one-thousandth the speed of Achilles, Achilles graciously allows the hapless creature a head start. According to Zeno, this was a blunder because for every interval of space Achilles covers, the tortoise will have covered one more interval. How can an arrow be shot from a bow? After all, the arrow can occupy only a space of its own size. When in this space, the arrow must be at rest. Something at rest cannot be in motion. Therefore, the arrow is always at rest. Many of Zeno's paradoxes went unresolved until Newtonian mechanics.*

One might delude oneself that life was easier in ancient Greece because there was so much less organized knowledge, so much less to know. But this misses the point, which is that the Greeks were creating science and had nothing to fall back on. Empedocles, Parmenides, Zeno, Plato, and Aristotle, to mention just a few, were astronauts of the mind, charting completely unknown territory, and they set the course of science for thousands of years. Some of the problems they posed were not resolved until the nineteenth and twentieth centuries A.D. and are still discussed today.

In the fifth century B.C., Parmenides of Elea, a student of Zeno, argued for a unitary or monistic view in which change is an illusion because the only reality is "the One" possessing neither qualities of differentiation nor any change of movement. Needless to say, this particular attempt to reduce all processes to a single one caused a crisis in Greek

---

* The Achilles and tortoise paradox arose essentially because of Zeno's refusal to grant the notion of an infinite number of time or spatial intervals in finite intervals of space and time. Concerning the arrow paradox, according to Newtonian mechanics it is possible for an object to be instantaneously at rest but accelerating. Such a notion requires integral and differential calculus, which deals with infinitesimal limiting procedures.

philosophy. At least Thales's "the One," water, was tangible. Among the immediate replies were those of the Sicilian Empedocles of Agrigento, who proposed that change is the result of combinations among four real elements: earth, air, water, and fire.

A more far-reaching proposal that included portions of Parmenides' "nothing" was made by Leucippus and his pupil Democritus. They suggested that all changes we see about us result from invisible entities incapable of further subdivision. These atoms are "being," while the spaces between them are "not being," or in Parmenides' terms, "nothing," the vacuum. Perhaps Zeno's paradoxes of division—such as Achilles and the tortoise—gave Leucippus the idea for suggesting a level of being that could not be further subdivided. Zeno's paradoxes convinced Leucippus that physical division differs fundamentally from mathematical division, which can go on forever.

According to the stunning proposal of atomism by Leucippus and Democritus, the tapestry of change is woven from the stuff of immutability. The writings of the Greek atomists are rich with statements that would not, and could not, be considered seriously for over a thousand years; one example from Democritus is, "Nothing can be created out of nothing, nor can it be destroyed and returned to nothing." In part this is a statement of conservation of matter, which was a guiding principle until Einstein's mass-energy equivalence replaced it. It is not mass that is conserved, but energy.

In order to explain macroscopic phenomena such as why salt dissolves in water but sand does not, Leucippus and Democritus postulated that atoms come in an infinite variety of shapes. They assumed that differences in shape between sand and salt atoms determined their ability to fit (i.e., dissolve) into a water atom. This explanation leaves something to be desired. Another quasi-scientific criticism of ancient atomism follows: If phenomena in the world of sense perceptions result from processes involving invisible particles whose motions are in principle understandable from mathematics, then atomism is materialistic and so atheistic, too. It leaves no place for the hand of God. Such was part of the case against Galileo's physics as well, which rested on the existence of atoms and a vacuum.

To conclude our discussion of Greek atomism, it is appropriate to say a few more words on the modern concept of "nothing"—the vacuum. A criticism of Greek atomism, and of Greek science generally, concerns its notion of "nothing." In contradiction with Democritus, it is not the case that "nothing can be created out of nothing." From quantum theory we know that the vacuum is not "nothing" at all. Rather it is seething with particles and antiparticles being created and destroyed. This is a striking feature of quantum mechanics.

One of the most fascinating and important features of quantum matter is fluctuations. One example is the random creation and annihilation of a particle with its antiparticle. Some theoretical physicists today have entered the realm of extreme speculation by proposing that our cosmos is the result of a fluctuation in which energy suddenly is created out of "nothing." This particular fluctuation, called the Big Bang, occurred some 10 to 20 billion years ago. Roughly speaking, some energy was converted into mass through Einstein's energy-mass equivalence, and the rest into the attraction in gravitational processes that hold certain parts of the universe together. In this way, the energy created sums up to zero: The part that goes into mass is positive, and the gravitational energy is counted as negative.

The probability for this random fluctuation is, of course, very small, but the point is that it can occur. We might say that from "nothing," "nothing" is created. As the physicist Alan Guth, who proposed many of these fascinating cosmological ideas, writes: "It is tempting to speculate that the entire universe evolved from literally nothing. The recent developments in cosmology strongly suggest that the universe may be the ultimate free lunch."

Galileo developed a different cosmology, one that is similar to today's widely accepted view—which is no surprise since he was one of the founders of modern science. In the *Dialog* of 1632, Galileo wrote that whereas God created the universe and its architecture, the job of the scientist is to explore its fabric. And basic to the fabric of the universe are atoms.

Doubts About Atomic Reality

Whereas atomism was considered taboo in Galileo's era in Italy because its materialism carried the stigma of atheism, Newton felt it natural to extend his theory of motion, down to the atoms constituting matter, without fear of religious persecution. But the atomism of Galileo, Newton, and others, such as Descartes, shared the sins of ambiguity and vagueness. Their atoms were either rigid point particles or endowed with particular shapes for no other reason than to explain certain phenomena. An important step out of this situation was taken by Immanuel Kant, whom we have met earlier.

As a young man, a focus of Kant's activities was theoretical physics Taking a stab at an elaborated atomic theory, he offered a view in which atoms are point particles that affect each other through fields of force—a marvelous premonition of modern field theory. Invisible particles responsible for phenomena were folded into some of the earlier theories of electricity; however, other theories based on analogy with fluids also did the job and were more intuitive because of analogies with, for example, water flow.

Meanwhile, the discipline of modern chemistry began to emerge with its own notion of an atom, which differed from that of, for example, Kant. Key figures in the emergence of chemistry from alchemy, such as the seventeenth-century English scientist Robert Boyle and the eighteenth-century French chemist Antoine Lavoisier, advocated the reality of indivisible elements. In the hands of John Dalton, in the early nineteenth century, these entities became what we refer to today as chemical elements constituting the periodic table. Dalton's "chemical philosophy" came equipped with a system of ideographs for elucidating how chemical elements combine (Figure 1).

Dalton placed great importance on visual representation, even constructing wooden models in order to illustrate how atoms combine. His symbols played an important role in aiding chemistry to visualize

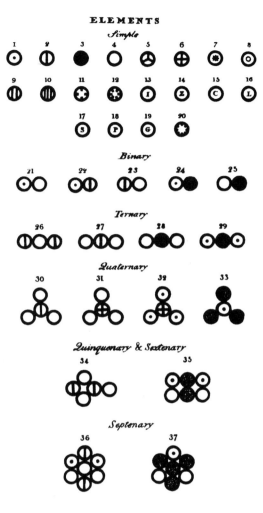

FIGURE 1.

Dalton's ideographs from his *New System of Chemical Philosophy*. Each numbered ideograph represents atoms (first three rows) and the remaining, compound atoms. Dalton's list is not shown.

complex chemical reactions. As the historian of chemistry William Brock writes:

> As in mathematics, chemistry could advance only to a certain degree without an adequate symbolism for its deeper

study. . . . Dalton completed a revolution in the language of chemistry.*

The reality of the chemical atom was another matter. Some scientists considered atoms merely auxiliary devices for performing calculations. In other words, the final result is independent of whether or not chemical atoms were used in the calculations. As William Prout, one of the pioneers of chemistry, put it in 1815, the atomic theory

> has been very analogous to that in which I believe most botanists now consider the Linnean system; namely, as a conventional artifice, exceedingly convenient for many purposes but which does not represent nature.

Among the stunning results to emerge from Dalton's view, with its accompanying visual representation, was the proposal offered in 1811 by a physics professor at the University of Turin, Amedeo Avogadro, for a number of atoms in a standard amount of gas (one gram molecular weight). This number turned out to be huge, 6×10^{23} atoms in a gas per gram molecular weight. The principal point is that if Avogadro's number were infinite, then the substratum of our world would be a continuum. This makes sense intuitively because the limiting case of an infinite number of particles is a continuous distribution of matter, one that can be chopped up forever.† However, Avogadro's work was not taken seriously until the 1860s, following developments in gas theory by physicists like Maxwell and Boltzmann. The historian of science

* Starting in 1813 the Swedish chemist Jakob Berzelius formulated a chemical notation in which capital letters of the alphabet represent elements. By the 1830s, it developed into the notation we use today, in which, for example, H denotes hydrogen and K, potassium. Much to Dalton's chagrin, in 1837 the British Association for the Advancement of Science persuaded British chemists to adopt Berzelius's system. One predominant reason was the additional printing expense for Dalton's symbols. A great advantage of Berzelius's notation is that it enabled chemists to express reactions in equations.

† Readers familiar with the concept of entropy will recall that if Avogadro's number were infinite, then Boltzmann's constant would be zero, and a definition of entropy in terms of probable molecular arrangements would be meaningless. In addition, Planck's constant would be zero.

Stephen Brush concluded from a detailed study of this episode that "no matter how appealing a new world view may be, it cannot by itself bring about a revolution if it is not relevant to the problems that a generation considers crucial." In Avogadro's day, the prevalent research efforts in gas theory concerned the subtle fluid caloric and ideas on the affinities of chemical atoms championed by giants such as Dalton.

The physicist's atom was altogether different from the chemist's. The former was supposed to be the ultimate constituent of matter and so the building block even of the chemical atom. Chemists relegated atoms to the realm of metaphysics (pejoratively understood). This began to change with the advent of the gas theories of Clausius, Maxwell, and Boltzmann, which permitted physicists to investigate actual properties of atoms, such as size. One result of their research was estimates of Avogadro's number. In the 1870s, the successes of gas theory, combined with advances in structural chemistry that utilized actual models of how atoms combine, led most chemists to drop the distinction between the putatively auxiliary chemical atom and the purportedly real one of the physicists.

Yet, beginning in the 1880s, there were formidable holdouts— among them were Mach, Planck, and Poincaré—against *hypotheses* on the atomic structure of matter. On the physics side were outstanding problems in gas theory such as those concerning the properties of rarefied gases, sound propagation, and heat conduction and diffusion in gases and liquids. In retrospect, and to many physicists of that era, these were side issues to be tackled somewhere down the line (as, in fact, they were in the second decade of the twentieth century, when the Maxwell–Boltzmann gas theory was generalized with proper mathematical methods of approximation).

On the other hand, philosophically inclined physicists pointed out that since many results of gas theory could be obtained without atoms, then such directly unobservable entities should be excluded right from the start. In their favor, it did seem as if certain phenomena required different atomic models with no theoretical basis. The other side of the debate centered about the recurrence paradox, which we discussed in Chapter 4. The crux of that issue was that physicists believed that irreversibility ought to be a law of nature and not something that is only highly probable, as advocated by Maxwell and Boltzmann with their atomic-based views of entropy. In other words, such physicists asked why the second law of thermodynamics is universal and yet open to violations. If Maxwell and Boltzmann prevailed, then the atomic hypothesis would have to be taken seriously: Ultimately, probability and atomism go hand-in-hand.

In summary, whereas most scientists agreed by the end of the nineteenth century that matter is comprised of atoms, they were concerned over hypotheses made about atoms and over the defects atomic theory showed for certain problems. Certain philosophically minded scientists, on the other handed, reacted against the possibility of an atomic basis for science. Instead, they preferred theories that kept a close tie with experimental data. The best-known philosopher of this antiatomist group was Mach.

POSITIVE THINKING

The positive approach to physics, or positivism, was launched in the 1880s by the Viennese physicist-philosopher-psychologist Ernst Mach. The name "positivism" had been coined some 50 years earlier by the French philosopher Auguste Comte to denote positive progress toward a science devoid of theology and metaphysics. In Mach's era, "positivism" had come to mean a critical approach to science. Mach adopted this attitude because of a malaise that was spreading over physics. The promise of Newton and his followers of a mechanical foundation for all of physics was falling short. The phenomena of optics, electromagnetism, and thermodynamics seemed not to be reducible to Newtonian mechanics. As if this were not enough, truly "deep" questions, such as

"What is the nature of force, of velocity, of the mind?" and so on, seemed unapproachable with current physics.

The influential German physiologist and philosopher Emil du Bois-Reymond concluded in his 1872 book *On the Limits of the Knowledge of Nature* that certain phenomena, despite their importance, would forever remain a mystery. But du Bois-Reymond did not stop here, writing *"igno-ramus et ignorabimus* [we do not know and we shall never know]." *Ignora-bimus* became the slogan of an antiscience movement and of defeatism in science itself. Shortly thereafter, *ignorabimus* was replaced by the pungent slogan "bankruptcy of science." After all, science seemed to offer no solutions to the deep and pressing questions of life. In the face of this criticism, a good many scientists in the latter part of the nineteenth century simply went right on working. But for some, a reply had to made.

A MACH ATTACK

Mach was the man on the white horse who went right for the jugular in attacking what he rightly took to be the very root of the malaise—unwarranted assumptions in the foundations of Newtonian mechanics. Einstein was right on the mark when he assessed Mach's "greatness in his incorruptible skepticism and independence." In his youth, Mach had flirted with Berkeley's idealism and Kant's metaphysics. But Mach's experimental results in psychology led him to conclude that the study of sensations is fundamental to science. Reading Charles Darwin's *Origin of Species* convinced Mach that the most economical scientific theory is the one most likely to survive competition with others. Such a theory is one stripped to its barest bones—namely, whittled down to sense perceptions.

The psychologist-philosopher Gustav Fechner's mathematical equation relating sensation and stimulus struck Mach as a result that placed psychology on a scientific basis. Actual measurements could be made.

Mach pushed this empiricist view to its limits by claiming that serious scientists ought to concern themselves only with laboratory data reducible to the senses. "All intellection," Mach wrote, "starts from sense perceptions and returns to them. . . . From sensations and their conjunctions arise concepts," which are higher abstract faculties for reasoning. But not so fast, he cautions, because there is an intermediary, namely, intuition, which Mach regarded at its most basic level to be common sense.

These results served as a basis for Mach's hard-hitting criticism of such idle metaphysical baggage of mechanics as Newton's absolute space and time, which cannot be measured and so have no place in physics. He was no less outspoken about "artificial atomic theories," with atoms being merely "things of thought" because they cannot "be perceived by the senses." He concluded that the goal of science is to describe experimental data as economically as possible. The role of equations is to summarize in the most economical way a great amount of experimental data.

In addition, the hitherto assumed primacy of Newtonian mechanics was just a historical happenstance. It just happened to be the first subject developed mathematically. Actually, Mach stressed, biology ought to be the science to which all phenomena should be reduced, due to the central role sense perceptions play in both the origins and experimental methods of science. The 45-year-old philosopher-scientist developed these views in his widely read book of 1883, *Science of Mechanics*, which went through 16 editions in Mach's lifetime.

Later in life, Mach recalled the initial resistance to his ideas. But soon the converts multiplied. Following Mach's lead, the German philosopher-physicist Heinrich Hertz succinctly put the crux of Mach's preaching as follows: A new rule has been added to scientific inquiry—serious scientists reserve the right to declare certain questions "illegitimate." On Hertz's list were the "What is the nature of   ?" questions. Perhaps at some future time these questions can be taken off the shelf. This is what happened to such taboo topics circa 1900 as "What is the nature of the mind?", "Why are we here?", and "What is consciousness?"

By advocating strong dependence on experimental data, Mach returned the moral high ground to scientists while opening the floodgates for a new generation of scientific "philosophizers," or philosophers of science. While Mach's positive message succeeded, variations on it abounded. Consider, for example, Hertz's book *Principles of Mechanics*, published in 1892, two years before his death at age 37. Hertz dedicated the book to Mach and then went on to set up a version of mechanics based on organizing principles accepted as axiomatic, which is most un-Machian. Yet in applied areas of his research, Hertz toed the positivist line in opting, for example, for a nonatomistic interpretation of cathode rays. The weight of the support that Hertz and Mach lent to antiatomism cost Walter Kaufmann a Nobel prize. Kaufmann knew better the next time around. Kaufmann's colleague, Max Abraham, although an atomist, specifically developed certain mathematical techniques that in his view provided the most "economical expression" for his electron theory. These techniques are still used today, reflecting a beneficial aspect of Mach's positivism.

Poincaré, as well, spun a philosophical system based on empirical data—"experiment is the sole source of truth." But Poincaré emphasized the importance of mathematical generalizations. The higher the level of generalization, the more difficult it is to falsify a theory. Like Hertz, but unlike Mach, Poincaré attributed a major role to mathematics as being something more than a mere instrument for economy of thought. Poincaré eventually threw in his lot with those seeking a unified theory of matter based on electromagnetic theory and not biology. He became a believer in the reality of electrons, even though they are unobservable.

Planck Converts to Atomism

Max Planck was another important convert to atomism. His biographer John Heilbron relates Planck's quest for absolute laws of nature to his being the "exemplar of the classical physicist [who] came from a line of pastors, scholars, and jurists." This is a reasonable background to Planck's early attraction to thermodynamics, whose laws claimed a timeless universality over all of physics. Heilbron writes, "This doctrine spoke to the lawyer and the theologian in Planck." In 1879, at age 21, Planck completed his doctoral dissertation, in which he explored the two known laws of thermodynamics, with emphasis on the concept of entropy. In particular, he stressed the capacity of thermodynamics to solve problems without the atomic hypothesis, which, at best, he deemed to be an auxiliary device to expedite calculations. Over the next 20 years, he would change this view.

What made the difference was Planck's research in extending thermodynamics to problems in chemistry, for which successful results had already been obtained using gas theory. The research path decisive to Planck was his application of thermodynamics to problems of radiation. Planck was convinced that the entropy of radiant energy in a cavity inside matter should increase irreversibly, with no possibility of slipping backward. But Boltzmann pointed out to him the elementary fact of life that Maxwell's equations for radiation are time reversible and so cannot escape a determined reversibility, just like Newton's mechanics. The recurrence paradox struck again. Thus, Planck was forced to admit that in order to explain the measured properties of cavity radiation, additional hypotheses were needed which brought with them the randomness intrinsic of Maxwell and Boltzmann's gas theory. Among the results of Planck's monumental theory of cavity radiation, which he published in 1900, were his calculations of Avogadro's number, the charge on a *single* electron, and the first value for Planck's constant.

Planck's radiation theory was a key argument toward accepting the atomic hypothesis and the notion of irreversibility as a probabilistic concept—anything that one can conceive of happening can happen. What emerged full-fledged was the link between discontinuity, probability, and atoms.

In light of these results, we might think that Planck's work went far to convince Boltzmann that the atomistic hypothesis had become completely credible. Two years earlier, in 1898, Boltzmann had his doubts about this: "In my opinion it would be a great tragedy for science if the theory of gases were temporarily thrown into oblivion because of a momentary hostile attitude toward it." The real tragedy, writes Stephen Brush, was that Boltzmann "somehow failed to realize that the new discoveries in radiation and atomic physics occurring at the turn of the century were going to vindicate his own theories." No doubt Boltzmann's depressed frame of mind over the possible fate of his life's work was a contributing factor to his suicide in 1906, at age 62.

With all the zeal of the converted, in 1908 Planck lambasted Mach's philosophical view, arguing for a scientific realism. Planck conceived of science as progressing primarily because of an increasing lack of dependence on sense perceptions, a realization that had not come easily to him. After all, continued Planck, an atom is as real as a planet: We speak of how much an atom weighs, yet no one has actually ever weighed one. We do this similarly for the weight of the moon, and surely we concede that the moon exists.

On the other hand, Mach continued to advocate sense perceptions as the basis for all of science. The Viennese physicist Stefan Meyer told how, in 1903, he convinced Mach to look into an evacuated tube that displayed scintillations made by positively charged rays called α-rays. According to Meyer, Mach was astounded and changed his "entire world-picture," accepting the reality of atoms. Meyer misunderstood: Mach remained an antiatomist. No doubt Mach realized that what he saw were not atoms but complex interactions between the α-rays and the phosphorous-coated glass surface of the tube.

In Planck's view, the universality of such constants of nature as Planck's constant implies the universal character of science that holds for all cultures—"earthly and nonhuman ones," too. Part and parcel of this advance had been the reality of atoms. And none of this scientific progress had anything to do with notions of economy. In what Mach may have taken to be the most cutting of Planck's criticisms, Planck accused Mach of writing false history in his *Science and Mechanics* because none of the great scientists Mach discusses were at all concerned with describing experimental data with the fewest and least abstract hypotheses: Economy of thought has nothing to do with real science.

While the tone of Mach's rejoinder to Planck is one of sarcasm and self-justification, it contains an interesting retrospective of his views, as written by someone in his seventies. With regard to Planck's case for atomism, Mach writes: "I cannot conceal my dislike for hypothetico-fictitious physics." Instead, he continues to claim that sense impressions as the source of all experiences ought not to be abandoned so quickly for the atomic hypothesis. He dismissed Planck's speculations on the universality of certain constants of nature as "almost ludicrous." Mach's defense of intellectual economy was to claim its central role in biological and psychological processes. Finally, against the atomic hypothesis in general and its claim—supported by the vast majority of physicists in 1909—that it is the route to a unified theory of electromagnetism and mechanics, Mach writes:

> If belief in the reality of atoms is so important to you, I cut myself off from the physicist's mode of thinking, I do not wish to be a true physicist. . . . I denounce with thanks the communion of the faithful. I prefer freedom of thought.

Planck's subsequent reply is ad hominem in tone, vociferously defending scientists' rights to let their imagination roam free, beyond their sense perceptions: "A physicist who wishes to advance his science must be a realist, not an economist."

Despite Boltzmann's suicide and the confrontations between atomists and antiatomists, which included such titans as Mach and Planck, the early 1900s were "business as usual" to most scientists. In 1908, the scientific and philosophical sides of the debate over atoms came together, with experimental proof for the existence of atoms.

ATOMIC VINDICATION

Proof for the existence of atoms came from an unexpected quarter. Among Einstein's three classic papers from his *Annus Mirabilis* of 1905 is the one on Brownian motion, which is the erratic motion of, for example, dust particles in air. The source of the energy that kept this motion going incessantly was a mystery. Physicists wondered if in this sort of process energy was not conserved. Einstein clarified the situation by assuming that the dust particles collide with each other as if they were diffusing through a liquid. But the most amazing result is that Einstein was able to relate Avogadro's number to observed zig-zag paths of macroscopic particles. In 1908, the French experimentalist Jean Perrin performed measurements of Avogadro's number based on further developed versions of Einstein's theory. The number obtained agreed with other means for calculating Avogadro's number—means also conceived of mostly by Einstein.

These other avenues were from such apparently unrelated areas as thermodynamics, the flow of incompressible fluids, and the way light is scattered in the atmosphere to make the sky blue. That such different methods yield the same result for a single entity is called convergence. Scientists take convergence very seriously, as do most philosophers. This is especially so when experiments are performed by a redoubtable researcher like Perrin, who concluded that "it will henceforth be difficult to defend by rational arguments a hostile attitude to molecular hypotheses."

During Mach's lifetime, Poincaré felt it incumbent to change his view on the reality of atoms. Although in 1902 Poincaré considered atoms to be metaphors, two years later he believed otherwise and went on to make lasting contributions to electron physics and quantum theory. True philosopher-scientists resonate with scientific developments. They do not always succeed: For example, Poincaré never agreed with special relativity, and similarly Einstein with quantum theory, and Heisenberg with post–World War II elementary particle physics.

It was astounding and unexpected that unquestionable proof for the existence of atoms came from measurements of *macroscopic* quantities. But to his dying day, Mach would not concede the existence of atoms. His dogmatic adherence to antiatomism in the face of evidence to the contrary is a damning verdict against his methodology of combining philosophy and physics. Yet Mach's critical outlook had a widespread effect on fields other than physics, for example, psychology, particularly through William James. Gerald Holton, who has studied this episode in some depth, quotes James as writing in 1902 "of his attempt to give his students at Harvard 'a description of the construction of the world built up of pure experiences related to each other in various ways'." The behaviorist school in psychology begun by John B. Watson in 1915 and honed by B. F. Skinner is a direct outgrowth of Mach's emphasis upon sense perceptions.

Mach's influence was also seen in the thrust of research in atomic physics from 1923 to 1925 in which the programmatic intent was to ground the theoretical structure only on quantities that are measurable. Recently, the historian of science Peter Galison has argued that the aim of Mach's philosophical heirs, the Vienna Circle, during the 1920s and 1930s was to integrate science with philosophy, art, architecture, and social values. From 1929 through 1941, the Vienna Circle organized a series of congresses under such titles as "International Congress for the Unity of Science," with the aim of wooing scientists away from what these philosophers took to be Kantian metaphysics. They did not generally succeed. Developments in philosophy and history would bring their program entirely to a halt.

Before proceeding with our discussion of the reality of atoms, let us bring the state of physics circa 1900 to a happy close. The *fin de siècle* mood of failure in science was dispelled in no small part with Mach's philosophical refresher, which advocated that science is the only legitimate pursuit of objective knowledge and should proceed with utmost critical acuity. Then there were several astounding discoveries. On November 8, 1895, Konrad Röntgen discovered x-rays. Like Hertz's discovery of electromagnetic waves, x-rays were quickly exploited for industrial and medical uses; in the Spring of 1896, the British expeditionary force to the Sudan went equipped with an x-ray machine. In March 1897, J. J. Thomson discovered the electron, in 1896, Henri Becquerel radioactivity, followed up in 1898 with the Curies' discovery of radium. That same year Lord Rayleigh and William Ramsay unraveled the noble gases. Physics ran headlong into the twentieth century with high hopes for incredibly new and far-reaching generalizations of Maxwell's electromagnetic theory. This was not to be. The root of all problems would be atoms, whose existence was assumed proven in 1908 to nearly every scientist, and some philosophers.

ON MACH'S PHILOSOPHICAL HEIRS

Mach died in 1916 at the age of 78. His immediate philosophical heirs never fully understood the atomistic position in relation to theories such as quantum mechanics. They should have known better, because many of them were scientists. The reason for this problem lay in part in their misinterpretation of another of Einstein's great papers of 1905, "On the electrodynamics of moving bodies." In this paper Einstein set out what became known as the special theory of relativity. Part of Einstein's message was that if you want to break loose from theories containing entities that seem to be unmeasurable—like the ether—you should take measurement seriously. Einstein meticulously defined by means of at least in-principle macroscopic measurements such basic quantities in a physical theory as space, time, simultaneity, and mass.

Ignoring the results of Einstein's Brownian motion research and Perrin's experiments, a *Kaffeeklatsch* of philosophers started to meet in Vienna in about 1912 to tease out and systematize what they took to be Einstein's message in relativity theory. They concluded it was living proof of positivism. Einstein never agreed.

Most of the Vienna Circle were well versed in physics. Declaring themselves to be Einstein's amanuensis, they developed a "logical positivism," as two of their members, A. E. Blumberg and Herbert Feigl, referred to it in 1931. The "logical" component advocates the usefulness of mathematics to axiomatize already well-developed scientific theories, in addition to a sharp distinction between the logic of discovery (there is none and so the genesis of scientific theories is irrelevant to philosophers of science) and the context of justification (which is the purview of the philosopher of science *qua* logician). The "positivist" or "empirical" component is based on the belief that all science is grounded in experience and so there is a distinction between theoretical terms (terms introduced by a theory, such as the noun "electron") and observational terms (terms from experiment, such as the predicate "is deviated by an externally imposed magnetic field"). Theoretical terms are interpreted through observational ones in ways such as: "An electron is deviated by an externally imposed magnetic field." Consequently, theoretical entities like an electron change according to what theory holds sway at the moment. (In the case above, electromagnetic theory was in vogue.) Logical positivists assumed it possible to find a theory-neutral language for observational terms. The upshot is that since entities like "electron" are tied to particular theories, then there is relativism in their meaning and so electrons do not exist in any absolute way, that is, apart from a theory about them. (Sometimes logical positivism is referred to as logical empiricism.)

The problem is that logical positivism bears little relation to actual scientific practice. For example, crucial to Einstein's thought experiments were organizing principles, such as the principle of relativity, the acceptance of which led him to discover phenomena beyond sense impressions, such as the relativity of time. The sterility of logical positivism became clear in the 1950s. It degenerated into attempts at finding a theory-neutral language for discussing observations and at formulating an inductive probabilistic logic to calculate the degree of confirmation of hypotheses in terms of the experimental evidence supporting them. The desired result would have been statements like "a certain scientific theory is 85% correct"—not very useful. Indeed, they never achieved anything useful regarding actual scientific research. One wonders why the impossibility of their task was not abundantly clear to everyone right from the start. Be that as it may, logical positivism never adequately addressed many fundamental philosophical problems (e.g., the underdetermination thesis that had been pointed out earlier by philosopher-scientists such as Duhem and Poincaré, among others). The futility of it all was finally revealed by proper historical investigations into how science developed, which turned out to be in a manner diametrically opposite to logical positivism: There is no distinction between theory and observation, because all observations and measurements are theory-laden. Consequently, there can be no neutral observation language and no inductive probabilistic logic that is useful for scientists. This point was forcefully emphasized in the late 1950s by Paul Feyerabend, Norwood Hanson, and Thomas S. Kuhn.

BOREDOM RELIEVED, MOMENTARILY

Boredom was relieved by the appearance in 1962 of Kuhn's view of science progressing by revolutions. Although, as I will discuss in Chapter 7, I disagree with Kuhn's general conclusions, he performed a great service for historical and philosophical studies of science. The opening sentence of his classic book *The Structure of Scientific Revolutions* tells it all:

> History, if viewed as a repository for more than anecdote or chronology, could produce a decisive transformation in the image of science by which we are now possessed.

Kuhn's message is *take the history of science seriously* for studying its development, especially so if one wants to study its philosophical aspects. Another key point in Kuhn's historical scenario was emphasis on the context of discovery. Kuhn spoke the language of Gestalt psychology and Piagetian psychology, introducing such buzzwords as "Gestalt switch."

He writes of the moment of changeover from one theory to another as inarticulable. Kuhn appeared to abolish emphasis on axiomatization and methods of an inductive logic with all its dryness, and he pointed out the importance of a scientific community in doing research, thereby starting a new field of study called sociology of science.

Kuhn's scenario goes as follows: Over a period of time, scientists work on a theory, or *paradigm*, accepted by the *scientific community*. *Puzzles* arise that become *anomalies* and by consensus provoke a *crisis situation*. (The emphasized words have become registered trademarks for Kuhn's view.) A breakthrough occurs, causing scientists to see the problem situation differently. The scales fall from their eyes, and they see the world through another sort of spectacles. A discontinuous change (Gestalt switch) has taken place to a higher-level theory. According to Kuhn, when a scientific revolution occurs, the meaning of all scientific terms and their relationship to the world about us radically differ and derive their meaning from the new paradigm. There is no projection of scientific terms between new and old paradigms, and so different paradigms are *incommensurable*. If Einstein and Newton could meet, there would be definite limits to their scientific communications. Kuhn's book galvanized everyone. For a time it was the most widely used book in American universities. Every discipline sought its paradigm. But the message belied the medium.

Let us use Kuhn's terminology for scientific progress to discuss developments in atomic physics from 1913 to 1925 (Figure 2). The response to puzzles took the form of different versions of Bohr's theory, including one that had no visual imagery. But by 1925 it became clear that the theory could provide no resolution to certain puzzles that became anomalies. This opinion, in addition to the belief that a crisis situation had developed, was made by the major domos of the subject, such as Bohr, Born, and Heisenberg. In June of 1925, Heisenberg formulated the quantum mechanics, which became the new paradigm. A scientific revolution had occurred, accompanied by a Gestalt switch because scientists "saw" the world through the new set of spectacles, quantum me-

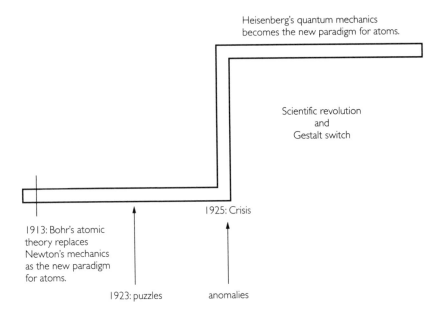

Heisenberg's quantum mechanics
becomes the new paradigm for atoms.

Scientific revolution
and
Gestalt switch

1925: Crisis

1913: Bohr's atomic
theory replaces
Newton's mechanics
as the new paradigm
for atoms.

1923: puzzles   anomalies

FIGURE 2.

An illustration of how Kuhn's view of scientific progress applies to
developments in atomic physics during 1913 to 1925.

chanics. The old texts were discarded. But this does not mean that the
old concepts were immediately thrown out, too.

The development of quantum mechanics during 1913 to 1925 is
the best example of Kuhn's scenario. As we have already discussed,
however, particularly in Chapter 1, this is not the way developments
occurred. In Chapter 7 we will discuss scientific progress in some detail.

As the spokesperson for pragmatism William James would have
asked, "What is the cash payoff of Kuhn's view of scientific progress?"
Let's take another concrete example. According to Kuhn, the thing
called the "electron" differs from theory to theory, and so J. J. Thomson's electron of 1897 is not the same entity as Bohr's planetary electron
of 1913 or the one in quantum mechanics in 1927, or in quantum electrodynamics of 1948, and so on. Scientific theories are not a means for
understanding an entire world, because all meaning is relative. What the

external world is depends on what theory you have in mind. Since theories have no common measure or meaning among their basic terminology, then, as the philosopher of science Paul Feyerabend stated, following the songwriter Cole Porter, "anything goes." Since all meaning is relative, science cannot offer any absolute truths whatsoever. The world is asunder. There is no unification of the sciences. Feyerabend and Kuhn presaged the postmodernist philosophy of the 1980s.

But in the end, it turned out that, as we have noted for the electron, what Kuhn proposed is essentially a rewrapped version of logical positivism. Incommensurability is an inherent part of logical positivism in which supposedly theoretical terms like "electron" are defined relative to a particular theory. The only difference between Kuhn and logical positivism was the moment of irrationality in theory change, on which Kuhn never elaborated, anyway. Historical studies show that scientists grasp wildly at continuities between old and new theories in order to further explore and expand emergent concepts into another possible world. We will look into this in some detail in Chapter 7. I use the term "possible world" as follows: Physical theories are interpreted as providing an entrance into realms of nature beyond our sense perceptions. These realms are possible worlds. Whether a physical theory tells us something about a real world depends on its agreement with relevant extant experimental data, its capacity to predict, and perhaps its ability to unify apparently disparate fields. In 1632, for example, Galileo tried to unify terrestrial and extraterrestrial phenomena within a theoretical framework based on a world in which there are vacuums, and it produced such predictions as all bodies fall at the same acceleration regardless of their weight. Part of this theoretical package was that the universe is sun-centered, for which Galileo claimed to have empirical as well as rational evidence.

Having been educated and working as a physicist before focusing my attention entirely on matters of a historical and philosophical nature, I can seriously entertain only scientific realism. This is the case for every scientist. Only scientific realism can satisfy our curiosity about how the world is constituted. Scientific theories are our means for exploring possible worlds. Before proceeding, we must be more precise about what we mean by scientific realism.

SCIENTIFIC REALISM

I define "scientific realism" along lines advocated by the philosophers Richard Boyd, William Newton-Smith, and Hilary Putnam,* which I will call (R1) to (R4):

(R1) Scientific theories are approximately true, and more recent theories are closer to being true than previous theories in the same domain. Consequently, science is a means for learning about absolute truth, universal constants of nature, and universal laws of nature, such as those of thermodynamics.

(R2) Observational and theoretical terms in theories actually refer to entities with an assumed ontological basis even though these entities are neither *directly* observable nor *directly* measurable, such as electrons and atoms. When we say that elementary particles have an ontological basis, we mean that they actually exist. "Ontology" is a term from metaphysics meaning the reality status of something.† I emphasize the word "directly" because we never directly measure or perceive anything. Even our most mundane representations are products of our perceptual and cognitive systems. How we represent the world depends on how much we know about it. Realists believe that science can teach us something about the unobservable from a combination of measurement, physical theory, theory of meaning, metaphor, and cognitive science.

---

* Putnam later changed his mind on many of these points.

† Metaphysics is the very foundation of philosophy. It deals with the nature of things, that is, existence, reality, and how we come to know these concepts. The goal of metaphysics is the ultimate ontology, which includes science and values, with no *testable* criteria. In this book we restrict ourselves to the problem of a scientific ontology in which existence is empirically testable.

(R3) Successive theories are formulated in a way that preserves most of the referents of earlier theories, which are their limiting cases.

(R4) Ultimately, it is the success of theories formulated according to (R1), (R2), and (R3) that provides empirical confirmation for realism. By "success" I mean not only explaining experimental data within the theory's domain of applicability, but also making bold and unexpected predictions, in addition to the theory's power to unify apparently unconnected phenomena. The success factor is essential.

There are deep and contentious concepts within realism such as truth and the convergence of scientific theories to the actual physical reality underlying natural phenomena. Scientific progress will be discussed in some detail in this chapter, and in Chapters 6 and 7 as well. Here we focus on comparing realism with its antithesis, antirealism.

Among the contentious notions contained in statements (R1) to (R4) is that of truth. Two concepts of truth are used here. The truth of a scientific theory is one of approximation whose measure is the theory's success in inference to the best explanation. We set out a detailed example of this type of reasoning in Chapter 2 with the double-slit experiment and the wave theory of light. The truth of a theory is necessarily linked with theories being closer and closer approximations to an underlying physical reality. For example, we say that Newtonian mechanics has been superseded in the atomic domain by quantum mechanics and in the domain of high speeds by special relativity. From this we conclude that Newtonian mechanics has a radius of applicability, not that Newtonian mechanics is wrong or has been disconfirmed. Newton's theory is approximately true. Newtonian mechanics is used to build bridges and automobiles, and we don't blame accidents on it.

In Chapter 3 we discussed the complexity of testing a scientific theory. Consider again Newton's laws of motion, which in their pristine form apply to the entire universe. In order to use them, we must add certain auxiliary statements. For example, in order to calculate how fast it takes a dropped stone to reach the ground, we must add to Newton's theory the auxiliary statement that we restrict the situation to the interaction between two bodies—Earth and stone. We neglect interactions between the stone and you and I and the moon, sun, Mars, Venus, and so on. Then we find the "usual" result of distance varying as the square of the time of fall. The problem would have been insoluble without this restriction. If the solution to a problem is unsatisfactory, we may want to alter the auxiliary statements before claiming that Newton's laws are in trouble. What are actually tested in an experiment are a scientific theory and its *auxiliary* statements. Falsification and verification in actual science are sophisticated processes.

SCIENTIFIC ANTIREALISM

Antirealists claim that there are no such things as electrons or any other indirectly measurable entity. Of key importance to any antirealist is the underdetermination thesis, which is logically unbeatable. Recall that the underdetermination thesis states that there are in principle an infinite number of theories that can describe any set of experimental data. Consequently, with what right can anyone claim that one of these theories is not only *the* correct one, but that its unobservable quantities are physically real? Therefore, electrons, for example, are just fictions useful for formulating theories but should disappear in the final analysis. Antirealists claim, as did Mach, that electrons are not real since we cannot observe them at first hand, but planets are real because we can see them.

For antirealists, scientific theories are useful only for categorizing and describing scientific data. Antirealists have a very simple definition for truth: true = correct, as verified by experiment. Although experimental verification, or success, is important for every scientific theory, scientific realism seeks explanations for a theory's success which involve postulation of unobservable entities. Scientific progress is linked, some antirealist philosophers propose, merely to better problem-solving capability. In Kuhn's view, the direction of science does not depend on any fixed goal. Rather, scientific theories emerge that by consensus provide more satisfactory solutions. Regularities in nature are supposed to be incorporated into scientific theories, but need not be explained by them. Just as Mach desired, scientific theories should be formulated from laboratory data and so contain only directly measurable entities. Their goal should not be to explain regularities such as atomic electrons.

Except for notions such as reference (to be discussed in detail in Chapter 7), scientific progress toward physical reality, and an ontology for unobserved entities, most antirealist frameworks such as Kuhn's satisfy the same stringent scientific standards as realist ones. The terms "rational" and "objective" can be applied to all scientific theories, according to certain philosophical guidelines. By "science as a model of rationality," philosophers mean that science has a goal and a way to compare competing theories. One standard way is by examining how competing theories fare in the face of experimental data. We have already noted cases in which this is not straightforward. For example, there is logically no such thing as a crucial experiment, as we discussed in Chapter 2 with regard to the nature of light. Yet sometimes scientists *agree* that an experiment is crucial, as was the case for the Michelson-Morley experiment. On the other hand, scientists can make any number of changes in a theory to help it to agree with data. In addition, there are guidelines that can go beyond science, as we witnessed in the Schrödinger-Heisenberg debate. In short, there is no single method by which competing theories are compared. Objectivity and rationality are rescued by the fact that, in time, a consensus arises as to the preferred theory. Philosophically, the goal depends on whether you are a realist or antirealist.

Kuhn's scenario was somewhat unique at first, because it did not allow competing theories to be compared. Competing theories are framed in languages that Kuhn assumes are not translatable to each other and are not chosen on rational grounds. He drew back from this untenable position by giving five factors that provide "the shared basis for theory practice," as he wrote in 1973, eleven years after the publication of *Structure of Scientific Revolutions*. A scientific theory should be accurate within its domain, be consistent, have broad scope, be simple, and be fruitful (i.e., successful). These notions are part of statements (R1)-(R4) and contain the nonrational aspect of scientific progress because, for example, judgments of simplicity are not always articulable, at least at first. Kuhn has never clarified how these five shared notions can be folded in with the Gestalt switch and its inherent incommensurability, to which he still adheres.

Be that as it may, what will survive Kuhn's scenario is his emphasis on objectivity being something decided to some extent by scientific communities. So, for example, notions of explanatory power and simplicity are often set by standards formed by a scientific community. As part of his reply to claims that his view of scientific progress is one merely of "mob psychology" in which scientists en masse decide that a change is needed—as the Hungarian philosopher Imre Lakatos once accused—Kuhn has argued that objectivity in his framework can be achieved with scientific reasoning and need not be merely a matter of taste. We emerge then with the following conclusion about electrons: According to Kuhn, they are objectively real relative to a paradigm, but they have no independent ontology.

THE SOCIAL SIDE OF SCIENCE

Whether or not a theory is based on scientific realism, there will always be culturally relative elements to it. In Chapter 3 we explored certain cases of this situation. We recall Galileo's not drawing from his thought experiments the proper conclusion that motion in a perpetual straight line is the one on which to base a new theory of motion. His reasons were rooted essentially in his religious belief that the universe cannot be infinite in extent. Indeed, cultural relativism enters science through the creative act itself. Because although most of the day-to-day activities of scientists are rational (i.e., they follow a scientific method of investigation), the act of creation is irrational. We have noted some culturally relative elements in early twentieth-century atomic physics research which entered through the concept of *Anschauung*. Nor are the views of scientists to the structure and meaning of scientific theories capable of being pinned down in any rationalistic manner. Whereas all scientists are realists in the sense of believing in the existence of atoms, their view on the theories with which they deal verges on a specie of instrumentalism, in which theories are approximations that function like instruments for calculating the properties of electrons. Other scientists proceed entirely along the scientific realist's route and deem theories as a bridge between the unseen world of elementary particles and our world of sense perceptions: Theories are probes into the unknown.

This view fundamentally differs from a currently popular philosophical view that atomic entities do exist, but the approximations necessary for calculations prevent us from knowing everything about them. But I believe fundamental theories are gradually better approximations with which we can actually get at things like "real" electrons and "real" hydrogen atoms. Another problem is that the logical probability of a universal theory being valid is zero: Essentially, universal theories are so wide in their scope that they are impossible to use.

Another view is held by the so-called social constructivists, who take to its extreme Kuhn's message of the importance to scientific research of the social community of scientists. The fundamental tenet of social constructivism is that science is a construct thought up by groups of scientists. So, for example, at the giant elementary particle physics laboratory at CERN, experiments are carried out by large teams of scientists sometimes numbering in the hundreds. Science is made by committees. Consensus involves deciding what is background and what are meaningful data. Careers are often at stake. Nobel prizes can be part of the payoff. This is a reasonable mindset. After all, who does not work for recognition? Moreover, it must be borne in mind that elementary particle physics experiments are extremely expensive. Consequently, to spend millions of dollars testing a possibly defective theory will not do. Social constructivists claim that under these communal pressures, experimentalists often choose a combination of methods for data analysis for the purpose of verifying a currently fashionable theory. Nature is a social construction.

What can we make of antirealist schemes? They have little appeal to the practicing scientist except to underline an important social component of science, as did Kuhn. Besides the postulated existence of elementary particles, what completely separates antirealism from realism are the realist's contentions that there is absolute truth and science is universal. If we ever communicate with an alien culture, we will do so with electromagnetic signals, and we will find that the aliens have the same *universal* constants as we do: the charge of the electron, the velocity of light, and Planck's constant. To the scientific realist, science is transcultural and gender-free. Science is the only knowledge system today that is capable of delivering absolute truth. Hand-in-hand with science's universality, and the awaited absolute truth, is that the science we use today—which was created by Thales and his successors—is the only knowledge of its kind. Science was created once and has since undergone many changes. Scientists of all cultures around the world perform research with methods rooted in those proposed in ancient Greece. To claim, for example, that the ritual rites natives in a rain forest perform in rare dry periods to summon down rain from the sky should also be called a science is ludicrous indeed. Before looking into the background from which such statements have emerged, let us further discuss antirealism and scientific practice.

ANTIREALISM AND SCIENTIFIC PRACTICE

We begin once again by taking a bow to the underdetermination thesis, which is, in principle, unbeatable. But what does the underdetermination thesis have to do with actual research in the physical sciences? While there may be an infinity of possible theories that can deal with a given set of empirical data, in the physical sciences there are guidelines for singling out a very small group of possible theories. Among these metaphysical assumptions are symmetry principles, conservation laws, and principles of high restriction such as the principle of relativity, and unification.

However, with what right can we insist on the validity of these statements without experimental or theoretical proof? What gives us the right, for example, to assume the *absolute* validity of conservation laws? One very good reason is that without them we would be unable to formulate any scientific theories. At first sight, this may seem like a circular argument. But it is not so, if we assume that conservation laws and other metaphysical statements such as the principle of relativity constitute part of a transcendental argument in which these statements are conditions for scientific theories. As we have seen, certain of these statements are chosen often, with little scientific basis. But if they produce successful theories that make predictions beyond their original intent, then the validity of these statements is enhanced.

Only with the greatest temerity have scientists cast a critical eye on a conservation law. In this sense, scientists are conservative. But surprises are always in store. Einstein's mass-energy equivalence replaced Lavoisier's law of conservation of mass, for over a century a centerpiece in chemistry. In general, scientists believe that conservation laws have a meaning deeper than merely for purposes of cataloging data. This belief is supported, for example, by the connection between conservation laws and symmetry principles as telling us about the fabric of nature. As long as scientists are not total instrumentalists (which none are) or complete Kuhnians (ditto), we can say that the equations of physical theory have ontological content because they tell us something about physical reality.

For example, in huge particle accelerators, swarms of electrons collide with multitudes of protons. Data are taken from instruments that indicate electrical pulses from the remnants of these collisions or provide photographs from bubble chambers.* Such laws of nature as the

---

* A bubble chamber consists of a contained liquid at the proper temperature and pressure so that a charged particle passing through leaves a trail not unlike the vapor trail of an airplane flying overhead.

conservation of energy, momentum, and electric charge are brought to bear on analyzing these data. In this manner, with highest chutzpah, scientists focus their attention downward into the interaction between just two elementary particles. The bubble chamber photograph in Figure 3(a) is particularly spectacular because it contains all of the presently known processes in elementary particle physics. Figure 3(b) is generated by a computer that has been programmed with such conservation laws as those of energy and momentum conservation.

This said, one must wonder about the origin of current interest in antirealism. Perhaps it is rooted in literary movements such as postmodernism.

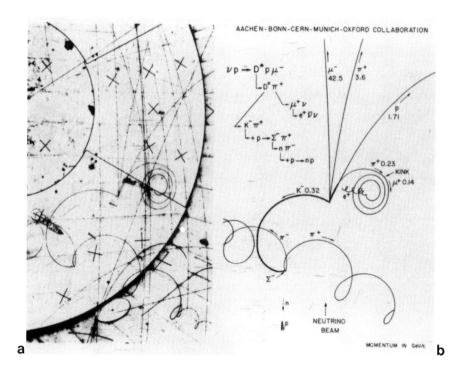

a   b

FIGURE 3.

(a) A bubble chamber photograph of a collision between a neutrino and a proton. (b) A computer tracing of the events in (a) which is executed with software containing such conservation laws as energy and momentum.

THE CHIC OF ANTIREALISM: POSTMODERNISM

Since I intend to be extremely critical about postmodernism, let me begin by saying that some antirealist philosophers have no social ax to grind and delve instead into serious philosophy, whatever is our opinion of it. There has always been a certain chic in antirealism, the cachet prominent literati lent to it. Today it goes under the rubric of postmodernism. (Today's postmodernism really ought to be called postmodernism II. Postmodernism I occurred during the late eighteenth and early nineteenth centuries, and its slogans were pretty much the same as today's.) Postmodernism is a rebellion against the Enlightenment program of the search for universal knowledge and a rational order in the world based on methods and successes of science. Postmodernism claims that this ideological system, which stands behind the culture and material practices of Western society, is bankrupt, having led to famine, sexism, wars, and environmental disasters as a result of pollution through technological excesses fueled by the quest for monetary gain. Postmodernists claim that humankind has lost its oneness with nature and no longer celebrates the individual self with his hopes, dreams, and aspirations. All of this has been sacrificed at the altar of science and technology.

Postmodernist commentators promise to lead us out of this morass. Yet, what they offer is a politics of despair and confusion: The world cannot be understood as a unity; it is whatever you want it to be—anything goes. Since there is no absolute truth, then any system of knowledge, be it political, scientific, literary, or artistic, *is* true. Art has no aesthetics. Politics is the message.

If there are *post*modernists, there must have been modernists. In the context of the late eighteenth and early nineteenth centuries, modernists are defined as supporters of the Newtonian clockwork universe. This was the basis of the Enlightenment, the Age of Rationalism. Critics of the modernists advocated a chaotic, nondeterministic, complex view of the world, which is usually referred to as Romanticism because it is the opposite of the supposedly dry rationalistic view. The most important postmodernist = Romanticist thinkers in postmodernism I were William Blake, George Gordon (Lord Byron), John Keats, and Percy Bysshe Shelley in England, and in Germany that literary giant, Johann Wolfgang von Goethe, Germany's Shakespeare. Names to be reckoned with. What scientist now or then could outdo them in prose?

These men believed they were defending the creative impulse that they deemed the cold rationalism of Newtonian science had attacked. We can see this in Shelley's 1821 reply to Thomas Love Peacock's claim that poetry was rightly in decline because most people were turning instead to more rational ways of understanding the world. With eloquence and sincerity in "A Defense of Poetry," Shelley writes that "reason is to imagination as the instrument to the agent, as the body to the spirit, as the shadow to the substance." Shelley makes it abundantly clear that he was responding to the adverse effect on society of the modernism of Newtonian science, namely, the Industrial Revolution:

> The rich have become richer, and the poor have become poorer; and the vessel of state is driven between the Scylla and Charybdis of anarchy and despotism. Such are the effects which must ever flow from an unmitigated exercise of the calculating faculty.

In his poetry, Shelley was even blunter:

> Knowledge is not happiness, and science
> But an exchange of ignorance for that
> Which is another kind of ignorance.

In a similar vein, Blake writes that "Art is the Tree of Life. . . . Science is the Tree of Death." And Keats adds,

> Do not all charms fly
> At the mere touch of cold philosophy? . . .
> In the dull catalogue of common things . . .
> Philosophy will clip an Angel's wings.

Goethe, writing much like the "nineties man," criticizes scientific thoughts:

> Unless you feel it, you will never achieve it.
> If it doesn't flow from your soul
> With natural easy power,
> Your listeners will not believe it. . . .
> How can anyone cling to such trash
> Keep any hope in one's head?

In 1774 Goethe published the most widely read novel of the eighteenth century, *The Sorrows of Young Werther*. The central character, a young man named Werther, is driven to suicide by the crassness of a rationalistic materialistic world, none of whose denizen understands his call to return to a philosophy near to nature. The book contributed to a wave of suicides throughout Europe, referred to as *Wertherfever*. Young men emulating Werther's suicide went out into the woods and shot themselves. The situation got so out of hand that in a later edition Goethe added a foreword asking the reader not to take Werther seriously, adding, "Be a man and don't follow me."

Goethe's "back to nature" theme can be traced to his sojourn in Italy, which sparked his romantic views to sublime excesses. The color and serenity of the Italian countryside, with its classical beauty, brought his *Sturm und Drang* period to a close. He obsessed over the classic purity of the color white. With the aid of a prism, Newton had demonstrated that sunlight, or white light, is comprised of six primary colors. To Goethe, Newton had tortured nature. Rather, the virginal purity of white light had been transferred into darker realms by the "deeds and afflictions of light," which it suffers by passing through and over surfaces. Goethe's theory of color was developed in outrage against the rationalism of Newton's mechanics.

Goethe strove to produce a theory of color in which the human subject, the "I," was of the essence, believing that "optical illusion is optical truth." As in *The Sorrows of Young Werther*, in science Goethe tried to dissolve the boundary between subject and object. He argued that colors are generated in the eye by a combination of the vicissitudes of white light and its clash with the substantial being called darkness. Goethe saw polarities everywhere: attraction and repulsion being the two basic forces; and contraction and expansion being central to plant growth. Along with Mayer and Oersted, Goethe was a supreme practitioner of Nature Philosophy.

Notwithstanding its mystical trimmings, Goethe's theory of light was part of the curriculum at German universities from about 1826 to 1831. With ideas of complementary colors, Goethe did indeed make a contribution to the subject. Rudolf Arnheim points out that Goethe divided Newton's six primary colors into "intertwining pairs of complementaries." Despite his immense contributions to literature, Goethe considered his scientific work to be more important. In a letter of 1829, shortly before he died, Goethe wrote that "As for what I have done as a poet, I take no pride in it whatever. . . . I am the only person in my century who knows the truth in the difficult science of colors."

We can describe Goethe's bringing of the "I" back into science in more modern terms. In reply to your beloved's question about the colors in a beautiful sunset, should you reply that the blue sky is due to the scattering of long wavelength light in the atmosphere and that yellow light is a wave phenomenon definable objectively (i.e., rationally) as waves of wavelength 5,896 Angstroms? On the romantic front, this will get you nowhere. Goethe suggests that you reply in personalized terms in a way more complex, to be sure, than electromagnetic theory and spectroscopy, but more introspective, and without torturing nature. Thus, in a Romanticist vein, one would object to imagining a tree to be nothing more than a water pump whose appendages derive their colors from determined chemical reactions like photosynthesis. Consider a story about the British astronomer Arthur Stanley Eddington, who discovered that the stars shine because of nuclear reactions in their interior. Soon after, he was staring at the starry evening sky with his girlfriend, who commented romantically on how beautiful the shining stars are. Eddington replied, "Yes, and I'm the only man who knows why they shine," which he then proceeded to explain.

This tale of a reactionary cycle in the early eighteenth century has a happy ending. The notion of cycles of history in the physical sciences in the eighteenth and nineteenth centuries has been most thoroughly studied by Stephen Brush, who has concluded that a Realist period counter to Romanticism lasted from approximately 1835 until 1870. This period coincides with the development of gas theory, with its emphasis on materialism, and its goal of explaining macroscopic processes in terms of unobservable atoms. This era witnessed the rise of such realist writers as Fyodor Dostoyevsky, Gustav Flaubert, Thomas Hardy, and Leo Tolstoy. Its painting style is characteristic of Goya, and its philosophy that of August Comte, Karl Marx, and Friedrich Engels. In contrast to the previous era, the trend became simplicity in explanation and rationalism of the calculational sort considered characteristic of science. Great advances in science occurred, such as Darwinian evolution, Maxwell's electromagnetic theory, two laws of thermodynamics, and the rise of germ theories in biology, which were somewhat analogous to the atomic theories in physics.

With the emergence of the second law of thermodynamics, this era effectively ended, because its success encouraged biologists to seek similar materialist mechanist theories. Darwin's theory of natural selection claimed to provide a mechanism for evolution of the human species from lower forms of life. Thomas Huxley, one of most influential popularizers of Darwinian evolution, lectured in 1868 about the possibility of reducing physiology to the laws of physics, and with it such manifestations of consciousness as feelings.

Although most scientists were somewhat cautious about making such pronouncements, Huxley's opinion and those of other influential popularizers offended public opinion and opened up a counterrealist movement that lasted into the first decade of the twentieth century. It is one of pessimistic neo-romanticism with a *fin de siècle* aura of decadence in life and the failure of political institutions. The new age's music was that of Mahler and Bruckner. One of its main philosophical voices was Friedrich Nietzsche's. This age was marked by rampant misinterpretations of science by such figures as the historian Henry Adams and Nietzsche, which we discussed in Chapter 4. But, as we shall see in Chapter 10, during this era there was great excitement in art, with sources in neo-Romanticism, science, and technology. As we have noted earlier in this chapter, there was also a great deal of activity in philosophy, particularly Ernst Mach's positivism. So, in the main, the *fin de siècle* mood was for the most part in the minds of the literati, as it was in the transition era between the eighteenth and nineteenth centuries.

In summary, the identifiable postmodernism of the late eighteenth and early nineteenth centuries—postmodernism I—had payoffs in science and literature. Among them are Mayer's discovery of the first law of thermodynamics and Oersted's generation of magnetism by electricity. The sublime antiscience rantings and ravings of Byron, Keats, and Shelley aside, Goethe carried out interesting scientific work: He produced a theory of complementary colors, and he discovered the intermaxillary bone in humans. At least Goethe knew whereof he spoke, which is more than we can say about today's postmodernist critics of science.

Two indicators of the backlash against the increasingly important role of science in post–World War II society were C. P. Snow's 1959 lecture "The Two Cultures" and F. R. Leavis's critique of it. While Snow attempted to educate the lay public on the importance of a place in society for science, Levis railed against it.

The Two Cultures

The reader has no doubt noticed an asymmetry in the dialectic between humanists and scientists: Criticism flows only one way, from the humanists to the scientists. The chemist and writer C. P. Snow pointed this out forcefully in his famous 1959 Rede Lecture at Cambridge University,

"The Two Cultures." From personal experience he described how often scientists know something about the humanities, sometimes in depth, rather than vice versa. Snow's description is worth quoting in full.

> A good many times I have been present at gatherings of people who, by the standards of the traditional culture, are thought highly educated and who have with considerable gusto been expressing their incredulity at the illiteracy of scientists. Once or twice I have been provoked and have asked the company how many of them could describe the Second Law of Thermodynamics. The response was cold: it was also negative. Yet I was asking something which is about the scientific equivalent of: *Have you read a work of Shakespeare's?* [emphasis in original]

On the other hand, some scientists are woefully unaware of the humanities. Needless to say, both extremes are deplorable and cry out for improvements in educational systems. Snow set up as straw men the extremes of the two cultures and argued that this will not do in the new postwar world of science and technology.

Unfortunately, when Snow's essay was published, it was misinterpreted by the well-known vitriolic Cambridge literary critic F. R. Leavis. Leavis's response was so vicious that his publisher showed it to Snow before publication to inquire whether Snow would sue. Being a scientist, and so used to fighting for his opinion, Snow declined. After accusing Snow of exhibiting "an utter lack of intellectual distinction and an embarrassing vulgarity of style. . . . He doesn't know what he means, and he doesn't know he doesn't know," Leavis emphasizes science's depersonalization of society and Snow's omission of the fact that it is built upon "a prior human achievement of collaborative creation, a more basic work of the mind of man."

Commonalities begin to emerge from antiscience criticism: Science's critics understand, and too often know, nothing about the origins, practice, substance, and goals of science, even though some of these critics have an undergraduate degree in a science. They caricature science as placing humankind in the center of a dehumanized and godless universe, while portraying the practice of science as an uncultured activity whose ultimate products debase everyday life. Yet these same critics write their essays on computers, whose basis is Newtonian mechanics and quantum theory. They ride to work in state-of-the-art automobiles while speaking on mobile telephones and are quick to visit their physicians for the latest medical care, which may even involve CAT scans and the like. But they deride scientists and technologists for not policing their products and for forcing God out of their deliberations. As Paul Gross and Norman Levitt put it in their hard-hitting critique of postmodernism, *Higher Superstition*, "For every bomb there is a vaccine; for every ICBM, a CAT scanner." In a perverse way they are lost souls seeking the scientist king to lead them out of an intellectual wilderness.

This brings us to postmodernism II.

SOME VERY MODERN POSTMODERNISTS

Proscience and antiscience attitudes have gone through phases with one or the other in the ascendant. Today we are in an antiscience phase almost as deep as the one that confronted Galileo. The only difference is that scientists are not persecuted as part of any state policy, although this did occur somewhat during the McCarthy era in the 1950s. Yet in some ways today's antiintellectualism is even worse than in McCarthy's day. Today more people know about astrology and mysticism than about astronomy and how science can, in fact, dispel many a mystery. A recent issue of the *New Scientist* states that

> Witches, druids, ecofeminists, and other modern pagans are steadily growing in numbers and influence and represent a new threat to 'rational' public policies on scientific issues— not to mention funding for research.

Unique to today's attack on science is that many of its critics are on the faculties of leading universities. It is an attack on objectivity itself.

The social constructivists, known in a more general guise as cultural constructivists, go beyond antirealism, claiming that since science is a social construction, it possesses all the defects of society such as sexism, an issue that happens to be vigorously addressed particularly in science departments and without any urging from antiscience proponents. But the cultural constructivists go further, claiming, for example, that Newtonian science would have been different had Newton been a woman: It would have better described the world as closer in touch with nature, which, compared to men, women are. Cultural constructivists assert that science is culturally biased toward Western society because we regard science as first formulated in Greece. If science were formulated elsewhere, it would be different. They believe that other cultures should be able to stake their claim of having a knowledge system comparable to what we have in science, for example, the superstitious beliefs of primitive tribes. As the philosopher of science Alan Chalmers writes in his book *Science and Fabrication*, "I do not think that [the cause of the cultural constructivists] is helped by construals of science as a capitalist male conspiracy or as indistinguishable from black magic and voodoo."

Postmodernism runs counter to the entire fabric of science with its claim to universality and objectivity. A recent spirited rejoinder can be found in the book of Gross and Levitt, *Higher Superstition*, which I summarize here in a highly condensed manner as it pertains to questions of scientific realism. Basically, everyone will agree that there is no masculine or feminine science, but good science and bad science, to paraphrase Stephen Jay Gould. Nor is there any concrete evidence "that some nontrivial, long-lasting outcomes of science *are* in error because of contextual values, and that science done in a new context—feminism—can be both important and correct," write Gross and Levitt with emphasis. Earlier in this chapter, we addressed social constructivism, which is the basis of feminist science. Interestingly, certain female commentators blanch at the image of woman having some sort of mystical relationship with nature, which harkens back to a quaint sexism.

Claims aside, science as we know it has produced spectacular results spanning the gamut from the medical sciences, to space vehicles, to an understanding of the universe. Postmodernists produce little concrete support for their case. Instead they harp on cultural relativism and the harm science has caused the environment. The environment has been damaged, but executives of oil companies, for example, are to blame rather than scientists.

In arguing against science, postmodernists display an abysmal lack of scientific acumen. For example, postmodernists sometimes reach into recent chaos theory in order to support a claim that deterministic Enlightenment science has spent itself. According to chaos theory, small changes in initial conditions can build up and cause certain mechanical systems to deviate strongly from what would have been expected according to strict Newtonian causality. Besides the fact that any research on chaos theory begins with good old-fashioned Newtonian physics, scientists feel these so-called nonlinear effects provide deeper insights into Newtonian physics. Quantum mechanics is also touted as a harbinger of the end of deterministic physics and of reliable knowledge because of Heisenberg's uncertainty principle. Although quantum mechanics' intrinsic probabilities do represent a rupture with classical physics, with proper understanding of the uncertainty principle, quantum mechanics nevertheless provides reliable information about the physical world, in addition to making definite predictions, as we discussed in Chapter 2.

Statesmen have recently gotten into the act. Of some note is Václav Havel, the poet, playwright, and first president of the Czech Republic. During his presidency, Havel wrote two influential editorial columns in the *New York Times*, published March 1, 1992 and July 8, 1994. To him the evils of the twentieth century resided not in the whims of men like Hitler and Stalin but in "rational cognitive thinking," "depersonalized objectivity," and the "cult of objectivity." According to Havel, Communism was the "perverse extreme" of science. "Traditional science, with its usual coolness, can describe the different ways we might destroy ourselves. [Science] is impersonal, inhuman, [it] kills God."

Among those affected by Havel's columns was George E. Brown, a long-time member of the House of Representatives and chairperson of the U.S. Congressional Committee on Science, Space, and Technology. Inspired by Havel, in 1992 Brown addressed the American Association for the Advancement of Science with a talk entitled, "The Objectivity Crisis: Rethinking the Role of Science in Society." Brown emphasized the need to return to "other types of cognition and experience" than the objective ones offered by science. Spurred on by Havel, Brown advocated the importance of science's not playing a central role in society. A member of Brown's select panel even seriously proposed that the membership panels judging scientific research proposals be opened to include "a homeless person [and] a member of an urban gang." Because of the powerful position Brown occupied, no immediate objections were made. By 1994, however, as a result of mulling over objections, Brown reversed himself and wrote in *Physics Today*, in a vein similar to C. P. Snow, that it is

> a moral imperative to enlist science and technology in a campaign for a more productive and humane society in which all Americans can enjoy the benefits of an improved quality of life.

As Gerald Holton has pointed out, Representative Brown was one of the few policy makers to protest a recent suggestion that "federal spending for 'curiosity-driven' research be cut back in favor of supposedly quick-payoff 'strategic research'." The intent here was to put a lid on basic research, relegating science strictly to the control of nature, one of the faces of antirealism.

We might ask where politicians come into all this, especially someone like Havel who admittedly knows essentially nothing about the subject he is criticizing. I think it important to concede that science works effectively within well-defined boundaries and is *one* way we have of understanding our world. Galileo understood this when he wrote that "The Bible tells one how to go to Heaven, but not how the heavens go." Nevertheless, Descartes, Galileo, Kepler, Newton, and other giants of the Age of Rationalism believed science to be a means for picking the mind of God, a point glossed over by postmodernists. Today we know better, despite the hype of such luminaries as Stephen Hawking, who concluded his best-selling *A Brief History of Time* with the statement: "Once we have the theory of everything we will know the mind of God."

Back to Reality

Sensing the antiscience sentiments of his era, in 1902 Henri Poincaré, the completor of Newtonian mechanics and the discoverer of chaos, wrote of "Science for Its Own Sake." He recalled such antiscience literati as Leo Tolstoy, who accused scientists of picking and choosing experimental data almost at random. Poincaré responded that the choice of what are good data is not at all straightforward: "Certain facts are more interesting than others." The scientist is not the fact-gathering machine that Francis Bacon described in the seventeenth century. Poincaré continues in an impassioned manner:

> It is only through science and art that civilization is of value. Some have wondered at this formula: science for its own sake; and yet it is as good as life for its own sake, if life is only misery; and even as happiness for its own sake, if we do not believe that all pleasures are of the same quality, if we do not wish to admit that the goal of civilization is to furnish alcohol to people who love to drink. . . . All that is not thought is pure nothingness; since we can think only thought and all the words we use to speak of things can express only thoughts, to say there is something other than thought is therefore an affirmation which can have no mean-

ing. And yet—strange contradiction for those who believe in time—geologic history shows us that life is only a short episode between two eternities of death, and that, even in this episode, conscious thought has lasted and will last only a moment. Thought is only a gleam in the midst of a long night. But it is this gleam which is everything.

So, what is the meaning of science? Put another way, what is scientific progress? Although curiosity may have killed the cat, curiosity is the fulcrum about which vital societies function. In a closed society, "science for its own sake" is virtually nonexistent. Such was the case in Galileo's Italy, Hitler's Germany, Stalin's U.S.S.R., and almost in America during the McCarthy purges, which caught no Communists but ruined the lives of thousands. Scientists are by nature curious critters. Einstein once described himself as someone who asks childlike questions.

So let's be curious and inquire into the meaning of science. If we reply as a positivist, then the meaning and goal of science become trivialized: The goal of science is merely to catalog empirical data, or even perhaps to describe but not explain. The positivist—the antirealist—is not curious about regularities in nature. The antirealist accepts them and then folds them into schemes for describing nature, while refusing to grant any depth to mathematical formulation. In stark contrast, the realist assumes that the structure of the universe is mathematical. The next chapter examines how such an assumption could hold.

6

THE
REASONABLE
EFFECTIVENESS
OF
MATHEMATICS
IN
PHYSICS

This chapter's title is a play on the title of a lecture the physicist Eugene Wigner presented in 1959, "The Unreasonable Effectiveness of Mathematics in Physics." It is a wonderful and often-quoted title for a delightful paper in which Wigner describes his awe at the

> enormous usefulness of mathematics in the natural sciences [which] is something bordering on the mysterious . . . there is no rational explanation for it.

The nature of the relation between mathematics and physics is a deep problem with which philosophers and scientists have wrestled through the ages. Whereas mathematics is exact, physics is not, because it ultimately refers to measurements, which may entail errors. What right, then, have we to express the laws of a contingent subject such as physics in exact mathematics? This chapter explores this issue in ways that avoid Wigner's attraction for the mysterious in the following manner.

Philosophers of mathematics worry about whether the entities of mathematics really exist waiting to be plucked out of the air, like our knowledge of real electrons, which are invisible to the naked eye. A positive reply puts you in with the Platonists, who argue for a world of perfect geometrical figures. A negative reply puts you into the intuition-ist or constructivist camp, according to which mathematics is a con-struction of the human mind. Rather than arguing about these philo-sophical positions, I will explore them as they pertain to physics. By "Platonism in physics," I mean exploring a world beyond appearances with theories based on the existence of unchanging physical laws that are expressible in mathematics. We have already encountered some of these laws in the principle of relativity and the conservation of energy.

MATHEMATICS AND PHYSICS

Wigner was a pioneer in the use of symmetry principles in quantum physics, for which he was awarded the Nobel prize in 1963. In his 1959 lecture he referred to the "miracle of the appropriateness of the lan-guage of mathematics" for physics research. (Wigner loved to provoke, so we assume he was using the word "miracle" in a secular sense.) He was not alone in his astonishment. Although mathematics is framed in exact terms, it describes very well the contingent or inexact data taken from the world we live in. Everyone who has solved a physics problem marvels at the predictive and explanatory power of mathematics.

Consider the following example of Newtonian causality. Hold a stone six feet from the floor, drop it, and record the time it takes to hit the floor. Now, plug the stone's initial conditions into Newton's second law (that is, the stone's initial distance from the floor—six feet—and its initial velocity, zero). After certain mathematical manipulations, out comes an equation relating the stone's height from the floor with the time at which it was at that height. Next calculate the time it takes for the stone to hit the floor. Lo and behold, it is very close to the one obtained from experiment. This is no miracle, because this equation, as well as the method used to deduce it, applies to many other phenomena.

Is mathematics simply a tool for calculation or is there something deeper going on here? Might the fabric of the universe be mathematical at its most fundamental level? We know that mathematics is more than merely a tool. Let's focus on the underlying structure of the universe being mathematical. But we must be somewhat wary here, because arguing for a mathematical fabric is not necessarily an argument for a Platonic realism in mathematics or physics. The added ingredient to a strictly mathematical argument is the *existence* of unchanging laws of nature expressible in mathematics.

Mathematics often provides us with a rich variety of possibilities, some of which we reject on the basis of occurrences in the natural world. For example, calculating the relationship between the stone's distance from the floor and the time through which it falls yields two solutions for the time, one positive, one negative. We discard the negative one under the assumption, or boundary condition, that negative times make no sense. But we don't always discard negative quantities. A famous example occurred in 1928 when P. A. M. Dirac retained the negative energy solution of his relativistic equation describing the electron.* At first this result was misunderstood and led many physicists to reject Dirac's theory out of hand. Heisenberg called it the "saddest chapter in

---

* According to relativity, a particle's energy, momentum, and rest mass are related through a quadratic equation.

modern physics." But it turned out that Dirac predicted a wholly unexpected class of matter—antimatter.

PYTHAGORAS'S PREESTABLISHED HARMONY

Maybe Newton had Pythagoras in mind when he described himself as "a child playing on the seashore of knowledge." Perhaps it was contemplation of the tiny indivisible grains of sand that gave Pythagoras the notion of the reality and centrality of integers and led him to give them prime importance in his description of terrestrial and extraterrestrial phenomena. Pythagoras understood the import of his making a connection between sensory aspects of our world and the realm of numbers.

He reinforced this connection with experiments on musical instruments in which he discovered that the laws of harmony depend on integers. For example, to play tones an octave apart requires strings with lengths in the ratio of 1:2; a fifth requires a ratio of 2:3, and a fourth, 3:4. Pythagoras boldly generalized these results to a theory of the universe in which the heavenly bodies revolve about the Earth in the most perfect of all geometrical figures, the circle, at distances varying as the ratio of integers. This supposedly produced the most harmonious configuration, giving rise to a harmony of the spheres.

As with every innovative but not well-developed notion, others went on to mold it around their own philosophical views. Plato emphasized the geometrical aspects of Pythagoreanism while demeaning its cry for experimental data. Over the portals of his Academy in Athens, he inscribed, "Let No One Enter Who Is Ignorant of Geometry." In the seventh book of the *Republic*, Plato has Socrates say,

> As you will know, the students of harmony make the same sort
> of mistake as the astronomers; they waste their time in measuring audible concords and sounds one against the other. [They]
> tease and torture the strings, racking them on the pegs.

This resembles what Goethe complained about over 2,000 years later. For Plato, insights are all that are necessary to gain access to an ideal world of perfect geometrical figures. He conceded that from time to time empirical input is helpful.

This extreme mathematical rationalism surfaced again in the writings of Descartes, but with an important difference. Descartes supported connecting mathematics with the sensory world. This was possible by a momentous decision, made first by Galileo and then more emphatically by Descartes himself, to distinguish between living and dead matter. Galileo and Descartes assumed that rocks, fluids, and other inert matter can be reduced to bare primary qualities such as position, velocity and the forces acting on them. All of these properties can be described with mathematics.

Plato's student Aristotle moved in the opposite direction to his mentor by overemphasizing the sensory aspects of the world. Aristotle strove to explain phenomena such as motion with hypotheses based on the behavior of four basic elements (earth, water, air, and fire). Although he succeeded admirably, Aristotle, like Plato, missed the essential connection between mathematics and terrestrial phenomena.

Shmuel Sambursky, an Israeli physicist and historian of science, summarized Pythagoras's influence through the ages as follows: "Scientific daring, poetical depth and religious fervor combined to provide this theory with such a powerful appeal that it continued to fascinate thinkers right up to modern times." What Pythagoras provided was less a theory than a program for understanding the world about us through numbers, with the goal of realizing the preestablished harmony of the cosmos. Plato's variation on Pythagoras's theme was to replace number with geometry. Galileo's Pythagoreanism is unquestioned, as his musings on the properties of integers and their squares clearly demonstrate. Galileo insisted that "the book of nature is written in mathematics."

For Leibnitz, Pythagoras's preestablished harmony provides the means to combine the Greek organismic universe with aspects of Newtonian science. Leibnitz accomplished this by granting almost mystical importance to conservation laws of motion such as momentum and energy (for mechanics), which seemed to pick out a single path from among many possibilities with no external influences. Much as Pythagoras would have said, Leibnitz wrote that "Music is the arithmetic of the soul, which counts without its being aware of it."

When Bohr's atomic theory first appeared, physicists waxed Pythagorean over the appearance of integral quantum numbers. The otherwise pragmatic physicist Max Born wrote in 1923 of the "magic" that, according to Bohr's atomic theory, the visual imagery of the solar system applies to atoms, too.

NEWTON, EINSTEIN, AND MINKOWSKI'S DREAM

When he began to write the *Principia,* Newton wondered about the relation between the ethereal world of mathematics and the contingent one of physics, writing in his preface: "It comes to pass that mechanics is so distinguished from geometry that what is perfectly accurate is called geometrical; what is less so, is mechanical." In a similar vein, Einstein wrote in his 1921 essay "Geometry and Experience": "As far as the propositions of mathematics refer to reality, they are not certain; and as far as they are certain they do not refer to reality."

As theoretical physics became more abstract, the need arose for more and more abstruse mathematics, which always seemed to be at hand, having been formulated by mathematicians with no particular use in mind. Quaternions were available for James Clerk Maxwell in the 1860s, non-Euclidean geometry and the tensor calculus for Einstein in 1912. The German mathematician Hermann Minkowski voiced frustration over having to live in the reflected glory of physics. Based on previous work of Henri Poincaré, in 1907 the 43-year-old Minkowski formulated a geometrical representation of special relativity that would be especially important later on for Einstein's development of general relativity. Although by this time Minkowski's research assured him lasting fame among mathematicians, his wider renown among physicists and philosophers resulted from the linking of his name with his erstwhile former student at the Swiss Polytechnic Institute, Zurich. As Minkowski wrote in a letter circa 1908, "Oh, that Einstein, always cutting lectures—I really would not have believed him capable of it."

Minkowski's great realization in physics is that the velocity of light links space *and* time into a four-dimensional *space-time*, thereby taking physics beyond the three-dimensional world of Euclidean geometry:

> Such a premonition would have been an extraordinary triumph for pure mathematics. Well, mathematics, though it now can display only arm-chair wit, has the satisfaction of being wise after the event.

Even if mathematics could never do better than this, Minkowski was adamant that it held the power "to grasp forthwith the far-reaching consequences of such a metamorphosis of our concept of nature."

Since we cannot imagine more than three dimensions, Minkowski instead offered two-dimensional cuts of four-dimensional space-time, with one spatial and one time dimension. Each point in space-time has physical connotation because its coordinates signify spatial position at a time read from one of many synchronized clocks (Figure 1).

Minkowski's accomplishment was a giant step toward realizing Descartes' program of geometrizing our view of the physical world. Minkowski's views were well known, due in no small measure to his imaginative prose:

> Henceforth space by itself, and time by itself, are doomed to fade away into mere shadows, and only a kind of union of the two will preserve an independent reality,

he wrote in 1908 in his widely read lecture "Space and Time." The union of space and time is the collection of points in Minkowski's space-time diagrams. The cost of this union is increased abstraction. Realizing this, Minkowski assuages everyone's temperaments at the painful "abandon-

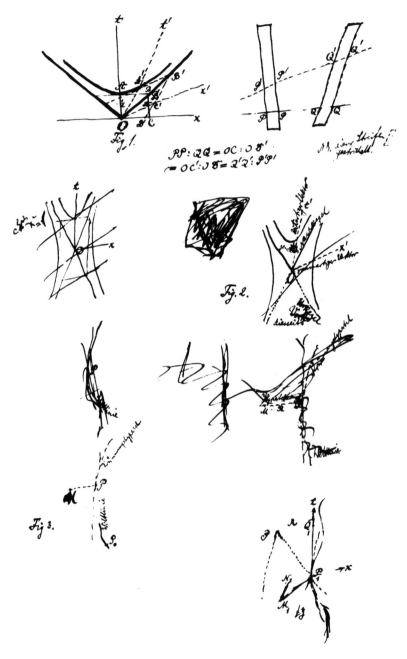

FIGURE 1.

Minkowski's sketches of space-time diagrams for his address "Space and Time," presented in 1908.

ment of old-fashioned views," a necessary prerequisite to glimpse the "idea of a preestablished harmony between pure mathematics and physics."

Minkowski's Platonism is of the sort held by many scientists, which differs from that of Plato himself. The difference lies in Minkowski's emphasis on experimental data—"The views of space and time which I wish to lay before you have sprung from the soil of experimental physics, and therein lies their strength"—and so in the relation between mathematics and our world. Like Plato, twentieth-century scientists do not consider atoms to be unchanging geometrical figures that exist in an ideal realm. Sadly, Minkowski died, in 1909 at the age of 45, before he could develop his philosophical views.

Today the border between pure mathematics and physics is almost totally blurred. Such incredibly abstract theories as string theories are formulated in 10 dimensions, where physicists themselves have generated the necessary mathematics. An indicator of this blurring of boundaries is that in 1990 the physicist Edward Witten of the Institute for Advanced Study, Princeton, was awarded the Fields Medal, equivalent to a Nobel prize in mathematics. Witten was recognized for developing new mathematics specifically for the purpose of physics research. This lends additional weight for assuming that the underlying framework of the universe is mathematical. To explore this point further, we turn to the cognitive status of science.

KANT, GEOMETRY, AND THE COGNITIVE STATUS OF SCIENCE

Among the philosopher-scientists who concerned themselves with the relation between geometry and physics was Immanuel Kant. Just like Descartes, who in his famous experience in the *poêle* (well-heated room) realized the necessity to doubt all existing knowledge and begin anew—"*Cogito ergo sum*"—Kant received a devastating philosophical awakening from the resounding message of the British philosopher David Hume. Hume argued persuasively that necessary laws of nature cannot be obtained from experimental data. Hume's reasoning was straightforward, the result profound: No matter how many experiments or measurements scientists make in order to establish a particular claim, the next experiment or measurement may disprove it. Perhaps they overlooked an experiment whose result contradicts the hypothesis under examination. In other words, verification can never suffice for a theory's acceptance. Philosophers refer to this as the problem of induction. Newton broached it in the *Principia* by inquiring into the relation between laws of physics, which are formulated in universal exact mathematical terms, and empirical data, which are necessarily particular and inexact. How can a scientist raise a hypothesis to the status of an exact law of nature beyond experimental reproach?

Kant interpreted Hume's problem as throwing doubt on Newton's purportedly exact laws of motion, particularly Newton's concept of causality in his mechanics: Inserting a system's initial conditions into Newton's laws of motion produces an analytical expression capable of predicting with, in principle, perfect accuracy the system's subsequent motion in space and time. Since all humans ultimately consist of atoms (in the vague sense of atoms in Newton's time), then by knowing the atoms' initial conditions (initial position and velocity), we can predict the course of our life. Debates about whether such a causal framework means the end of our belief in free will abounded in the eighteenth century. Of essence to us here is how Kant dealt with what seemed to him as the undermining of Newtonian physics. Being a scientist, Kant appreciated that the very core of the Newtonian edifice was at stake.*

---

* Kant was well versed in Newtonian physics. Besides Kant's knowledge of terrestrial mechanics, in his book of 1755, *Universal Natural History and Theory of the Heavens*, Kant made cosmological conjectures not confirmed until the advent of powerful twentieth-century observational techniques. For example, he conjectured that our solar system is one of a very great number of stars comprising a galaxy. Moreover, he suggested that the nebulous stars are galactic systems external to our own galaxy, and that the universe is made up of many such galaxies.

Kant asked himself whether the incredible success of Newton's physics indicated that it was more than merely another scientific theory. Kant believed that Newton had discovered properties of the mind as well. As a result, in his response to Hume, Kant considered the exact status of the law of causality. He started with the unarguable fact that, by definition, causality relates the time order of events separated in space.

In Kant's era, two poles of philosophical thought divided the intellectual world: empiricism and rationalism. Empiricists, such as Hume, believed in the primacy of sense perceptions and empirical data. Hume was also a nativist because he assumed that data are processed by a mind that has no innate knowledge. (Innate knowledge is knowledge possessed at birth.) According to Hume, we are born with a *tabula rasa*, or clean slate, and learn, at first, by association, that is, by bumbling about the world. This bumbling never stops: We learn new concepts by analogy, association, resemblance, etc., with familiar ones. As a consequence, our concepts (e.g., causality) and our universal statements (e.g., physical laws) cannot have the mark of necessity. This is another statement of the problem of induction. Consequently, in Hume's framework, exact laws of nature can never be obtained. Hume believed that Newton just had a lucky day in discovering a theory that describes experimental data extremely well.

On the other extreme of the philosophical pole were the rationalists, among them Descartes. He believed in the ability of the human mind to divine true and necessary laws of nature through intuition of "clear and distinct ideas" and based mathematical deduction from them.* Experiments serve only as a check on or an illustration of whether the scientist is proceeding correctly.

---

* Consider the following illustration of a "clear and distinct idea" given by Descartes. Before we place an oar in the water, we note that it is straight. But after we put that oar in the water, the part of it that is under water seems to be bent. We have a "clear idea" of what is going on here because we know intuitively that nothing happened to the oar. Using the physical laws, we can deduce how light is refracted when going from air into water, thereby gaining a "distinct idea" of the phenomenon of why the part of the oar in the water seems to be bent.

To see how Kant dug himself out of this seeming quagmire requires examining language and logic as understood in the late eighteenth century.

THE "RIGHT" SORT OF STATEMENT

During Kant's lifetime, the line dividing the precise world of mathematics and the imprecise world of physics was made abundantly clear by the only two possible types of logical statements. These statements, which date back to the time of Aristotle, have the grammatical structure of a sentence, with a subject and a predicate.

One is the *analytic a priori* statement, where "analytic" means "mathematical" and "a priori," "prior to experience." An example is

All bodies are extended.

The predicate ("are extended") is contained already in the subject ("all bodies") for the following reason: The term "are" can also be read as an equals sign, in which case the term "body" is synonymous with "extension": "extension" is part of the concept of "body." In this sense, analytic a priori statements do not advance our knowledge of nature.

The other possible statement for Hume was the *synthetic a posteriori* statement, where "synthetic" means that the predicate is not contained in the subject and so the connection between them is made by experiment. "A posteriori" means "after the fact." So a *synthetic a posteriori* statement is valid after the fact, as proven by experiment. For example, consider the statement

A body has gravity.

Here the relation that connects predicate ("has gravity") to the subject ("body") is empirical evidence, because only by observations and experiment (as Newton did) can we discover that there is a force (gravity) that makes bodies fall down.

But neither statement says anything about how exact laws of science are obtained. As Kant asked in his monumental work of 1781, *The Critique of Pure Reason*, how can our mind possess necessary laws of nature, such as Newton's, that are not given to us directly by induction from experience and yet have empirical consequences?* To paraphrase Kant from *The Critique of Pure Reason*: How are judgments that are synthetic *and* a priori possible?

To this problem Hume's analysis of causality provided a clue: As we have seen, according to Hume, in the synthetic a posteriori statement there is no direct connection between subject and predicate except experience. The synthetic a posteriori statement "a body has gravity" is based on our past experiences and on our habit or custom of believing that the future will be the same as the past. But due to the problem of induction, experience cannot provide us with necessity. Consequently, according to Hume, the foundations of Newtonian science are merely psychological: Newton's science can never be necessary.

Kant, on the other hand, believed that a hallmark of science is necessity. His reply to Hume's challenge was to combine the best of the opposite polarities of available statements into a new kind of logical statement, the *synthetic a priori* statement. The law of causality exists in the mind before all else (a priori) and has empirical consequences (synthetic). But it is not a priori in the sense of being innate. For Kant, the law of causality is a priori because it is generated from deeper properties of the mind.

---

* As the Kant scholar Norman Kemp Smith writes: "Seldom, in the history of literature, has a work been more conscientiously and deliberately thought out, or more hastily thrown together, than *The Critique of Pure Reason*." Kant, himself, readily admitted this, for example, in a letter to Moses Mendelssohn of August 16, 1783, from which I quote the first sentence, which is also indicative of the book's dense prose style:

> [Though the *Critique* is] the outcome of reflection which had occupied me for a period of at least twelve years, I brought it to completion in the greatest haste within some four to five months, giving the closest attention to the content, but with little thought of the exposition or of rendering it easy of comprehension, by the reader— a decision which I have never regretted, since otherwise, had I any longer delayed, and sought to give it a more popular form, the work would probably never have been completed at all.

Kant reasoned that cause and effect refer, after all, to the time order of spatially separated occurrences (events). Our ordinary concepts of space and time are Newtonian in kind. So, the Newtonian notions of space and time are taken as organizing principles, that is, principles whose role is to organize into our knowledge the potpourri of perceptions that enter our mind. In Kant's system of philosophy, space and time are regarded as *synthetic a priori* intuitions. This means that instead of being born with a *tabula rasa* [clean slate], we are born with *synthetic a priori* intuitions of space and time. These intuitions are hard-wired into the mind in order for us to organize experiences. Kant reasoned as follows: Representations of space and time must be prior to our sensations because we can neither perceive anything nor give an order to our sensations if we remove space and time.

In essence, Kant considered the astounding implications of Newtonian mechanics as extending beyond physics to the workings of the mind. He sought a cognitive basis for Newton's physics, concluding that Newtonian space and time are not empirical a posteriori concepts based on psychology (as Hume claimed), but the a priori bases (intuitions) on which our very psychological perception of the world depends as a condition or presupposition. Other principles of understanding, such as the law of causality in Newton's science, are grounded on these bases for organizing perceptions into knowledge. This is what is meant by the "cognitive status of science." Figure 2 compares and contrasts the views of Hume and Kant.

In *The Critique of Pure Reason*, Kant argued for the *apodictic* certainty of three-dimensional Euclidean geometry ("apodictic" means "necessarily true"). Consequently, accepting Newtonian space means that we must accept as well that three-dimensional Euclidean geometry is the only possible geometry for the investigation of nature. This commonsensical result was shattered in the 1820s with the discovery of non-Euclidean geometries, simultaneously and independently by Janos Bolyai, a Hungarian officer in the Austrian army, and Nikolai Ivanovitch Lobachevskii, on the faculty of the obscure Kazan University in southern Russia.

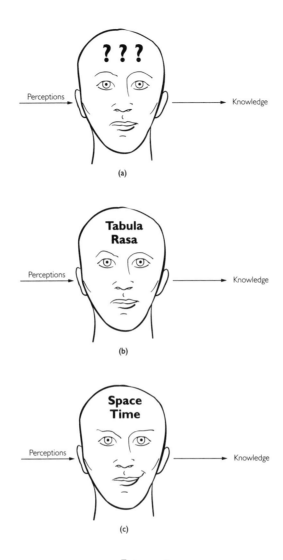

FIGURE 2

Part (a) is a schematic representation of a general problem faced by cognitive science and science: the problem of how perceptions, or empirical data, are turned into knowledge. Figuratively speaking, in part (a), perceptions enter the brain and somehow (question marks) knowledge emerges. Part (b) encapsulates and greatly simplifies Hume's view in which we are born with a *tabula rasa* (clean slate), and so we learn by experience that certain reactions to perceptions are good and some are to be avoided. The face in (c) has a sly grin because he has hard-wired into his mind the Kantian *synthetic a priori* intuitions of space and time and so can construct necessary knowledge of the world, which includes an exact Newtonian science.

Our discussion of the relation between mathematics and physics has led us to explore the connection between science and cognition, in particular Kant's proposal of *synthetic a priori* intuitions. It is important to continue this line of discussion because it serves as an introduction to the nature of scientific discovery, the concept of scientific progress, the notion of visual imagery, and the relation between art and science. These subjects form the core of this book. Our study now proceeds with the origins of non-Euclidean geometry and then the cognitive theory offered by Henri Poincaré as a replacement for Kant's.

NON-EUCLIDEAN GEOMETRIES AND EUCLIDEAN PRISONS

Can there be non-Euclidean geometries? This problem had been bandied about ever since Euclid's codification of geometry in about 300 B.C. All inquiries focused on Euclid's fifth postulate, or axiom, which can be stated in two equivalent forms:

1. Consider a straight line and a point outside of this line; only one line can be drawn through that point which is parallel to the already drawn line.

2. The sum of the interior angles of a triangle is exactly 180° (see Figure 3).

Of all the postulates of Euclid's geometry, only the fifth invites experimental testing. Yet the postulates of a purely mathematical system are not supposed to do this. Euclid's fifth postulate invites you to walk along two supposedly parallel lines and check whether they remain parallel infinitely or, equivalently, to measure the interior angles of a triangle and add them up. The first test is clearly impossible to conduct, and the second cannot give exactly 180° because of unavoidable measurement inaccuracies. These are among the reasons why mathematicians tried for 2,000 years to prove that Euclid's fifth postulate is not a postu-

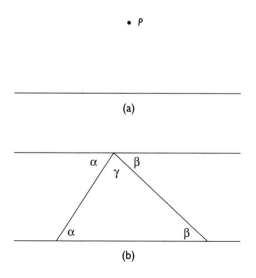

FIGURE 3.

Equivalent statements of Euclid's fifth postulate. The statement in
(a) is that through the point P only one line can be drawn parallel
to the straight line indicated in (a). The equivalent statement in (b)
is that the sum of the interior angles of a triangle add up to 180°,
that is, $\alpha + \beta + \gamma = 180°$. The proof of (b) follows from the construction in the figure, which depends on the statement that alternate interior angles of parallel lines are equal.

late at all but a statement that can be deduced from other postulates of
Euclidean geometry. Bolyai and Lobachevskii finally succeeded in the
1820s.

   While serving in the Austrian army, Janos Bolyai was kept *au
courant* on non-Euclidean geometry by his father, Farkas. Farkas had
been at the University of Göttingen with the great mathematician Karl
Friedrich Gauss, who had been keenly interested in the problem of constructing non-Euclidean geometries, a term he invented. This became
Janos's lifelong passion as well, but not before Farkas warned his son
that

   You must not attempt this approach to parallels. I know this
   way to its very end. I have traversed this bottomless night,

which extinguished all light and joy of my life. I entreat you,
leave the science of parallels alone.

In the end, father and son decided to risk all together.

Like others before them, they began by seeking inconsistencies in
alternative geometries. In 1823 Janos realized the error in this procedure and took the direct approach of formulating a geometry with a
new version of the parallel postulate: Through a point outside of a
straight line, many parallel lines can be drawn. This geometry is not at
all as intuitive as Euclid's. On November 3, 1823, Janos emphatically
announced to his father that he had *"created a new universe from nothing."*
Overjoyed at his son's success, Farkas replied with the wisdom of a vet-
eran researcher:

> Every scientific struggle is just a serious war, in which I can-
> not say when peace will arrive. Thus we ought to conquer
> when we are able, since advantage is to the first comer.

Yet Janos delayed publication until 1831 because of doubts about the
consistency of the new geometry, doubts that his father finally convinced him to put aside. Far away, at Kazan University, Nikolai
Ivanovitch Lobachevskii had been publishing on "imaginary geometries" since 1826 in the obscure Russian-language journal the *Kazan Messenger.* In 1837 he began to publish in more widely circulated German
mathematical journals. Bolyai's and Lobachevskii's results are startlingly
similar.

Bolyai and Lobachevskii liberated geometry from Euclid's fifth
postulate. Fittingly, many years later, in praise of Lobachevskii, the English mathematician William K. Clifford wrote, "what Copernicus was to
Ptolemy, that was Lobachevskii to Euclid." Yet it would take almost 30
years before the world of mathematics grasped the importance of non-
Euclidean geometries. A major problem was determining whether non-
Euclidean geometries possessed any inconsistencies. In 1868 the Italian
mathematician Enrico Beltrami demonstrated that the two-dimensional

version of Bolyai's and Lobachevskii's non-Euclidean geometry could be mapped onto discs in Euclidean space. Since the geometry of Euclidean space—Euclidean geometry—was assumed to be consistent simply because no inconsistencies had emerged in over a thousand years, non-Euclidean geometry must be consistent as well.*

So it was that mathematics escaped from its Euclidean–Kantian prison. Since the 1840s mathematicians have devised many non-Euclidean geometries, such as geometries defined on spheres and ellipses in any number of dimensions. Liberation was further underscored in the 1860s by Hermann von Helmholtz's demonstrations of visualizable non-Euclidean worlds. A familiar example is a parabolic sideview mirror on a car.

An interesting historical question arises here for which no definitive reply has been given. I have developed the history of non-Euclidean geometry in the standard manner: any geometry for which the sum of the interior angles of a triangle is not 180° is not Euclidean. Yet the sum of the interior angles of a triangle drawn on a sphere is greater than 180°. This means that spherical geometry is non-Euclidean. But spherical geometry had been used for navigational purposes before Bolyai and Lobachevskii formulated a non-Euclidean geometry. The question arises, then, why no one realized that there could be non-Euclidean geometries before Bolyai and Lobachevskii. Jeremy Gray, a historian of mathematics, offers the following conjecture. The problem of concern to everyone, from Euclid to Bolyai and Lobachevskii, was the problem of parallels, and not lines on curved surfaces. Everyone believed that if two lines through a common point are extended indefinitely they will never meet again, thereby enclosing an area. Only after Bolyai and Lobachevskii showed that a geometry could be formulated without the parallel postulate, followed in the 1840s by the German mathematician Bernhard Riemann's results on the geometry of *space*, was spherical geometry referred to as non-Euclidean.

---

* The consistency of Euclidean geometry can be proven using methods of modern logic developed since the 1930s.

Very basic and nagging questions continued to arise. Many people wondered why the Greeks dealt only with what was called Euclidean geometry if so many non-Euclidean geometries are possible. People also questioned why only Euclidean geometry applies to our world. These questions relate to the connection between mathematics and physics. Among the most fascinating and far-reaching responses are those of Henri Poincaré. In fact, they are part of the basic literature of cognitive psychology.

POINCARÉ, THE ORIGINS OF GEOMETRY, AND THE BOUNDARIES OF THOUGHT

In the late nineteenth and early twentieth centuries, the most developed view on the relation between mathematics and physics was Poincaré's, which he formed through research in pure mathematics coupled with a keen interest in psychology. Struck by the necessity to use non-Euclidean geometry for his work on differential equations in 1881* and then for another, apparently disparate discipline, the theory of numbers, Poincaré decided to explore the subject in some detail. In an 1887 paper "On the fundamental hypotheses of geometry," Poincaré explored a result obtained in the early 1870s by the moody Norwegian mathematician Marius Sophus Lie, with whom Poincaré had been in close correspondence. Lie had demonstrated that the motions of bodies in a plane are the generators of continuous groups of transformations, which became known as Lie groups. By this I mean the following: Lie groups are the mathematical description of how objects are transported from one point to another in an infinite number of infinitesimal steps. This derivation made such an enormous impression on Poincaré that he concluded his mathematical paper with a lengthy philosophical section on the origins of geometry. Poincaré realized that he could now offer a response to the

* To be discussed in Chapter 9.

problem of how it is that we can conceive intellectually of non-Euclidean geometries, yet only Euclidean geometry applies to our world. For this explanation Poincaré moved from science and mathematics per se into the realm of thinking and cognition. Poincaré's view of the origins of geometry is referred to in many texts on cognitive psychology and has influenced such major psychologists as Jean Piaget.

Children push and pull objects. They are curious and move things around. Certain objects retain their shape when their position is altered. We call them solid objects and notice that only positions of solid objects relative to other solid objects have any objective meaning. Imagine placing a ball on a particular spot on a table. Call this spot A. Now move the ball to another spot, B. The ball can also be moved from A to B along some indirect path, but the end result is the same as moving it along a straight line (Figure 4).

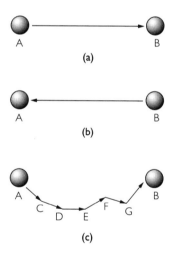

FIGURE 4

Imagine displacing a ball from a position on a table which we call A to another position B as in (a). In (b) the displacement is done in reverse and along the same line. Then (b) is called the inverse operation to the one in (a). Combining (a) and (b) is the same as having left the ball alone at A. In (c) the displacement from A to B is accomplished by a series of displacements. The sum of the movement from A to B in (c) is the same as the one in (a).

Poincaré noticed that this collection of displacements of the ball from position A to position B has three very interesting properties. Each displacement has an *inverse* because the object can be moved back to its original location. The sum of a displacement and its inverse is equivalent, or *identical*, to leaving the object unmoved at A. Lastly, any displacement from A to B is a member of the *set* of displacements taking an object from A to B. Mathematicians call any set of quantities with these three properties a group. Poincaré's example examines the group of displacements in three dimensions.

Let us pause at this point to present, as Poincaré did, a striking illustration of the difference between mathematics and physics. This is given by contrasting what he called mathematical and representative spaces.

Mathematical Space	Representative Space
homogeneous	not homogeneous
isotropic	not isotropic
three-dimensional	not necessarily three-dimensional
infinite in extent	not infinite in extent

Mathematical space is the space of axiomatic Euclidean geometry with its perfect geometrical figures. It is homogenous because no one point in it is more important than any other. It is isotropic because its properties are the same in every direction. Owing to its homogeneity and isotropy it is infinite in extent. Since the geometry of mathematical space is Euclidean, it is three-dimensional. The space we live in is representative space, and its properties are diametrically opposite to those of mathematical space. Owing to the physiology of our eyes, for example, representative space is neither homogeneous nor isotropic, nor is it infinite in extent, simply because we have restrictions on our vision. Nor is representative space necessarily three-dimensional. As Poincaré deftly points out, since our retinas are essentially two-dimensional planes and we have two eyes, why do we not see in four dimensions? The reason is that the muscles connecting our eyes work in such a way as to accommodate signals on both retinas into a three-dimensional display.

Yet we notice that the representative space of our world is, to a good approximation, homogeneous and isotropic. For example, in our daily lives we notice that objects generally retain their shape when they are moved from one position to another. Among exceptions might be moving a candle over a surface hot enough to melt it. On the scientific side of things, the failed ether-drift experiments of the 1880s and 1890s offered high-precision evidence for the isotropy of space for light propagation. On the other hand, as we explored in Chapter 1, representative space was not supposed to possess this property for light. Poincaré and other physicists puzzled over this result.

At this point, on the basis of his purely mathematical results drawn from Figure 4, Poincaré wondered if the next step was to take this group and generate Euclidean geometry. But this is not the Euclidean geometry used in mathematics. Rather, this is the *rough Euclidean geometry*, based on the properties of solid objects in the world about us, which are not the perfect solids of axiomatic Euclidean geometry. This led to three questions:

1. How can mathematicians discover axiomatic Euclidean geometry?
2. What is the relation between axiomatic and rough Euclidean geometry?
3. More generally, what is the relation between mathematics (formulated in a world of perfect objects—in mathematical space) and physics (which applies to our imperfect world—in representative space, and yet is formulated with mathematics)?

This was precisely what bothered Newton in the *Principia*. Poincaré was the first to address it squarely. For this purpose he turned to the only portion of Kant's philosophy of science that he felt had survived the crushing discovery of non-Euclidean geometries: *synthetic a priori* organizing principles. As mentioned earlier, for Kant these are the Newtonian notions of space and time, which inexorably lead only to Euclidean geometry.

As an alternative to Kant's *synthetic a priori* statements of Newtonian space and time, Poincaré proposed assuming that hard-wired (innate) into the structure of our mind (a priori) is the notion of continuous groups of transformation which have consequences (are synthetic) for our world. This gives humans the intellectual capacity to invent any type of geometry. We can imagine each group of transformations to be like a pigeonhole. Our mind surveys the collection of the objects' displacements: These are our sense perceptions. We realize that these displacements can best be organized by the mathematical group of displacements in three dimensions, which generates a rough three-dimensional Euclidean geometry.

But axiomatic geometry deals only with the displacements of ideal solids, which can be imagined to occur in very tiny or infinitesimal steps. In this way, using Lie groups, Poincaré proceeded to deduce non-Euclidean geometries in his 1887 paper. Consequently, Poincaré's next step was the mental leap to the creation of ideal solids. Imagining the displacements of ideal solids to occur infinitesimally requires that the total displacement consist of an infinite number of infinitesimal displacements. Then we pass to the limit of continuous motion. For this purpose Poincaré proposed a second *synthetic a priori* organizing principle, the principle of mathematical induction. This principle enables us to move in one fell swoop from a series of infinitesimally small steps to a single finite step. Whereas finite displacements of objects sufficed for constructing the rough geometry, summing up infinitesimal displacements plays this role for the axiomatic geometry (Figure 5). In Poincaré's view, once the proper geometry has been determined, then so is the nature of mathematical and representative spaces. In this case these spaces are three-dimensional and Euclidean.

This is the geometry based on ideal solids, among which are triangles whose interior angles add up to exactly 180°. With this framework, Poincaré could reply to a problem that had bothered mathematicians, including Bolyai and Lobachevskii, for over 2,000 years—whether geometries that deny Euclid's fifth postulate are mutually compatible.

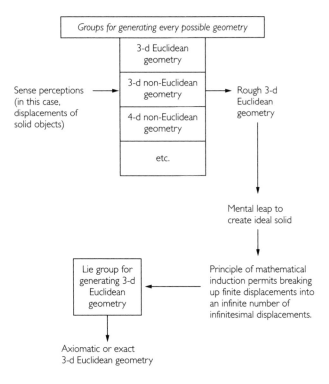

FIGURE 5.

This schematic shows Poincaré's notion for the origins of three-dimensional (3-d) Euclidean geometry. Displacements of solid objects are fed into bins hard-wired into the brain. The mind "surveys" them and realizes that sense can be made of them only with a rough three-dimensional Euclidean geometry. What occurs next is the mental leap to create the ideal solid. The principle of mathematical induction next comes into play to break up finite displacements into an infinite number of infinitesimal ones, which are fed back into the Lie group for generating three-dimensional Euclidean geometry. The axiomatic or exact three-dimensional Euclidean geometry emerges.

Poincaré believed they are, because he could not see how mathematical groups could be incompatible with each other. His reply goes deeper than Beltrami's proof because it goes beyond mathematics into the mind's organization. Why then do we choose Euclidean geometry over all other possibilities? Not because it is truer than any other geometry, but because we are guided to it by experience. In Poincaré's terminology, three-dimensional Euclidean geometry is more "convenient." As Poincaré himself summed up the situation in his 1887 paper:

> The chosen group is only more convenient than the others and we can no more say that Euclid's geometry is true and Lobachevskii's is false than we could say that Cartesian coordinates are true and polar coordinates false.

In order to more fully explore this statement we must move to the problem of whether and how geometries are testable.

TESTING GEOMETRY

Beltrami's proof that non-Euclidean geometries in two dimensions can be mapped into Euclidean geometry led Poincaré to conclude that any experimental result on what sort of geometry pertains to our world can be interpreted according to any geometry one wants. It is like translating between German and French using a dictionary. For example, the notion of a straight line in Euclidean geometry translates into an arc on the surface of a sphere in three dimensions.

Poincaré pursued this point by emphasizing more clearly than anyone before him that a theory of physics, such as Newton's mechanics or Maxwell's electromagnetism, consists of an axiomatic geometry *and* laws of physics. With this combination, the scientist makes predictions. If the predictions contradict theory, then the scientist can alter the axiomatic geometry or the laws of physics, or both. Every theory of physics thus far had been based on three-dimensional Euclidean geometry, which is also the one we use every day. Consequently, everyone assumed that Euclidean geometry is the only proper geometry in which to formulate physical theories. If a theory's prediction came out false, scientists changed the theory but maintained the axiomatic geometry. Poincaré was the first to point out that another option existed—you could change the axiomatic geometry *and* keep the theory. Poincaré's view of science in which there is the freedom to set up many different theories for any set of experimental data is known as "conventionalism." It is another statement of the underdetermination thesis.

Consider a concrete example. In 1829 Lobachevskii proposed accurate measurements of stellar parallax as a means to detect whether the interior angles of huge triangles might not add up to 180°. (Needless to say, the measurements would have had to be incredibly accurate, because the sum could never be exactly 180°, anyhow.) Twenty years later, the mathematically inclined German astronomer Friedrich Bessel, the measurer of stellar parallax, was unable to conclude one way or the other from his data (Figure 6).

However, if Bessel had measured the interior angles of the triangle (whose vertices are as described in Figure 6) and found their sum to be 170° or 200°, then according to Poincaré's conventionalism, scientists would have three possible moves:

1. Assume that the physics is correct and change to a non-Euclidean geometry.

2. Maintain Euclidean geometry and change the theory of how light rays propagate. For instance, the theory could be changed to assume that only to a very good approximation do light rays travel in a straight line. (Actually, their trajectories curve slightly.)

3. Alter both the geometry and the physics.

Poincaré and all other scientists circa 1900 opted for the second choice. Thus far nobody had proposed a theory of physics that depended on any geometry other than three-dimensional Euclidean geometry. This would have to await Einstein's general theory of relativity in 1915. In fact, general relativity was the first theory for which the geometry in which it was expressed really mattered.

But Bessel was testing the rough geometry, and Poincaré believed that there is no connection between the ideal objects of axiomatic Euclidean geometry and the worldly objects of the rough geometry. Con-

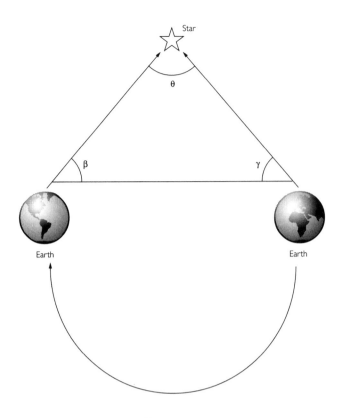

FIGURE 6.

Consider the measurement of a star's position from an observatory on the Earth. The first measurement is made when the Earth is on the right. The angle γ is made by the measuring instrument with respect to the line drawn from the Earth to its position six months later, when the Earth has moved to a position on the left. The angle β is defined similarly to γ. The angle θ is twice the angle of stellar parallax, which is *measured* by noting the movement of the star in question against the background of fixed stars. The distances are exaggerated because stars are so far from the Earth that the lines to them drawn from the Earth are almost perpendicular to the line connecting the Earth's two positions. Even if the sides of the triangle defined by the Earth in its two positions and the star are curved, the vertices that define the angles β and γ are to an excellent approximation straight lines. The reason is the measuring instrument is so incredibly small relative to the line connecting star and Earth. Adding up the three angles gives a measure of the flatness of this huge triangle. The stellar parallax of the star 61 Cygni was measured in 1838 by Friedrich Bessel and turned out to be vanishingly small, a mere 0.3 seconds of an arc. Bessel added the three interior angles θ, β, and γ and found their sum to be very close to 180°.

sequently, he concluded that axiomatic Euclidean geometry cannot be tested experimentally; neither, therefore, can the nature of space. Poincaré took space to be flaccid, taking whatever form one wishes to give it depending on what geometry is chosen. In a 1921 essay aptly entitled "Geometry and experience," Einstein agreed with Poincaré *sub speciae aeternitatus* (in the eternal sense). But, continued Einstein, Poincaré had missed the point, which is that the rough geometry is the important one for physics.

The physicist Einstein was even more Platonist than the mathematician Poincaré. For Einstein, "only daring speculation can lead us further and not accumulation of facts." Empirical data were still important, yet for Einstein thought experiments were of the essence for going beyond sense perceptions. Einstein was as clear as he could be on how to bridge the gap between empirical data and the exact statements of theory:

> The creative principle resides in mathematics. In a certain sense, therefore, I hold it true that pure thought can grasp reality, as the ancients dreamed.

Einstein's Platonic tendencies are also obvious in his address marking the sixtieth birthday of Max Planck, a fellow Platonist:

> The supreme task of the physicist is to arrive at those universal elementary laws from which the cosmos can be built up by pure deduction. There is no logical path to these laws; only intuition resting on sympathetic understanding of experience, can reach them. [T]he world of phenomena uniquely determines the theoretical system, in spite of the fact that there is no logical bridge between phenomena and their theoretical principles; this is what Leibnitz described so happily as a "pre-established harmony."

Belief in a preestablished harmony in nature was a guiding theme in Einstein's research, and mathematics was the means for its discovery.

But nowhere more dramatic than in atomic physics was mathematics not only a guiding principle, but also a key to physical reality.

Twentieth-Century Atomism, the Platonic Turn

In 1913 most physicists thought that although electrodynamics would require a major overhaul, mechanics needed less. This hope, or programme, rested on the visual imagery of atoms as minuscule solar systems. By 1923 this imagery had been jettisoned and mathematics became the guide. Soon it became the key to the atomic realm, as emphasized particularly by Heisenberg in the 1927 uncertainty principle paper.

Without a doubt, quantum mechanics is one of the most successful scientific theories ever formulated. In 1928, as a result of its marriage with relativity in Dirac's electron theory, many physicists began to believe it was *the* fundamental theory of nature, needing only some further tinkering to finish off all other fields. Bohr's assistant Léon Rosenfeld recollected a colleague's comment to the effect that

> In a couple of years we shall have cleared up electrodynamics; another couple of years for the nuclei, and physics will be finished. We shall then turn to biology.

Pronouncements like this are made every now and then when things are going well. It was said at the turn of the century, just before relativity and atomic physics upset the apple cart. Recently Stephen Hawking proclaimed the end of physics yet again—with more chutzpah than in previous eras because there is presently no theory agreed to by consensus among physicists to back him up.

What happened circa 1930 is that the roof fell in. Physicists realized that the relativistic extension of quantum mechanics to include the interaction between light and electrons (quantum electrodynamics) produced infinite results for quantities we know are finite, such as the electron's charge and mass. Therefore, quantum electrodynamics was deemed to be fundamentally flawed, and no alternative was in sight.

This created a quandary. One cannot begin totally anew. Just as Galilean–Newtonian theory grew out of Aristotle's physics and Bohr's atomic theory developed from Galilean–Newtonian theory, scientists in 1930 looked to the established nonrelativistic quantum mechanics, and the methods by which it had been formulated, for a guide to a new theory of quantum electrodynamics. Heisenberg and Pauli, among others, took up "correspondence limit procedures" to move what they took to be trustworthy variants of the established nonrelativistic quantum physics into ever-smaller spatial domains of the order of the electron's radius. As discussed with reference to Figure 9 of Chapter 2, correspondence limit procedures had been successful during the period from 1913 to 1925, when they were directed at passing from classical physics to quantum mechanics and vice versa. After 1928, complex mathematical methods were devised principally by Heisenberg for the purpose of moving *between* quantum mechanics and quantum electrodynamics. Heisenberg, and others, believed that seeking proper correspondence limit procedures was the key to branch out from quantum mechanics.

An example of the methods and mathematics that evolved from this period is Heisenberg's exploration of what he called a "lattice world," in which space is divided into cells of the size of the electron's radius (about 10^{-13} cm). In this way he hoped to eliminate the notion of a point electron and, so too, the electron's infinite mass. Although the attempt failed, today we refer to such theories, in improved form, as "lattice gauge theories."

Over the years Heisenberg's lattice world became linked with his research on cosmic rays, esoteric nonlinear unified field theories, and then with his concern that quantum mechanics used certain quantities from classical mechanics that essentially treat particles as points. This is illegal in quantum mechanics, where a particle's boundaries are ill defined because of the wave/particle duality and, consequently, the uncertainty principle.

The spatial regions of interest in quantum electrodynamics have dimensions smaller than the electron's radius. Moreover, this is the region where the unwanted infinities appear in calculations of the electron's charge and mass. On this scale, wrote Heisenberg in 1938, there will be "completely new concepts which will find an analog neither in quantum theory nor in relativity theory." He constructed a universal length out of quantities that included Planck's constant h and the velocity of light c and assumed it to be of the order of the electron's radius. In Heisenberg's opinion, h and c are truly universal constants "because they are associated with invariance properties as well as designating the limits which are set to the application of intuitive concepts." As mentioned in Chapter 2, the velocity of light c leads to the relativity of simultaneity and Planck's constant h to the wave/particle duality. The universal length is part of Heisenberg's attempt to use in 1938 the successful methodology of 1925, tried again in 1932—moving across new frontiers requires correspondence limit procedures that serve as bridges.

Heisenberg's attempts at a viable theory of cosmic ray showers based on a universal length failed, as did his work on lattice theories. Nevertheless, these lines of research led him, in 1943, to attempt a theory of elementary particle interactions based only on measurable quantities, which he called the S-matrix theory. Emphasis on observables had been his strategy in 1925 when he formulated the quantum mechanics. Heisenberg was desperate and hoped that this strategy would work again. He was optimistic that the S-matrix theory would enable him to find an appropriate mathematical description of elementary particles which, in some as yet-unknown correspondence limit, goes over into the classical description. As one might expect, Heisenberg's next step after formulating the new theory of elementary particles would be to consider its "intuitive contents [*anschaulichen Inhalt*]." Once again, mathematics containing universal quantities is the key to what is intuitive in a world beyond appearances.

Of essence here is Heisenberg's eventual realization, starting with his work in nuclear physics in 1932, that the mathematics of quantum physics generates the theory's imagery. (The next chapter discusses this.) Quantum physics thus differs from classical physics, whose visual imagery is imposed on it, as discussed in Chapter 1. This search for proper visual imagery set Heisenberg on a journey from his mid-1920s positivistic stance to Platonism.

As a youth of 18, while standing guard duty against Communist sympathizers in Munich, Heisenberg read Plato's *Timaeus*. In his memoirs he vividly described the intellectual awakening he experienced while perched on a rooftop bathed in early morning sunshine. The description strongly evokes Goethe's account of a similar experience, brought on by the brilliant Italian sunlight. The passage that struck Heisenberg was Plato's assignment of four Platonic solids to the four basic elements: earth, air, water, and fire. Heisenberg recalled his uneasiness with such speculation, but nevertheless it fired his interest to understand more. By the 1930s he was intensely interested in interpreting the mathematics of quantum mechanics to represent a Platonic-like realm of elementary particles reachable through a universal length. In his autobiography of 1969 Heisenberg wrote, "Our elementary particles can be compared with the regular solids of Plato's *Timaeus*. They are the Archetypes of the Ideas of matter."

If the structure of the universe is indeed mathematical, then we ought to be able to reach it from the opposite direction, going from physics to mathematics.

CAN PHYSICS GENERATE MATHEMATICS?
MINKOWSKI IN REVERSE

An essential part of any argument for Platonic realism is that through physics we ought to be able to discover the mathematics pertinent to the problem under investigation. This has, in fact, occurred. During the early 1700s, in the course of investigating vibrations in stretched membranes and heavy chains, members of the eminent family of Swiss mathematicians, the Bernoullis (first James and then Daniel) produced solu-

tions in the form of functions later called Bessel functions after the astronomer Bessel. In 1824 Bessel developed this class of functions in
great detail for use in dynamical astronomy. Bessel functions have
proven to be indispensable over the whole range of physics. It turned
out, particularly in the hands of Henri Poincaré, that Bessel functions
are special cases of more general ones called automorphic functions,
which, in turn, are related to number theory.

Then there is Fourier, whom we discussed in Chapter 4. Fourier's
research on heat transport, published in 1822, led to his formulation of
what are called Fourier series. By the end of the nineteenth century,
Fourier series became an essential part of such pure mathematical investigations as the theory of functions (by Bernhard Riemann) and such es-
oterica as properties of the continuum of real numbers as explored by
Georg Cantor. Another example is Heisenberg's discovery of matrices
in his 1925 quantum mechanics paper (see in Chapter 2).

A more recent example is the American physicist Murray Gell-
Mann's discovery in 1961 of a classification system for elementary particles that has similarities to the periodic table of the elements. In the late
1950s, Gell-Mann tried to generalize the known result that the quantum mechanical operators for certain properties of elementary particles
do not commute with one another. We discussed this situation in Chapter 2 regarding position and momentum. Gell-Mann dealt with more esoteric entities such as "strangeness" and "isospin." He tried fruitlessly to
generalize these results, until 1961 when a mathematician pointed out
that Gell-Mann was in the process of rediscovering the means for generating Lie groups, on which a voluminous literature existed.

Gell-Mann quickly developed a classification scheme for elementary particles based on Lie groups, for which he was awarded the Nobel
prize for physics in 1969. Just as in the periodic table, the different elementary particles are assigned positions according to the slots left open,
only now they are slots in the various representations of an appropriate
Lie group, here called SU(3), which plays a central role in Gell-Man's
theory (Figure 7). As in the periodic table, empty slots predict new ele-

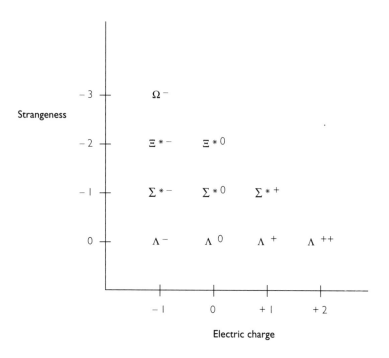

FIGURE 7.

This is the decuplet representation of the Lie group SU(3) for the set of 10 elementary particles called hyperons, which are the Λ's (lambdas), Σ's (sigmas), Ξ's (Xis), and Ω's (omegas). The vertical axis is scaled off with regard to values of the totally nonclassical property called "strangeness." The horizontal axis is marked according to electric charge. Asterisks indicate that the particle in question is an excited state of a more basic particle, which means that, for example, the Σ^{*-} is an excited state of the Σ^-. Each particle finds its place at a particular (x,y)-coordinate given by (strangeness, electric charge). I chose to display this representation because all other particles in it except the Ω^- had been found. The Ω^- slot was empty and so the particle's existence was a prediction of Gell-Mann's theory. It was found in 1964.

ments. Here the empty slot in the decuplet representation was the Ω^-, which was discovered in 1964 and considered as establishing Gell-Mann's theory. (There are instances where a single gold-plated event suffices for a theory's acceptance. Needless to say, other predictions of SU(3) were soon established as well.) But unlike the periodic table, every elementary particle in a particular representation has the same mass in a state of purest symmetry. Over the succeeding years, physicists proposed methods for how this exact symmetry is broken. One proposal is part of the successful electroweak theory.

SU(3)'s fundamental representation has three particles, which, at first sight, were so peculiar that Gell-Mann referred to them in a purely mathematical vein as "fictitious," then whimsically as "mathematical quarks," and finally as "quarks" (see Figure 8). The reason is that they have fractional electric charges, in contradiction to Millikan's experimental result that all electric charges in nature are integral multiples of the electron's charge. Gell-Mann used terms like "fictitious" and "mathematical" in 1966 to convey his view that quarks are trapped inside such nuclear constituents as protons and neutrons, a state of affairs which became known as "confinement." As Gell-Mann recalled in 1983, "Now why did I use that language? Probably because I dreaded philosophical discussions about whether particles could be considered real if they were permanently confined." Gell-Mann's original three quarks, in addition to three others postulated by 1974, have all been found (with their antiparticles). Millikan did the best he could with the theory available to him. The Lie groups of elementary particle physics are among the unchanging structures in a Platonic realism revealed through research in physics.

At a conference in 1983 on symmetries in physics, I asked Gell-Mann about his rediscovery of Lie groups. His reply was interesting:

> As I said, my views on this question [about reinventing mathematics] have changed somewhat; I still understand my old views, but I have different ones. Now, when I was younger, I thought that history showed us that a physicist did not really need to learn much mathematics. When it came to the crunch as in the case of Heisenberg, they would reinvent matrices, or whatever they needed. Or they would invent things for the first time, like Fourier who invented Fourier se-

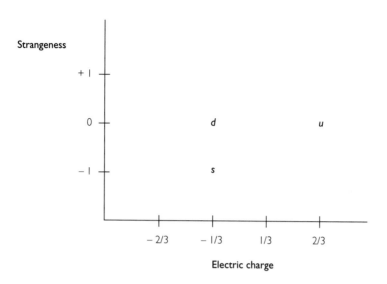

FIGURE 8.

This is the fundamental representation of SU(3), the triplet representation, postulated in 1964, where s, d, and u are quarks. The modern nomenclature for them is strange (s), down (d), and up (u). These quarks have been found experimentally by extremely indirect means because by hypothesis they never occur freely in nature. By 1974 three more quarks had been postulated. All of them have been discovered. According to SU(3) the decuplet representation in Figure 7 is "constructed" from the fundamental representation in this figure. Consequently, the particles in Figure 6 are comprised of quarks. Many physicists today believe that quarks are among the fundamental particles of nature, but they hold an open mind on this question.

ries. . . . Or if necessary they could order their mathematics delivered, like Einstein, who told Grossmann to produce for him the necessary mathematics, which Grossmann did. So, why bother? Besides I hated the style of people who knew this fancy mathematics and tried to impress their friends with it. I thought that a theoretical physicist should study the problems of physics, approaching them in a very pedestrian manner, and invent the mathematics as he needed it. . . . However, if mathematics can be taught differently, I think that these strictures do not necessarily apply. If very important, central parts of mathematics can be taught in a concrete manner, with non-trivial examples and an intuitive and computational approach. . . . I think it would be beneficial for physicists.

Gell-Mann's view on mathematics is close to Heisenberg's. In an interview with Thomas S. Kuhn in 1963, Heisenberg spoke at some length of his attraction for the "unclean mathematics" involved in mathematical schemes set up for specific "physical pictures," as for example in hydrodynamics and atomic physics. He considered this a "kind of sport":

> [This] way of representing something in nature by means of a nice mathematical scheme and then finding out something which you actually could see, well, that was quite a bit like my nature.

This rough-and-tumble approach was exactly what Heisenberg used in his groundbreaking papers; for example, in quantum mechanics in 1925 and nuclear physics in 1932.

Earlier this chapter showed how Lie groups appeared in Poincaré's theory of the origins of geometry. They are also of use in problems of classical physics. But not until quantum mechanics did anyone realize the depth of information they provided and the far-reaching consequences of their various representations. Lie groups form the basis both of a theory of geometry and of elementary particles. While the first result was obtained through mathematics, the second is from physics and argues for a mathematical structure of the universe which is realistic in the Platonic sense of there being unchanging laws and structures.

Heisenberg expressed it well in 1976, the year of his death:

> If we wish to compare the knowledge of modern particle physics with any of the old philosophies, the philosophy of

Plato appears to be the most adequate: The particles of modern particle physics are representations of symmetry groups, according to the quantum theory and to that extent they resemble the symmetrical bodies of the Platonic doctrine.

If a connection between the worlds of mathematics and physics is not assumed, then mathematics is just a means for cataloging experimental data. The emerging theories are then no more interesting than a telephone directory. Exciting things happen when we believe in a connection between mathematics and physics: We can explore worlds beyond our sense perceptions, such as the atomic realm.

While electron micrographs can be understood only with an atomistic theory, what we see in phenomena, ranging from the submicroscopic to the cosmos itself, is ultimately mediated by our physiology. We would like to have an intuitive version of how interactions among elementary particles manifest themselves in the macroscopic—that is, measurable—phenomena in the laboratory. For this purpose only a Platonic view of mathematics suffices. The Platonism for which this book argues is that the mathematics of physical theories generates the means for understanding worlds beyond sense perceptions. Among early successes were the wave interpretation of double-slit interference patterns and Maxwell's mechanical analogies for his research in electromagnetic theory. But argument by analogy sufficed here because the phenomena under study still bore a link, however tenuous, to phenomena from the world of sense perceptions.

Atoms are another story altogether, because they require a more abstract trope, the metaphor, which permits us, at least initially, to extend visualization and concepts of classical physics into the atomic realm. The next chapter examines the role metaphors play in scientific progress.

7

SCIENTIFIC
PROGRESS
AND
METAPHORS

Thus far we have explored
how scientific developments transform intuition and visual imagery. We discussed the importance of being
able to move between theories by means of universal
physical constants whose limiting values match what
we expect from perceptions. Two examples are the velocity of light becoming infinite, so that there is no relativity of time, and Planck's constant becoming zero,
thereby eliminating the wave/particle duality. We referred to this process as correspondence limit procedures. Behind all this is scientific realism, essential for
any understanding of the scientific enterprise. The per-

haps surprising ways in which scientists deal with data, and with each other, to achieve scientific progress have brought about an often highly public reappraisal of rationality.

No one can deny that science has progressed in the sense of producing more encompassing and exact means for explaining natural phenomena as well as a basis for technological advance. This is a strictly utilitarian view of scientific progress. A deeper view leads us to ask, scientific progress toward what? This is not just an academic inquiry. It goes to the very heart of the scientific enterprise. If there is no scientific progress in the sense of understanding nature, then science can serve only practical functions. This chapter takes up the concept of scientific progress and the crucial role of metaphors, which serve to extend our knowledge of poorly understood phenomena by means of those that are better understood.

THE PERVASIVENESS OF METAPHORS

Do we always mean what we say literally? If so, then what is a literal meaning? For example, I might say that "my wife is taking me to the cleaners." This *might* mean that she is driving me by car to the cleaners so that I may deliver some dirty clothes. But might it not also mean that my wife is taking all my money, and in nefarious ways? Consider the sentence, "George is an expert criminal lawyer." Is George a lawyer who specializes in criminal cases? Or is George an especially cunning rogue lawyer? Which is the literal meaning? The literal meaning of a word or expression is taken in the context of a sentence or in the tone of the person uttering it. My meaning would be clear from the context of my conversation. The meaning of words can be culture dependent, too. Any discussion of language is fraught with complexity. One reason is that we almost always express ourselves in tropes such as irony, hyperbole, idiom, oxymoron, and the most pervasive trope, metaphor.

Metaphors are a means for explaining a poorly understood entity in terms of something the reader or listener understands better. Metaphors are thus comparison statements. They are essential in teaching and popular science writing, where they sometimes occur ad nauseam. But choosing the proper metaphor is the essence of conveying a difficult or abstract thought or concept with pungency and style. Consider Polonius's advice to Laertes in Shakespeare's *Hamlet*, "Brevity is the soul of wit." Metaphors are pervasive.

How pervasive is made wonderfully clear by the linguists George Lakoff and Mark Johnson in their book *Metaphors We Live By*. Their treatment is well nigh exhaustive. They illustrate how we almost always express ourselves by conceptualizing one domain of entities in terms of another. This being the case, Lakoff and Johnson claim, there is no distinction between literal and metaphorical meanings, and I agree. But I disagree with this relativist conclusion as it applies to science. Scientific realism is the only proper philosophical and scientific view that does not trivialize science, its history, or the creative process. But if there are no purely literal statements, there is no scientific realism. How do we explain metaphor's clear importance in science and yet remain realists?

Among the points that will emerge in this chapter are:

1. Metaphors are an essential part of scientific creativity because they provide a means for seeking literal descriptions of the world about us. *These literal descriptions are scientific theories.*

2. There is a clear relation between metaphor and model. A model is an approximation to a theory. The comparison property of metaphors brings this distinction sharply into focus.

3. Metaphors underscore the continuity of theory change while ontology remains fixed. This is scientific realism. "Ontology" is a term from metaphysics meaning the reality status of something. For example, when we say that the concept of atoms has ontological content we mean that atoms are real entities.

THE INTERACTION VIEW OF METAPHOR

No adequate theory of metaphor exists. As the philosophers of science Janet Martin Soskice and Rom Harré have put it so well, "who, in any case, would be qualified to give it—philosophers, linguists, neurophysiologists?" Everyone agrees that a metaphor is a comparison statement, in which two subjects are compared. But so is a simile, in which the terms "as" or "like" usually play the role of connectors. Examples of similes are "She is like a block of ice," or "Man is like a wolf," where we can omit "like" without any misunderstanding. We recognize immediately the meaning of these sentences. One view of metaphor is called the comparison or substitution view, which regards metaphors as condensed similes. Consequently, the above examples of similes can also stand as comparison or substitution view metaphors. This is an example of the possible overlap between metaphor and simile.

Is there a way of distinguishing between metaphor and simile? The philosopher George A. Miller has suggested that "similes are less interesting only in that the terms of similitude are explicit and require less work from a reader." In the 1960s, the philosopher Max Black proposed a view of metaphor based on notions in I. A. Richards' classic *The Philosophy of Rhetoric*. Richards emphasized that the role of metaphors is more than just to relate words: they are meant to relate thoughts. Black's view is referred to as the interaction view and was formulated specifically to bring out the creative dimensions of metaphorical thought. Most important to Black was that in certain instances, "Metaphor *creates* the similarity" between two terms. This creation is obscured by the substitution theory, which, as Soskice and Harré write, "reduces metaphor to the status of a riddle or word game."

Black's interaction view of metaphor can be written as follows:

x acts *as if* it were a $\{y\}$,

where the instrument of metaphor—*as if*—relates the poorly understood primary subject x to the better-understood secondary subject y.

The curly brackets around *y* indicate that *y* stands for a collection of properties or statements, such as a scientific theory. Connections between the collection {*y*} and the primary subject *x* are usually not obvious and may not even hold, as could be the case in scientific research. Generally, *x* can also be replaced by {*x*}, a collection of properties or statements. The dissimilarity at first sight between primary and secondary subjects is referred to as the *tension* between them. The greater the tension, the greater is the creative powers of the metaphor.

We will see that metaphors in which tension is maximized involve nonpropositional, that is, nonlogical, reasoning, which often rests on visual imagery. The importance of metaphorical thought *emerges* from the study of actual scientific research.

METAPHORS AND MODELS OF SCIENTIFIC THOUGHT

As an example of creative scientific work being accomplished with an interaction metaphor, consider how, in the 1860s, James Clerk Maxwell arrived at his electromagnetic field equations along the following lines:

> The electromagnetic field behaves *as if* it were a collection of wheels, pulleys and fluids.

Here, the instrument of the metaphor—*as if*—signals a mapping from the secondary subject (the well-understood Newtonian mechanics of wheels, pulleys, and fluids, with its associated visual imagery with a means for translating mechanical terms into electromagnetic ones) toward exploring the poorly understood primary subject (electromagnetic field). Figure 1 is Maxwell's representation of a mechanical device that provides the mathematics for the metaphor's secondary subject. Here, according to Maxwell, "electrical current is represented" by the motion of small particles that have almost no mass and are spread out between neighboring vortices.

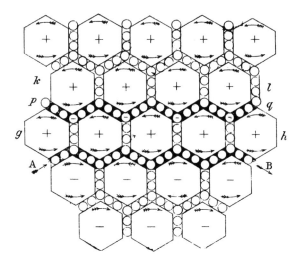

FIGURE 1.

James Clerk Maxwell's mechanical analogy for the flow of electric current. Basically, the hexagonal vortices spin counterclockwise, thereby causing particles between them to rotate as well as translate. The rotating particles represent electricity. (From Maxwell's "On physical lines of force." 1861.)

Although the speculative "as if" is essential here, as in every metaphor to be explored in this chapter, Maxwell's metaphor is a model. In Maxwell's case, for several reasons the tension is weak between the primary and secondary subjects. The secondary subject is not far removed from what was well known in the 1860s. In that era British scientists sought specifically to express their thinking in visual modes based on apparatuses that were, in principle, constructible and so could be actually experienced. Besides, no one believed that the electromagnetic field *really* has anything to do with wheels and pulleys. This is why Maxwell's metaphor is a model and his reasoning is referred to as analogical.

We will define a model as an interaction metaphor with less than maximal tension. Clearly, we cannot be so precise as to provide a quantitative scale for metaphorical tension. Rather, the notion of metaphori-

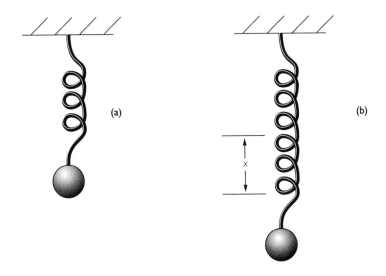

FIGURE 2.

(a) An object hangs by a spring. (b) The object has been pulled down a distance x. When released, it oscillates about the equilibrium position in (a). The periodic motion is called "simple harmonic motion."

cal tension pertains to how concrete the secondary subject is. However, this notion of concreteness itself will undergo severe transformations as science progresses. The philosophical literature contains many interesting analyses about how to define a model and what models mean. Models are approximations enabling scientists to grope their way toward a theory of some physical process, or they are simplified situations permitting application of a general theory.* Metaphor plays an essential role. As Black wrote, "Every metaphor is the tip of a submerged model."

One ubiquitous model is the simple harmonic oscillator, which was essential to Max Planck's research on cavity radiation (recall the discussion in Chapter 2). A simple harmonic oscillator is essentially an object attached to a spring (see Figure 2). Since Planck had already used

*Ameliorating circumstances concern the degree to which auxiliary statements are added to a general theory in order to make the theory applicable. This point has been discussed in Chapters 3 and 5.

thermodynamics to derive his law for cavity radiation, he could choose any model of matter for deriving this law from an atomistic viewpoint. The reason is that thermodynamics is independent of any assumptions on the constitution of matter. Consequently, Planck opted for the useful model of matter in which the electrons constituting matter can be thought of as tiny charged spheres attached to springs, "oscillator electrons." Most scientists choose this model because the motion of an object attached to a spring can be solved for exactly and quickly. Exact solvability is of the essence in the sciences.

Planck expressed this model with the metaphor of the kind:

> Cavity radiation can be explored by assuming that the electrons lining the walls of the radiation cavity behave *as if* they were charged particles on springs.

The secondary subject ("charged particles on springs"), with its well-known mechanical and electromagnetic properties, permitted Planck to explore the less well known properties of cavity radiation. Another reason this metaphor is a model is that no one seriously believed that the electrons constituting matter are attached to springs. Yet the model works because it led to a derivation of Planck's radiation law that brought into focus just how much this law violates classical physics.

During 1923 to 1925, harmonic oscillator electrons cropped up again in Bohr's atomic theory, providing the clue for Heisenberg's June 1925 breakthrough to quantum mechanics. Today oscillator electrons play a key role in the quantum theory of the electromagnetic field in which, by certain mathematical transformations,

> the electromagnetic field behaves *as if* it were a collection of harmonic oscillators.

It is instructive to compare this model with Maxwell's, in which the wheels and pulleys have been replaced by harmonic oscillators. Like Maxwell's and Planck's metaphor, this statement is also a model. Yet the ramifications of the harmonic oscillator representation run wide and deep, permitting advances in various parts of the so-called quantum theory of fields, where the mathematical representation of particles is once again set out in analogy with harmonic oscillators. Yet the harmonic oscillator representation has no ontological content.

So far we have begun to explore the role of metaphor as model, with some emphasis on the harmonic oscillator representation. We now turn to a case in which metaphor plays a more abstract role in formulating a scientific theory.

METAPHOR AND THE EMERGENCE OF SCIENTIFIC KNOWLEDGE

In his first paper on atomic theory in 1913, Bohr emphasized that although Newtonian mechanics is violated, its symbols permit visualization of an atom as a minuscule solar system. Bohr based all of his reasoning on the following visual metaphor:

The atom behaves *as if* it were a minuscule solar system.

The instrument of metaphor—*as if*—signals a mapping, or transference, from the secondary subject (classical celestial mechanics with its accompanying visual imagery, all of which is suitably altered by the introduction of Planck's constant), for the purpose of exploring the not yet well understood primary subject (the atom). Bohr employed a visual metaphor of an interaction sort, because primary and secondary subjects are webs of implications, each one having a visual imagery. Yet a measure of the abstractness of the secondary subject is that the folding together of Planck's constant with classical mechanics not only violates classical mechanics but also wreaks havoc with the primary subject regarding, for example, a visual description of the motion of atomic electrons. It was for such reasons that, in 1913, Bohr's atomic theory took scientists by surprise. One may say that the high degree of abstractness of the metaphor's secondary subject is a measure of high tension between the primary and secondary subjects.

Toward understanding the concept of scientific progress, Figure 3 depicts how Bohr's theory emerged from classical mechanics. This figure shows more than the historical development of Bohr's theory. It is a representation of the interplay between experimental and theoretical reasoning. Scientists examine information, which can come from either empirical data or from thought experiments. They realize a new way of organizing these data. (In this instance, Bohr was also influenced by the psychological research of William James. We will return to such instances of scientific discovery in Chapter 9.) Figure 3 is meant to depict more than a scenario in which scientists proceed in their research by proposing and then testing hypotheses. It is not meant to depict an anti-realist scenario in which higher level theories emerge only in order to describe more complex data. We are trying to do more than merely describe experimental data. More must be going on here. After all, so

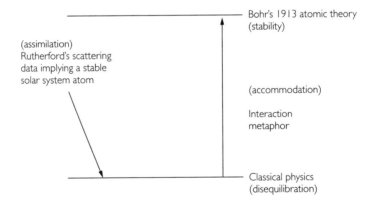

FIGURE 3.

Rutherford's data are considered by (assimilated to) classical physics, which is incapable of (disequilibrated by) them. Through an interaction metaphor, Bohr adjusted (accommodated) classical physics to these data with the emergence of a theory with a stable atom.

much is included in the figure: data, theory, metaphor, and scientific progress.

Cognitive science permits us to explore further the "deep structure" of Figure 3. As we began to see in Chapter 6, the connection between key problems in cognitive science and the history of scientific thought is not surprising (Figure 4). These two disciplines are confronted with the same fundamental problem: how information is transformed into knowledge. In both parts of Figure 4, the question mark is directly involved with metaphorical thought.

The Swiss psychologist Jean Piaget made it abundantly clear that the basic problem faced by psychology, philosophy, and science is how

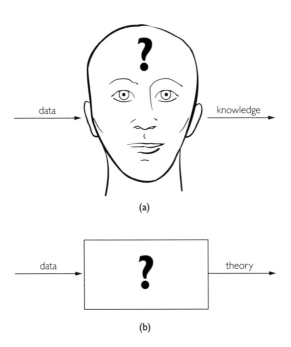

(a)

(b)

FIGURE 4.

Cognitive science and philosophy of science confront the same basic problem. (a) Cognitive science explores how the brain transforms incoming data into knowledge. (b) Philosophy of science studies the means by which data are transformed into theory.

knowledge emerges from sense perceptions or data. On this point, Piaget often noted his debt to conversations with Einstein and to hints from Poincaré's view on the roots of geometrical knowledge. Piaget wondered essentially about a paradox discussed by Plato in the *Meno:* How can new concepts emerge from ones already set into the brain? In other words, how can a system produce results that go far beyond the statements included in it? This is the problem of creativity.

Piaget's answer is the assimilation/accommodation process. Data are incorporated, or assimilated, into an already existing body of knowledge, which can be disequilibrated (Figure 5). In Figure 5 the lower level theory T_1 is thrown into confusion, or becomes disequilibrated, because of the assimilation of new information. Through reasoning involving metaphors, T_1 adjusts or accommodates itself to these new data, and a new and hopefully higher level of knowledge, T_2, emerges to provide a better understanding of the data in question. We shall see how this process of assimilation and accommodation emerges from actual case studies in scientific research.

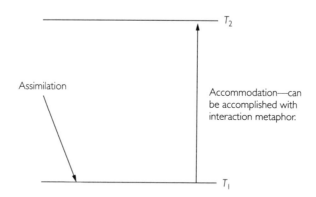

FIGURE 5.

Information is assimilated (incorporated) into an already existing level of knowledge, T_1, which can become disequilibrated. Accommodation (adjustment) can be accomplished with an interaction metaphor. The result is the emergence of a hopefully higher level of knowledge, T_2.

As the philosopher of science Richard Boyd has observed, metaphors are essential for "accommodation of language to the yet undiscovered causal features of the world." The emergence of Bohr's atomic theory in 1913 is one example.

That we learn new things by relating our intake of information to things already known may, at first sight, seem straightforward. This process is known as apperception. Apperception, or accommodation, embodies the concept of continuous change, and this is the same manner in which theories emerge one from another. The message of underlying continuity between theories was overlooked by the logical positivists and their close cousins, the supporters of Kuhn's notion of scientific revolutions. To their detriment, these groups paid scant attention to cognitive studies, focusing instead on analysis of language. This point is important for how we understand science.

MEANING AND ANTIREALISM

If we are going to talk about science as *actually* describing the world about us and not merely as an intellectual or social exercise, then we must be more precise about what we *mean* when we inquire how scientific theories inform us about, say, electrons and atoms. If science is an "anything goes" exercise, then all the ballyhoo about science and religion, and science and morals, is meaningless.

What do we *mean* by the term, or concept, "electron"? In philosophy, the term "electron" is called a *natural kind* because it is part of the fabric of the physical universe and requires a theory for its explication. So do lemons and pencils. But there is obviously something more fundamental about electrons. What is the reality status of the entity to which the term "electron" refers, and what role does this entity play in scientific theories?

In order to set our replies into their proper context, let us make a short detour into descriptivist or positivist theories of meaning. In these theories the meaning of the term "electron" is fixed by definition as follows. First, the electron's properties at a certain time in the history of science are denoted. These properties are then strung together with the word "and." This conjunction of the electron's properties is a definition or theory of what it is to be an electron. Consequently, the meaning of the natural kind electron is given by a theory that encapsulates all of the information we have about the term at a particular time. Meaning defines what sort of entities are electrons, that is, the term's extension or reference, and in this way the term itself, or referent. Let's take a concrete example.

Imagine that you are living in 1904 and you want to explain to someone what an electron is. According to positivist theories of meaning, you might say that the meaning of the term "electron" is given by the conjunction of the following predicates ("predicate" is used in its usual grammatical way, where the sentence's subject is "electron"):

(a) is a particle with negative charge,

(b) has a certain charge-to-mass ratio,

(c) has a certain radius,

(d) has a mass that depends on the electron's velocity in a way predicted by Max Abraham's electron theory.

Although the class of entities, or extension, of each of the predicates (properties) (a)–(d) is the same (electrons), the meanings of the predicates (a)–(c) and (d) differ. The reason is that, in 1904, properties (a)–(c) were accepted by everyone, whereas property (d) was predicted by Abraham's electron theory. As we know, (d) was in competition with the predicate

(d') has a mass that depends on the electron's velocity in a way predicted by Lorentz's electron theory.

According to positivist theories of meaning, we have two theories of the electron. Abraham's can be written as the conjunction of the predicates (a)–(d), which is (a) *and* (b) *and* (c) *and* (d) and whose referent is $(electron)_{Abraham}$. Then there is Lorentz's theory, which we can similarly write as (a) *and* (b) *and* (c) *and* (d'), whose referent is (electron)$_{Lorentz}$. Everyone accepted predicates (a)–(c), and so attention was focused on (d) and (d'). Since (a)–(c) are assumed to be correct, then if either of the predicates (d) or (d') is incorrect, either Abraham's or Lorentz's theory is incorrect, accordingly. The predicate (d) turned out to be incorrect and so, too, Abraham's theory. Consequently, the extension of the term "electron" in the *logically* possible theory of Abraham is empty. There are no Abraham-type electrons in nature.

So it turned out that the meaning of the term "electron" was to be taken as defined relative to Lorentz's theory. But then physicists tried to extend Lorentz's theory to spectroscopic data, which we call (s'). This new form of Lorentz's theory comprises (a)–(d'), as before, only now with conjunction of the predicate (s'), which is "radiates in a way suitable for describing atomic spectral lines." As Bohr showed, (s') is incorrect. Consequently, according to any positivist theory of meaning, Lorentz's theory suffered the same fate as Abraham's. The new electron had to be defined relative to Bohr's atomic theory. However, Lorentz's theory, patched up with special relativity, is still valid for nonatomic physics. But if we persist in using Lorentz's theory, then we'll have a problem if we talk about Lorentz's electron, which is supposedly no longer a viable concept. Actually, I should say that antirealists have a problem.

This view of meaning is antirealist because, by definition, the term "electron" is tied to a particular theory. In such a relativist world view, literally anything goes. Should a rival electron theory emerge that somehow or other fits available data, then there is another kind of electron. But this means that ontology changes whenever theory changes. If theory offers nothing but a succession of mutually exclusive ontologies, why bother with science? The core problem with positivist theories of meaning is that they are based on analysis of language, rather than going the more reasonable route of allowing language to be tempered by the furniture of the world, which includes tables, chairs, and electrons, too.

For realists the major problem is how to set up a theory of meaning in which natural kind terms are not defined in relation to any particular theory, that is, are not theory dependent.

LANGUAGE AND REALISM

The American philosopher Hilary Putnam has offered a theory of reference and meaning that can be adapted to scientific realism regarding elementary particles. Putnam begins with metaphysics rather than a theory of language. He calls his view the "causal theory of reference." One of his illustrative examples, the natural kind term "electricity," shows why. What this term *refers* to was established in the late eighteenth century by Benjamin Franklin, who pointed out that electricity is something that flows from clouds during lightning storms as well as from the static electricity machines that were all the rage. This is an "ostensive definition"—defining a word by pointing to the physical object or process it denotes. Subsequent scientific investigations have widened and deepened our knowledge of electricity. But we can always refer back to Franklin's hands-on demonstration for the referent of the term "electricity."

More sophisticated theories of electricity emerged, such as Maxwell's and its subsequent elaboration by Lorentz based on electrons. The purpose of these successive theories is to explore the natural kind term "electricity" whose referent was fixed long before by ostensive definition. The causal theory of reference defines natural kind terms in such a way that they can be transferred from one theory to the next. It accomplishes this by lifting natural kind terms out of theories. In this way, setting the reference of a natural kind term once and for all by an introducing act enables the term to be used in a succession of different theories.

A key point in causal theory is that at any moment the notion of electricity is given by an incomplete set of characteristics that Putnam calls "stereotypes." Most of us identify natural kinds by stereotypes and leave it up to experts for adjudication, if necessary. We see someone wearing a ring that we identify as gold because it looks like the stuff we have been taught is gold. In a pinch, however, only experts can tell. The natural kind term's set of stereotypes, or properties, is *enlarged* by scientific research. This is in contrast to positivist theories of meaning, wherein reference and meaning are fixed by definition within a particular theory. For positivists, the term "electron" is defined relative to a succession of theories, each replacing the other in a discontinuous manner. In realist theories, on the other hand, scientific research explores the essence or microstructure of natural kind terms (Figures 6, 7, and 8).

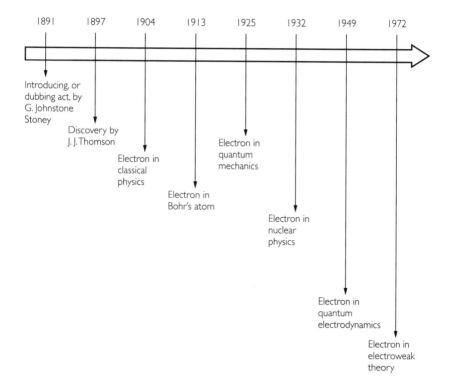

FIGURE 6.

The development of the electron from 1891 to 1972. It was dubbed by the English scientist G. Johnstone Stoney on the basis of experimental and theoretical results implying such an entity's existence.

Classical theory	Quantum mechanics	Quantum electrodynamics	Electroweak theory
— charge	— spin	— anomalous g-factor	— electron is fundamental
— mass	— radius is ambiguous		— interactions with neutrinos
— radius	— indistinguishability		

FIGURE 7.

This figure shows how scientific research adds to and sets restrictions on the electron's stereotypes, thereby adding to our knowledge of the electron's structure. Quantum mechanics informs us that in order to explain certain empirical data, in addition to having a more complete theory of the electron, the electron has a non-visualizable degree of freedom called spin. According to Heisenberg's uncertainty principle, the notion of a well-defined electron radius turns out to be ambiguous. Another surprise of quantum mechanics is the indistinguishability of electrons. This is counterintuitive according to classical physics in which we ought to be able to distinguish anything we want by, for example, painting certain molecules different colors. We learn from quantum electrodynamics that the interaction of the electron with its cloud of light quanta gives rise to the so-called anomalous g-factor. Electroweak theory (to be discussed later) reveals that the electron is truly fundamental and also informs us how it interacts with neutrinos. Throughout these theoretical and experimental explorations, the electron's charge and mass reappear, and the natural kind term can be causally related back to its dubbing in 1891 by J. Johnstone Stoney.

RELATIVISM IN LANGUAGE AND SCIENCE

Recall from Chapter 5 the discussion of Kuhn's view that old and new theories are so conceptually different that there is no common measure between their terms—a Gestalt switch has occurred (see Figure 2 of Chapter 5). The two sets of theories are said to be "incommensurable," and so theory change is supposed to be discontinuous. Can this be? What does this view mean for realism? The incommensurability of theories does not fit Putnam's causal theory of reference, or any other framework that assumes continuity of meaning based on an underlying reality. Since positivist theories of meaning are uniquely compatible

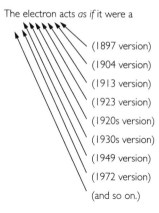

The electron acts *as if* it were a

(1897 version)
(1904 version)
(1913 version)
(1923 version)
(1920s version)
(1930s version)
(1949 version)
(1972 version)
(and so on.)

FIGURE 8.

According to the causal theory of reference and metaphors, the natural kind term "electron" is explored by successive scientific theories. The theories themselves emerged from metaphors. The 1897 version that probed the electron is classical electromagnetic theory applied to the electron as if it were a charged billiard ball. The 1904 version is Lorentz's theory of the deformable electron. The 1913 version is Bohr's atomic theory in which the electron became further removed from classical physics. The 1923 version is the harmonic oscillator one. The 1920s version was the one proposed by Heisenberg and Schrödinger attributing certain quantum mechanical properties to electrons. The 1930s and 1949 versions probed the electron as a primitively bare object surrounded by its covering of light quanta. The 1972 version explores the natural kind term "electron" with the electroweak theory. Other theories, yet to be formulated adequately, will further explore the electron.

with Kuhn's view, then incommensurability means antirealism, and relativism, too. After all, since incommensurable theories cannot be compared, then anything goes. Many different theories can be valid, but not necessarily true.

The forerunners of the scientific incommensurability advocated by Feyerabend and Kuhn were the linguists Edward Sapir and Benjamin S. Whorf. Stephen Pinker gives a succinct definition of the so-called Sapir–Whorf hypothesis of linguistic determinism and relativism in his 1994 book *The Language Instinct:*

Peoples' thoughts are determined by the categories made available by their language, and its weaker version, linguistic relativity, stating that differences among languages cause differences in the thoughts of speakers.

According to Sapir and Whorf, language is prior to thought and consequently primes our world view, or *Weltanschauung*. Whorf believed that his research among the Hopi people was proof positive. He claimed to have found that Hopi temporal and spatial terms differ from those in Western society. For example, their verb tense system has only a present tense and so Hopis have no concept of "time flowing." On this basis, Whorf concluded that Hopi is not translatable into English, and consequently, their world view is incommensurable with ours. He carried out similar studies on Apache people and had similar results.

The Sapir–Whorf hypothesis proposes that Apaches and Hopis think differently from us and so construct a different view of the world because they speak differently. This is a circular argument. No expert linguist would defend the untranslatability of Hopi syntax into that of English or any other European language. Also, language most certainly tempers our view of the world, but language cannot have any meaning without acting on the objects found in the world. In any case, Whorf's knowledge of Hopi (and Apache) turned out to be somewhat questionable. His thesis of linguistic determinism does not hold up under scrutiny.

Feyerabend and Kuhn tried to transfer Whorf's thesis of linguistic relativity into the history of scientific thought as conceptual relativism. According to conceptual relativism, different scientific theories are incommensurable and so noncommunicable in the same sense as people in different cultures with different languages, and so different world views, cannot communicate with one another. But what about dictionaries? Conceptual relativists reply that there is no unique translation dictionary, a result related to the underdetermination thesis. This view wants us to believe that Einstein could not explain relativity theory to

Newton, that Heisenberg and Bohr could not retain and separate the concepts of the new and old atomic physics—that no one could become bilingual and bicultural, too. Despite all the ink spilled over incommensurability, I have always thought that no one could ever believe this stuff. Besides historical studies to the contrary, there is the philosopher Donald Davidson's devastating critique:

> The dominant metaphor of conceptual relativism, that of differing points of view, seems to betray an underlying paradox. Different points of view make sense, but only if there is a common coordinate system on which to plot them; yet the existence of a common system belies the claim of dramatic incomparability.

In contrast to Kuhn's view, causal theory permits meaning to change while reference remains fixed. Another way of stating this is that causal theory permits continuous theory change while ontology remains the same. This view is consistent with scientific realism.

Positivist theories of meaning are inconsistent with the history of scientific thought and so offer but a caricature of scientific progress. Their conceptual relativism cuts off discourse on the value of science while placing science in the moral grab bag of abhorrent political philosophies.

METAPHORS, LANGUAGE, AND REALISM

The Bohr atom provides a concrete example of why it is natural to combine causal theory of reference with metaphors to formulate a view of scientific progress that is consistent with the history of scientific thought (Figure 9). The referent of the 1913 Bohr atom is fixed causally by its "dubbing" in Rutherford's 1911 experiments as an entity with a positively charged center and external electrons: This is the natural kind term "atom." So we may say that by metaphor, we explore the natural

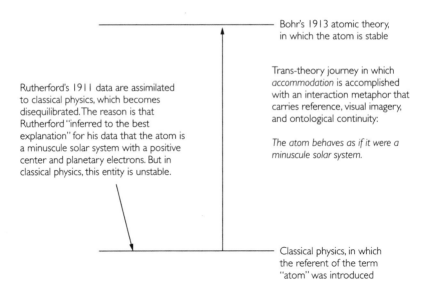

Rutherford's 1911 data are assimilated to classical physics, which becomes disequilibrated. The reason is that Rutherford "inferred to the best explanation" for his data that the atom is a minuscule solar system with a positive center and planetary electrons. But in classical physics, this entity is unstable.

Bohr's 1913 atomic theory, in which the atom is stable

Trans-theory journey in which *accommodation* is accomplished with an interaction metaphor that carries reference, visual imagery, and ontological continuity:

The atom behaves as if it were a minuscule solar system.

Classical physics, in which the referent of the term "atom" was introduced

FIGURE 9.

Example of how the interaction metaphor serves as the instrument of accommodation.

kind term "atom" in order to better understand its meaning (expand its set of stereotypes) and so, too, its essence (microstructure).

Combining metaphors with causal theory of reference opens up a view of scientific progress involving the notion of possible worlds. By definition, a possible world is a counterfactual situation consistent with logic. For instance, electrons *would have been* (electron)$_{Abraham}$ *if* their mass *had* behaved in the manner predicted by Abraham's theory. But in positivist theories of meaning, possible worlds are disjoint and so are inhabited by individuals who cannot communicate. This makes no sense. In causal theory of reference, on the other hand, we can make journeys between theories. Starting from our own world, we can use metaphor and correspondence principles to extend natural kind terms into another world by adding to the term's stereotypes. In this way scientists use scientific theories to explore the electron, which remains the same entity in Lorentz's theory, Bohr's theory, quantum mechanics, quantum electrodynamics, and so on (see Figures 7 and 8). Scientific theories are bridges to other worlds.

Thus, the causal theory of reference is no mere word game played by simply redefining "reference" in order to support realism. Scientific realism receives support from the history of scientific thought and makes us wary of any theory of reference or of language in disagreement with it. Since metaphors are one way of exploring physical reality, the next step is to combine metaphors with causal theory of reference.

AT A LOSS FOR WORDS

Consider a situation in which a natural kind term can be neither introduced ostensively (that is, by physical demonstration) nor even discussed with existing terminology. As Richard Boyd has suggested, catachresis is an essential function for metaphors in modern science. If reference cannot be set for such a natural kind term, then the term refers to nothing real. But if we assume that metaphor is the key to this situation, then the solution to the problem becomes crystal clear: Metaphors are a means for seeking literal descriptions of the world about us. These literal descriptions are scientific theories. Here is an example.

In classical physics there are two sorts of attractive forces: gravitational and electromagnetic. The principal problem nuclear physicists faced in 1932 was how to formulate a theory of an attractive nuclear force operating between a charged proton and the newly discovered neutron, an electrically neutral particle. This is an extremely nonclassical situation for which no language existed. (By "language" I mean the language of theoretical physics. But even ordinary language is problematic here because we are used to such phrases as "opposites attract while likes repel.") Nuclear physicists needed to enlarge their conceptual framework from what had become intuitive notions of electricity. They had to imagine how an electrically neutral object could attract a positively charged one.

In 1932, Werner Heisenberg saw a possibility, which he thought might also take care of the problem of β-decay. At this time, β-decay was known only as a process by which nuclei transform themselves by emitting an electron, coincident with a neutron's becoming a proton.* Heisenberg tackled these two problems by assuming them to be connected: the theoretical problem of how to describe the nuclear force, and the experimental puzzles posed by β-decay, such as whether the emitted electrons came from the nucleus itself. We may say that Heisenberg sought a unified theory of the strong and weak interactions. (As always, he was years ahead of his time. Although his theory of the nuclear force failed, it suggested the basis of modern elementary particle physics—forces transmitted by particles. *Sic gloria transit.*)

For this purpose, Heisenberg reached back to 1926, to one of his dazzling discoveries in quantum mechanics. In order to explain certain properties of the helium atom, he had postulated a force between the atom's two electrons that depended on their being indistinguishable.† Under this so-called exchange force, the indistinguishable electrons exchange places at a rapid rate. Consequently, no notion of visual imagery can be imposed on this *nonclassical* process, nor can one be generated by the theory's mathematics. On the other hand, the helium atom had resisted all attempts using the old Bohr theory to understand its properties.

The success of the exchange force in quantum mechanics led physicists to extend it to molecular physics. Of particular interest was another bane of the old Bohr theory in addition to the helium atom, the H_2^+ ion depicted in Figure 10(a). Bohr's theory pictures the H_2^+ ion as a planetary system consisting of an electron orbiting two protons. Despite attempts in 1921 by some of the strongest calculators, no reason-

---

*Some two decades later, β-decay was called a weak interaction in comparison with the strong interactions among particles such as neutrons and protons, and the electromagnetic interactions among charged particles.

†In classical physics we may follow particles by assuming that in principle we can color them as, say, red and blue. This is not permitted in quantum mechanics, where only certain sets of specifications are possible. Nor can we follow their trajectory by making position measurements, because any such experiment irretrievably disturbs the system.

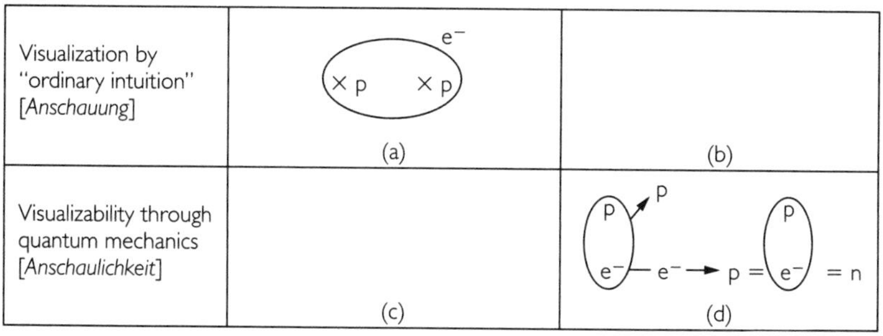

Visualization by "ordinary intuition" [Anschauung]	(a)	(b)
Visualizability through quantum mechanics [Anschaulichkeit]	(c)	(d)

FIGURE 10.

These two rows show the difference between visualization and visualizability. Frame (a) depicts the solar system H_2^+ ion, which is the visual imagery imposed on the mathematics of Bohr's atomic theory, where the p's denote protons about which an electron (e^-) rotates. But Bohr's theory could not produce proper stationary states for this entity. Frame (c) is empty because quantum mechanics gives no visual image of the exchange force. Frame (b) is empty because classical physics yields no visualization for the nuclear exchange force. Frame (d) is the depiction of Heisenberg's nuclear force, which is generated from the mathematics of his nuclear theory, where n is a neutron assumed to be a proton–electron bound state and e^- is the electron carrying the nuclear force. Heisenberg had no compunctions about employing spinless electrons, because he believed that quantum mechanics might not apply within the nucleus.

able H_2^+ ion under the Bohr planetary theory emerged. Instead, the species of H_2^+ ion that it produced were unstable. Heisenberg recalled this failure, in addition to the helium atom, as ominous.

On the other hand, according to quantum mechanics, the exchange force for the H_2^+ ion operates through the electron being swapped, or exchanged, between the two protons at the rate of 10^{12} times per second. Clearly, this process is unvisualizable, and so the box in Figure 10(c) is empty. Because the exchange force ensures the stability of the helium atom and the H_2^+ ion, in 1932 Heisenberg decided to take it inside the nucleus.

Had Heisenberg tried to visualize the nuclear exchange force generated by the mathematics of his nuclear theory, the image would have looked like the one in Figure 10(d). Heisenberg assumed that inside the nucleus the neutron is a compound object consisting of an electron and a proton. The attractive force, wrote Heisenberg in 1932, operates through the disintegration within the nucleus of a compound neutron, causing the "migration" of an electron over to a fundamental proton, which captures it. The proton–electron pair becomes a compound neutron. According to Heisenberg, the migratory electron is the carrier of the attractive nuclear force.

Yet in order for the compound neutron to exist inside the nucleus, the electron could not have certain of its by then accepted characteristics. It could not have any "spin."* Heisenberg was untroubled by this because he assumed that upon leaving the nucleus in β-decay the electron would instantly acquire its one-half unit of spin. At this time Heisenberg and Bohr were willing to entertain the notion that quantum mechanics was invalid inside the nucleus! This typified Heisenberg's free-for-all attitude toward theorizing. Like Einstein and Mozart, Heisenberg had the tools of his trade so firmly in hand that he could let his mind wander above merely technical matters. As the Belgian physicist Léon Rosenfeld put it, "A wonderful combination of profound intuition and formal virtuosity inspired Heisenberg to conceptions of striking brilliance."

In Heisenberg's nuclear exchange force, the neutron and proton do not merely exchange places. Rather, what was an *analogy* for the H_2^+ ion becomes a visualizable metaphor in the nuclear situation. The metaphor of motion is of the essence here:

The nuclear force acts *as if* a particle were exchanged.

The secondary subject ("particle exchanged") sets the reference for the primary subject ("nuclear force"). The unexpected and highly abstract concept of an exchanged particle carrying a force, depicted in Figure

---

*One should not attempt to visualize a particle's spin as if the particle were actually spinning like a child's top. Spin is not a visualizable degree of freedom like position.

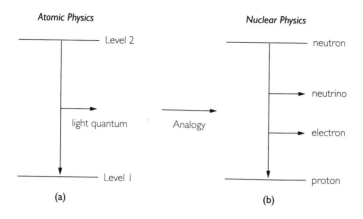

Atomic Physics   *Nuclear Physics*

FIGURE 11.

Fermi's "intuitive" thinking toward a theory of β-decay by analogy with the decay of an atom consisting of two levels, Level 1 and Level 2, which occurs by emission of a light quantum. In the nuclear case, a neutron decays into a proton with the emission of an electron and the newly postulated neutrino.*

10(d), allows exploration of the nuclear force using mathematics from Heisenberg's nuclear theory.

METAPHORS AND ANALOGIES IN PARTICLE PHYSICS

Alas, Heisenberg's nuclear theory was quickly found not to conform with available data on the strength of the nuclear force in light nuclei such as the deuteron, comprised of a proton and a neutron. But his play with analogies and metaphors generated by the mathematics of quantum mechanics was seen as the key to extending our intuition into the subatomic world. Consider the Italian physicist Enrico Fermi's "intuitive [*anschaulich*]" thinking, as it was called in 1934, toward a more viable theory of β-decay (Figure 11).

---

*It is noteworthy that Wolfgang Pauli postulated the neutrino in 1931 in order to preserve the law of energy conservation for β-decay. Most physicists accepted the existence of this particle even though it was not experimentally discovered until 1956, 25 years later. Its discoverer, Frederick Reines, did not receive a Nobel prize until 1995!

Two analogies are in operation here: depictive and descriptive. Fermi's depictive or visual analogy can be expressed as

> The β-decay of a neutron into a proton with the emission of a neutrino and electron *can be likened to* the decay of a two-level atomic system.

The descriptive analogy offers the means for representing the theory mathematically:

> The β-decay of a neutron into a proton with the emission of a neutrino and electron *can be likened* mathematically to the electromagnetic interaction.*

Compared with the secondary subject in Heisenberg's metaphor for probing the nuclear force, the secondary subjects in Fermi's interaction metaphors had been extensively studied by 1934 and subsequently were considered to be no more abstract than Maxwell's pulleys, wheels, and fluids or Planck's oscillator electrons. Although the quantum theory of radiation was still not on firm ground in 1934, everyone took for granted the basic mathematical form by which electrons interact with light. Consequently, in Fermi's interaction metaphors, the tension is not high between the primary and secondary subjects. For this reason we refer to Fermi's reasoning, as we did with Maxwell's for the electromagnetic field, as analogical reasoning and Fermi's metaphors as models.

Heisenberg's method of research can be seen in a proposal of the Soviet physicist Igor Tamm, made later in 1934. Tamm proposed to take Fermi's β-decay theory as the basis for a theory of the nuclear force. Tamm suggested that the nuclear force between neutrons and protons is *analogous* to the Coulomb force between two electrons, which is carried by light quanta. But Tamm's nuclear force turned out to be too weak to hold the nucleus together.

---

*Namely, a vector electron current coupled with the vector electromagnetic field.

By November of 1934 there was still no adequate theory of the nuclear force. In particular, Heisenberg's concept of the nuclear force being carried by an improper electron had been discarded. We may say that Heisenberg's nuclear electron turned out to have no referent. Nevertheless, his theory was the most mathematically developed. This was the point that interested the Japanese physicist Hideki Yukawa, who made the ultimate use of Heisenberg's research method of employing metaphors to extend intuition. In November of 1934, Yukawa returned to the mathematics of Heisenberg's theory. He replaced the almost arbitrary mathematical form for the nuclear force with one tailored to describe the exchange of the proper sort of particle, which Yukawa called a "heavy particle" and which has been known since 1939 as a "meson." What Yukawa accomplished can be explained by returning to Heisenberg's metaphor of the nuclear force being transmitted by a particle (Figure 12). The literal referent of the secondary subject—exchanged

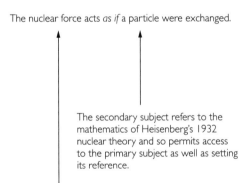

The nuclear force acts *as if* a particle were exchanged.

The secondary subject refers to the mathematics of Heisenberg's 1932 nuclear theory and so permits access to the primary subject as well as setting its reference.

The referent of the primary subject turned out to be identical to the secondary subject's referent when Heisenberg's 1932 nuclear theory is adjusted with a proper particle exchanged.

FIGURE 12.

Yukawa's use of Heisenberg's original metaphor for the nuclear force.

particle—turned out to be the literal referent of the primary subject and, astoundingly, the exchanged particle turned out to be physically real. Coincidentally, this established the proper terminology for the attractive force between charged and neutral particles (protons and neutrons) as due to particles being exchanged. This is a wonderful example of metaphor establishing catachresis. On the technical side, Yukawa had to adjust Heisenberg's nuclear theory for the exchange of a particle with appropriate properties. For this research, in 1949 Yukawa was awarded the Nobel prize.

DO SCIENTISTS ACTUALLY USE METAPHORS?

A cynic may ask whether I have contrived all of these uses of metaphor after reading primary and secondary sources. I hope this criticism has been dispelled; if not, let me offer two more instances of explicit use of metaphor.

The first is from another master of metaphor, Albert Einstein. One means by which Einstein introduced the light quantum in 1905 was via a metaphor of the form:

> Under certain circumstances light behaves *as if* it were comprised of particles.

Is this a new sort of metaphor? After all, in 1905 the referent of the thing called light was a wave. From scientific investigation, its stereotypes were: plane wave; can be polarized; propagates with a certain velocity in vacuum; can be reflected, refracted, and diffracted; and produces interference.

All of this being the case, what sort of particles comprise the secondary subject of Einstein's metaphor? For instance, how can light quanta produce interference? We have discussed the depth of this problem in earlier chapters. What an unusual metaphor this is, where the primary subject is better understood than the secondary one. Maybe this is a backwards interaction metaphor, a *rohpatem*? It took almost two decades of struggle before scientists agreed that the literal referent of the secondary subject is part of the primary subject. Since metaphors are all around us, perhaps the peculiar *form* of Einstein's metaphor *contributed* to the counterintuitivity that for so long prevented physicists from giving light quanta ontological status.

A PLATONIC INTERLUDE

Further discussion of the critical role metaphors play in scientific research requires focusing on a breakthrough in visual imagery Richard Feynman made in 1949. Feynman's diagrams are direct descendants of both Heisenberg's visual metaphors of the nuclear force as carried by particles and Yukawa's elaboration of Heisenberg's theory.

Since 1923 there had been no visual imagery in atomic physics. Schrödinger's attempt to bring back a form of classical visualization, suitably reinterpreted, had failed. At this time the emphasis switched from visual imagery to mathematical formalism because the latter provided a representation—first of Bohr's theory during 1923 to 1925, and then Heisenberg's quantum mechanics. Yet this representation offered no visual content. The term "representation" was taken to be synonymous with the mathematical formalism in which atomic theory was written. Feynman's diagrams relieved this uncomfortable situation.

Atomic physics changed drastically, both descriptively and depictively, in 1949 with the advent of two different quantum electrodynamics—the quantum theory of how electrons interact with light. The underdetermination thesis tells us this event should not be a surprise. The formulation proposed by Julian Schwinger and Sin-itoro Tomanaga has no imaginal content. Feynman's formulation is based on the visual imagery of visualizability, and not visualization; the distinction between these two forms of visual imagery is shown in Chapter 2. The difference can be understood by comparing different representations of the

Coulomb interaction between two electrons (Figure 13). The visual imagery in Figure 13(a) (visualization) is abstracted from phenomena that we have actually witnessed: Electrons are depicted as distinguishable billiard balls possessing electrical charge. This imagery, which was *imposed* on classical electromagnetic theory, turned out to be incorrect for use in the atomic domain, where electrons are simultaneously wave and particle.

Taking a hint from Plato via Heisenberg's uncertainty principle paper, which emphasizes the role of mathematics as a guide, physicists turned to the mathematical formalism of quantum mechanics to generate the proper visual imagery for electrons and light quanta: We would not have known how to draw Figure 13(b) without the mathematics of the quantum mechanics that *generates* it. This is visualizability. Details are unimportant. The essential point is that we can assume that the mathematics of quantum mechanics offers a glimpse of the subatomic world, where entities can be simultaneously continuous and discontinuous. Feynman diagrams *represent* interactions among elementary particles in a realist manner—there is ontological content in these diagrams. We must draw them with the usual figure and ground distinctions due to the

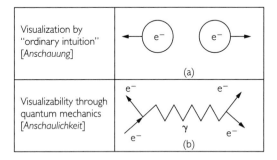

Visualization by "ordinary intuition" [*Anschauung*]	(a)
Visualizability through quantum mechanics [*Anschaulichkeit*]	(b)

FIGURE 13.

(a) The Coulomb force from elementary physics textbooks. (b) The Feynman diagram, which is the appropriate depiction of the Coulomb force, where two electrons interact by exchanging a light quantum (γ).

limitations of our senses. By this I mean the simple distinction between a well-defined structure set against a background of secondary importance. Today physicists visualize in Feynman diagrams. Chapters 8 and 10 elaborate further on Feynman diagrams and visual representations in physics.

ANOTHER NOBEL PRIZE METAPHOR

In the late 1960s, Abdus Salam and Steven Weinberg formulated a unified theory of the electromagnetic and weak interactions. Of this goal, wrote Salam in 1968,

> One of the recurrent dreams in elementary particle physics is that of a possible synthesis between electromagnetism and weak interaction. The idea has its origin in the following shared characteristics:
> 1. Both forces affect equally all forms of matter—leptons as well as hadrons.
> 2. Both are vector in character.
> 3. Both (individually) possess universal coupling strengths.

With only slight readjustment, Salam's metaphor for the so-called weak neutral current can be rewritten as*

> The weak neutral current in certain ways can be expressed formally *as if* it were the neutral current of electromagnetic theory.

Just like in the cases of Maxwell's and Fermi's metaphor, the electroweak metaphor's secondary subject is not as abstract as was the case in Bohr's metaphor. The reason is that in 1967, when the electroweak theory was first proposed, quantum electrodynamics had been deeply explored and

---

*The term "neutral current" is defined in Figure 14.

everyone took it for granted. Since degree of abstractness is a measure of the metaphor's tension, we refer to the basic metaphor of the electroweak theory as a model. Salam and Weinberg's accomplishment was to combine the analogical reasoning in the model with certain developments in quantum field theory in order to bring about unification, with the prediction of a new particle, the Z^0.

Let us explore the basis of these somewhat esoteric points further. Directly analogous features of electromagnetism and electroweak theory can be read off their lowest-order (or simplest) Feynman diagrams (Figure 14). The term "current" has been transformed in elementary particle physics from its everyday electrical meaning in which current flows because of a difference in polarity between the ends of a wire. In elementary particle physics, "current" still refers to the "flow" or motion of particles. It is expressed mathematically with quantities that represent the incoming and outgoing elementary particles as fields, analogous to the electromagnetic field. Currents in quantum field theory are generalized so that they can refer to the flow in space-time of neutral particles such as neutrinos.

The following excursion into some basics of Feynman diagrams is essential for getting the flavor of metaphorical reasoning in elementary particle physics. Currents are depicted in Feynman diagrams as the slanted arrows in Figures 14(a) and (b), representing incoming and outgoing particles. The currents in Figure 14 are neutral currents and so are connected with, or coupled to, neutral intermediate particles like the light quantum in Figure 14(a) and the Z^0 in Figure 14(c). This is necessary because electric charge must be conserved at every vertex (in Figures 14(b) and (d), the vertices are formed by the currents and the intermediate particles). The strength of the two couplings at the two vertices in Figure 14(a) is the familiar electric charge denoted by e, while the couplings at the vertices in Figure 14(c) are given by the electroweak charge, or coupling constant, g.

Now we can appreciate the power of Weinberg and Salam's reasoning. As it did for Fermi's β-decay theory, the electromagnetic theory

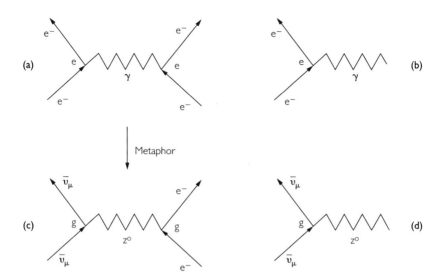

FIGURE 14.

The depictive form of analogical reasoning used in formulating the electroweak theory. (a) The Feynman diagram for the Coulomb force from Figure 13. According to quantum electrodynamics, the Coulomb or repulsive force between two electrons is carried by a light quantum (γ), where e is the electron's charge. Part (b) is the left-hand portion of (a) and is referred to as a "vertex." In this vertex the current of electrons is connected, or couples to, an intermediate light quantum. The strength of the coupling is given by the electron's charge e. Since there is an electron in and an electron out at this vertex, then the electron current is referred to as a neutral current.

We can write the Coulomb force as two electrons scattering from each other, as $e^- + e^- \rightarrow e^- + e^-$. By analogy, the weak interaction in (c) is the scattering of an electron (e^-) by a muon antineutrino $\bar{\nu}_\mu$. We can write this process as $e^- + \bar{\nu}_\mu \rightarrow e^- + \bar{\nu}_\mu$. This scattering process ought to be mediated by particles somewhat similar to the photon in (a), where g is the charge (coupling constant) for the electroweak force and the force's carrier is the Z^0 vector boson. An example of a weak neutral current is in (d) which is the vertex from the left-hand side of (c). Analogously to (b), the muon antineutrino current is called a "weak neutral current" and it couples to the Z^0 particle with a coupling strength given by g. The term "weak" is used here because the process depicted in (c) is a weak interaction, while the one in (a) is an electromagnetic interaction.

acts as a guide for the mathematical formulation of the electroweak theory. Only now Weinberg and Salam used the Feynman diagrams of quantum electrodynamics as a starting point, instead of an atom's energy levels. The depictive or visual analogy is read off from Figure 14(a) to produce Figure 14(c). The purely descriptive or mathematical analogy is in equating the coupling constants, or strengths, for the weak and electromagnetic interactions, which are g and e, respectively ($e = g \sin^2 \theta_w$, where θ_w is the Weinberg angle, which is a free parameter in the theory that is fit to experimental data). This is similar to Maxwell's equating the velocities of light and electricity and to Mayer's equating the units of heat and energy.

METAPHORS AND SCIENTIFIC PROGRESS

When we assume that all our utterances are metaphorical, the history of scientific thought takes on new and interesting turns. We have found that models are metaphors that can function like analogies. Sometimes metaphorical entities turn out to be physically real: particles transmitting forces, light quanta, and weak neutral currents. Being tools for scientific exploration, metaphors provide access into possible worlds that can become actual ones. Metaphors are a means for continuity in scientific progress.

The key point about metaphors is that metaphorical reasoning *emerges* from case studies in the history of scientific thought. So, the reply to the pragmatists' query—"What's the cash value of metaphors in history and philosophy of science?"—Metaphors play an essential role in a theory's formulation, as we saw with Maxwell, Planck, Einstein, Bohr, Heisenberg, Yukawa, and Salam and Weinberg. Consequently, metaphors are essential for exploring scientific progress.

We can condense our results thus far with Figures 15, 16, and 17. Figure 15 emphasizes the roles of metaphor and correspondence principles in the emergence of Bohr's theory from classical physics. This sequence becomes further understandable within the guidelines of a theory of reference. In Figure 16 the theoretical puzzle of the nuclear force wreaked havoc with quantum mechanics, which could not deal with an attractive force between charged and neutral particles; the experimental puzzle shook the foundations—disequilibrated—quantum mechanics because physicists had already ascertained that garden-variety electrons could not be nuclear constituents. Fermi's β-decay theory disequilibrated Heisenberg's 1932 nuclear force theory because its explanation for β-decay did not require any nuclear electrons. In Figure 17 quantum electrodynamics is disequilibrated by its extension to weak interactions because its mathematical properties are inappropriate for this purpose. All of these results can be illustrated with the assimilation/accommodation process. An essential ingredient for accommodation is metaphor.

The dynamics of scientific progress that emerge from historical studies turn out to be astonishingly similar to the assimilation/accommodation process offered by cognitive science to probe knowledge construction. This should not be surprising, because the problem of sci-

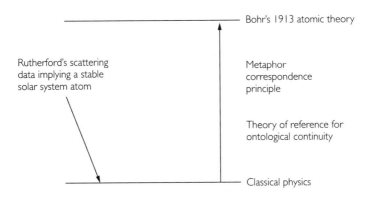

FIGURE 15.

A representation of the emergence of Bohr's 1913 atomic theory.

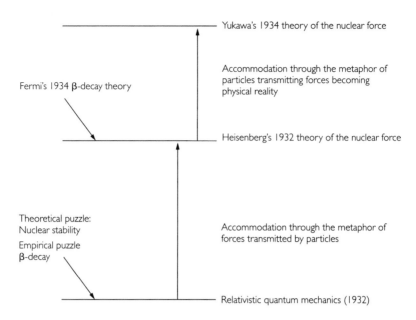

FIGURE 16.

A representation of the two-step process by which Yukawa's theory emerged from Heisenberg's work toward a nuclear force theory. Relativistic quantum mechanics became disequilibrated by Heisenberg's attempt at providing a unified theory of the nuclear force and the process of β-decay.

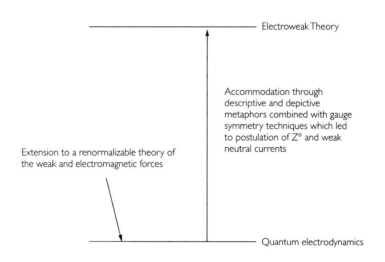

FIGURE 17.

The emergence of electroweak theory.

entific progress cuts right to the heart of the development of knowledge itself. Fundamental to any theory about how we construct knowledge is that we interact with the world about us. We take in, or *assimilate*, sense perceptions and act on them at first with levels of knowledge that are little more than primitive reflexes, such as responses to heat and cold. (Although more exotic knowledge is hard-wired into the mind; see Chapter 8.) These lower levels adjust, or *accommodate*, themselves to incoming sense perceptions by forming higher levels of knowledge (see Figure 5). The resulting higher level is in a state of relative equilibrium until new information is assimilated to it and the cycle of a punctuated upward spiral repeats as in Figure 18.

Figure 18 takes account of only the "winners" in the history of scientific thought. The path from Aristotle to quantum mechanics was not this direct. The many detours include subtle fluids like caloric, the ether of electromagnetic theories, the visual imagery or visualization of Galilean–Newtonian science, and the antiatomistic philosophies that spilled over into physics. Although these notions have either disappeared or undergone severe transformations, all were essential in their time. Theories may come and go, but only the fertile ones offer clues for how to proceed: Caloric and the ether were instrumental in setting the basis for mathematical theories of heat and electricity; the visual imagery of solar system atoms was essential for getting modern atomic theory off the ground using a quantized form of Newtonian mechanics with no firm basis whatsoever; and the antiatomistic philosophy of early positivism cleared the field of problems that physics in that era was incapable of handling. The hierarchy in Figure 18 is anchored in familiar soil and is a staircase ascending to the stars. Beginning with the earthbound physics of Aristotle, by degrees scientists have reached toward a unified theory of matter in motion. The history of scientific thought is not merely a graveyard of discarded and forgotten theories. Instead, certain of those theories contain concepts that enabled scientists to move onward to theories that are ever-better approximations to understanding the world.

The historian of ideas Herbert Butterfield coined the term "Whig history" to denote histories that in retrospect focus on the winners. Whig history aside, the tale told in Figure 18 is important because it indicates a persistent direction to scientific progress, toward increased abstraction of the concept of intuition and so toward knowledge of phenomena beyond sense perceptions. It is a direction as persistent as that of time's arrow.

It is interesting and informative to re-represent Figure 18 as Figure 19. The series of theories on the left are those that are closer to concepts of classical physics than those on the right, and consequently

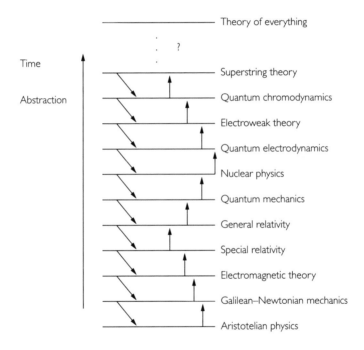

FIGURE 18.

An overview of the history of scientific thought, illustrating the development of physics from the view of assimilation/accommodation. The slanted arrows on the left represent the assimilation of information into a scientific theory that upsets the theory's foundations. The upward-pointing arrows on the right designate adjustments with the appearance of a higher-level theory that is able to handle the assimilated information in addition to covering a wider amount of phenomena than the lower-level theory.

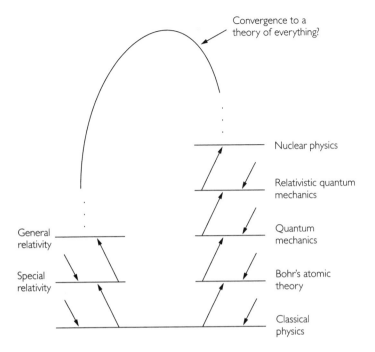

FIGURE 19.

This depicts the bifurcation with regard to conceptual transformations that occurred after classical physics, by which I mean Newtonian and Maxwellian physics. The sequence on the right-hand side is contained in Figures 15–17. The left-hand side refers to developments in which visual imagery is abstracted from imagery in the world of sense perceptions. The downward-pointing arrows denote assimilation and the upward ones accommodation.

their visual imagery is one of visualization rather than visualizability. Despite the bifurcation, cross connections exist in Figure 19 because, for example, there is a relativistic quantum mechanics. By "bifurcation" I mean that the new concepts of relativistic theories such as the relativity of simultaneity can be understood with the visual imagery abstracted from the world of sense perceptions (visualization, or *Anschauung*). But conceptual transformations in quantum theories cannot be understood in this way. Their visual imagery is generated by the mathematics in which the theories are expressed (visualizability, or *Anschaulichkeit*).

Figure 19 summarizes what we found earlier in this chapter in analyzing the emergence of Bohr's atomic theory from classical physics, quantum mechanics from Bohr's theory, nuclear physics from quantum mechanics, and then electroweak theory from quantum electrodynamics. The emergence of special relativity from classical mechanics occurred through a combination of thought experiments as well as data from actual experiments. General relativity emerged the same way, with more emphasis on thought experiments. These sorts of analyses are deep and complex. The principal point is that a scenario such as the one in Figure 19 emerges, having similarities with scenarios from cognitive science.

Figures 18 and 19 depict the goal of science as a "theory of everything," as it is known in today's parlance. This is, of course, a misnomer because it is not a theory of love, hate, or why Picasso's paintings are so interesting. It will explain interactions among elementary particles and the origins of the universe and so will be a magnificent intellectual achievement, still leaving much to be explained. But this book is written from the perspective of the end of the twentieth century. Who knows what surprises await us? One thing we can speculate on with a bit more confidence is why physics developed the way it did.

WHY DID PHYSICS DEVELOP AS IT DID?

Besides illustrating scientific progress, Figure 19 leads to the intriguing question of why physics developed the way that it has, instead of in some other direction. For this purpose, let us re-represent Figures 18 and 19 as Figure 20, which highlights the role of correspondence principles. The question marks indicate unknown correspondence limits, metaphors, and analogies toward new theories that are increasingly abstract and so extend our intuition and visual imagery.

Figures 19 and 20 drive home the realization that theorizing must begin somewhere, somehow. The ancient Greeks sought to understand

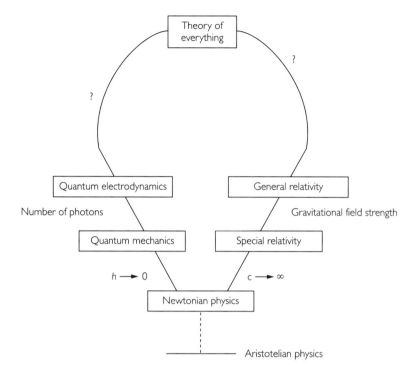

FIGURE 20.

This re-representation of Figure 19 emphasizes relevant correspondence principles. The question marks indicate presently unknown methods for achieving higher unifications. The correspondence principles concerning Planck's constant and the velocity of light were discussed in Chapter 2 (see Figure 9). Concerning the "number of photons," the larger the number of photons involved in any given process, the more the situation tends to be that of classical electromagnetic theory. For example, we can discuss ordinary FM radio transmission with classical electromagnetic theory because of the huge number of photons involved [of the order of 10^{17} per volume (which is the cube of the FM station's wavelength)]. Concerning "gravitational field strength," the weaker the gravitational field, the more the situation reverts back to ordinary Newtonian gravitational theory.

their world with a system of knowledge based on natural occurrences available to them. This is part of the accommodation process, which we also referred to as apperception. It is reasonable, then, that Aristotelian science came first, followed by Newtonian physics, and then special relativity, general relativity, quantum mechanics, and so on. Aristotle's "common-sense" intuition was simultaneously transformed by Galileo with the help of abstracting to possible worlds where there occurred phenomena beyond sense perceptions. Common-sense intuition had to be pushed to its limits before Galilean–Newtonian physics, with its transformed concept of intuition and its breathtaking abstractions, could become necessary. This new intuition would again be pushed to its limits and be altered again and again by the advent of relativity, quantum mechanics, nuclear physics, quantum electrodynamics, electroweak theory, and so on. Belief in the universality and objectivity of science means that although some alien civilization somewhere in the universe may start with something entirely different from Aristotelian physics, it would finally produce physical theories with foundations similar to ours. Almost certainly, its starting points would differ because the sense perceptions of its inhabitants would most likely be totally unlike ours.

HOW SCIENCE ADVANCES

Historical studies show that the history of scientific thought should not be regarded as a graveyard of discarded theories. Rather, progress *and* unification occur in sometimes surprising ways. Probing into how intuition is transformed, along with coincident changes of abstraction, has made it plausible to imagine how science progressed the way it did.

From analysis of actual cases of scientific research it has *emerged* that scientific advance occurs when including information into a scientific theory requires adjustment of the theory by such means as metaphor and correspondence principles, all of which underscore the continuity of scientific progress. In each case, both intuition and visual imagery were transformed. This process has parallels with the assimilation/accommodation model for the construction of knowledge. That there should be a relation between basic problems of philosophy of science and cognitive science is not surprising. Both disciplines focus on how knowledge is created from input information. As scientific theories become more abstract, the notions of intuition and of visual representation become more abstract as well. Scientific theories extend our intuition and visual imagery into realms beyond our daily experience. Accepting this conclusion means accepting a realist interpretation of science. In this way the notion of scientific progress, that is, of convergence of scientific theories, becomes meaningful. We cannot speculate exactly what that convergence is at present, nor can we speculate on what the character will be of increasingly more unified theories.

According to the underdetermination thesis, an infinity of theories can cover any data set. There seems to be no way of avoiding this. Yet only the theories that have included certain assumptions, such as the principle of relativity, have been able to survive, provide the basis for new science, and form connections with other theories through correspondence principles and metaphors. According to the "no miracles" argument for scientific realism, if science were not even approximately true, then these theories' successful predictions would be akin to miracles. The history of scientific thought provides evidence for the continuity of certain scientific concepts and assumptions, which, in turn, provide a panorama of science in which continuity emerges. Only in the light of a realist interpretation of scientific theories does such a panorama make any sense.

What if, in the far future, a scientific theory was formulated which had no need of electrons, or any of the other elementary particles we know of today? In this case the natural kind terms of today's scientific theories would suffer the same fate as caloric, and the ether of nineteenth-century electromagnetism. Hilary Putnam has referred to this situation as the disastrous "meta-induction." Antirealists tout it as a further argument against scientific realism: Why waste time on ephemeral entities? The realist philosopher Ernan McMullin takes the tack that the "history of science testifies to a substantial continuity in theoretical structures." If this were not the case, then we would not have been led to defend scientific realism in the first place. The realist can also reply that the science of the far future will be so unimaginably unlike ours that there may well be another sort of directly unmeasurable entities replacing electrons and perhaps even quarks. Yet we would not have been led to these other entities had not we not attempted to build theories based on electrons. Scientific realism is the philosophy of the optimist.

Thus far, we have explored the interplay between theory and experiment, intuition, and scientific progress. Since visual imagery has played a central role in all these topics, it is apropos that we probe this fascinating topic further.

8

VISUAL
IMAGERY
IN
SCIENTIFIC
THOUGHT

Why does visual imagery so often play a key role in scientific research? Exploring this question will take us into cognitive science and neurophysiology.* As the cognitive scientist Ned Block wrote in 1981:

> After fifty years of neglect during the hey-day of behaviorism, mental imagery is once again a topic of research in psychology. In-

---

*Cognitive science is a relatively new interdisciplinary subject that incorporates concepts from such subjects as cognitive psychology, computer science, neurophysiology, philosophy of science, and linguistics in its study of the human mind.

deed, with the emergence of a truly spectacular body of experiments, imagery is one of the hottest topics in cognitive research.

Among the hot topics in the field of imagery are: What are visual images? How do they originate in the mind? Is there a relation between how we process imagery from the world of perceptions and the visual imagery we imagine?* Do visual images play a causative role in thinking? In this chapter, we will explore these questions and see what light the history of science sheds on them. Chapters 9 and 10 continue this investigation by probing scientific creativity and the relations between art and science.

After setting the stage for the subject of visual imagery in cognitive science, we delve into the so-called imagery debate. One faction contends that visual images do not play an essential role in thinking; the other says they do. This leads us to consider such topics as: How does the brain extract information from the varying intensities of light entering the eye? How is this information stored, accessed, and processed? The complexity of these operations is staggering, especially since the scene before the eye is not necessarily what is on the retina. A circle is round only when viewed head on; it can be elliptical otherwise. Yet we "know" that we are viewing a circle. A vast amount of information goes into making sense of a retinal image, in the process creating pitfalls for vision theory in terms like "see."

After examining the imagery debate, we will look at how all of these issues are relevant to the history of science, in the notion of thought experiments in creative scientific thinking and the role visual imagery plays in scientific theories. We conclude the chapter with conjectures on what the history of science reveals about the mind's visual capacities.

---

*To avoid possible confusion in terminology, by "visual imagery" I mean the imagery of objects that are not actually present. This is to be distinguished from the imagery processed as a result of retinal stimulation.

Some Background to Mental Imagery

The general term "mental imagery" refers to imagery related to our five senses—visual, tactile, auditory, olfactory, and taste—without the stimulus being present. Mental imagery has been a central problem throughout the history of philosophy and psychology. Aristotle considered thought to be impossible without an image. Plato likened mental images to impressions made on wax tablets and stored away for future use. If our skulls were then opened up, might visual images fall out? A criticism against the usefulness and meaning of mental images is the notion of "in-my-head." On this point, one of mental imagery's more facetious critics, the Oxford philosopher Gilbert Ryle, wrote in his 1949 book *The Concept of Mind* that

> no one thinks that when a tune is running in my head, a surgeon could unearth a little orchestra buried in my skull or that a doctor by applying a stethoscope to my cranium could hear a muffled tune.

For its critics, mental imagery is a metaphor, in the term's most derogatory sense.

Early psychologists focused on the notion of the "idea," which they believed could be represented ultimately by images. Chiefly on the basis of introspective accounts, the founder of modern psychology, Wilhelm Wundt, at Leipzig in the late nineteenth century, assumed that images accompany all thought processes. (Introspectionists study the contents of consciousness by having subjects "introspect," or "look inside" their mind. This process is usually carried out with trained subjects under controlled conditions.) In the last decade of the nineteenth century, Wundt's ex-student Oswald Külpe and others of the Wurzburg school found that simple association tasks seem to require no visual imagery. This result shocked the psychological world, and the "imageless" thought controversy ensued. The Wundtians at Leipzig were unable to respond convincingly because they had no data on mental states other than reports based on introspections. In addition, there were sharp disagreements between the results of various laboratories and there was no theory to describe the data. These were the principal defects of introspectionist psychology that led to its decline in the early part of the twentieth century.

In the 1920s, the impasse was temporarily resolved by a dose of Ernst Mach's positivism: Declare the question of whether there are mental images to be illegitimate because they are inaccessible to laboratory demonstration. The attack was launched, against consciousness and mental imagery simultaneously, by John B. Watson, the founder of the positivistic school of psychology called behaviorism. Since in Watson's view only responses to stimuli are measurable, then no information can be inferred about internal events (the reader is invited to look again at Figure 4 of Chapter 7). In his 1928 book, *The Ways of Behaviorism*, Watson gave vent to his frank opinion about the "rubbish called consciousness—prove to me that you have auditory images, visual images. . . ." The similarity to Ryle's opinion is not accidental. The predominant research effort in psychology was studying behavior in its own right. By the 1960s, however, behaviorism had run its course due to its limitations for providing adequate explanations for language acquisition and perception, among other functions.

In the early 1970s, exciting new data were found which researchers believed could be interpreted as evidence for the mind processing visual images. The most widely cited of these experiments was performed in 1971 by Roger N. Shepard and colleagues at Stanford University. They showed subjects 1600 pairs of figures like those in Figure 1(a)–(c) and asked them whether pairs were congruent or not.

Shepard found that the larger the angular disparity in each pair, the longer the subjects took to respond. The relation between angular disparity and response time turned out to be linear, as in parts (d) and (e) of Figure 1. When the subjects were asked how they obtained their answers, they reported imagining rotating one figure until it became superposed on the other, just as they would have done manually. From

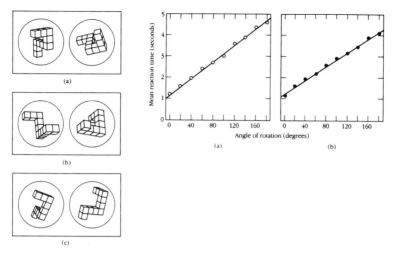

Figure 1.

On the left are shown stimuli used in 1971 by Shepard and Metzler for studying mental rotation. The two objects in (a) and (b) are identical. An 80° rotation in the plane of the paper will bring the ones in (a) into congruence. Similarly a depth rotation by 80° suffices for the ones in (b). The pair in (c) cannot be brought into congruence. The figures on the right indicate the mean time for determining whether two objects have the same three-dimensional shape as a function of the angular difference in their orientations. The plots are linear for pairs rotated in the picture plane (a) and by a rotation in depth (b). [From J. Metzler and R.N. Shepard, "Transformational studies of the internal representations of three-dimensional objects," in R.L. Soslo (ed.), *Theories of Cognitive Psychology: The Loyola Symposium*]

this, Shepard hypothesized a connection between imagery and perception. Many scientists considered this a real breakthrough in imagery research. These results inspired a series of experiments by Stephen M. Kosslyn at Johns Hopkins University and later Harvard.

In a typical Kosslyn scanning experiment, subjects are shown a map with sharply distinguishable features. The map is then removed. Subjects are asked how long it would take to move between various points on their now *imagined* map (Figure 2). The time for scanning the imagined maps turned out to be directly proportional to the distance

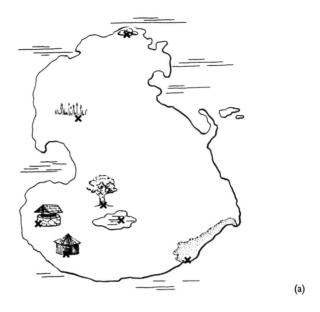

(a)

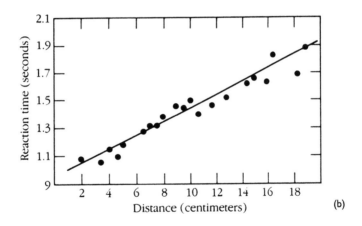

(b)

FIGURE 2.

In Figure 2(a) is a map with easily identifiable locations used by Kosslyn. After showing subjects the map, Kosslyn removed it and asked them to visualize it. He next requested that they scan between various points. Figure 2(b) indicates the time to scan between pairs of all points on the map that the subjects visually imaged.

between pairs of points. As Shepard had done for rotations, Kosslyn inferred a continuity in the mental processes involved in scanning, which is essential for the image–perception link. Similar brain processes, he claimed, were responsible for both externally and internally generated visual images.

In summary, on the basis of studying subjects' responses to stimuli, Shepard and Kosslyn proposed hypotheses on the brain's internal functions: There is a connection between the way we process visual imagery and the way we perceive a phenomenon such as rotation actually occurring. This is the scientific method that deeply repulsed the behaviorists. In the 1970s, there appeared a means of developing models for internal mental processing—the computer.

Visual Imagery and Computational Psychology: A Study in Architecture

Cognitive scientists differ on the degree to which thinking can be reduced to the manipulations of symbols in the mind, and so, too, on the fundamental nature of visual imagery. In order to develop these differences, let us discuss how cognitive scientists use computer science.

On the desk in front of me sits an antiquated Macintosh IIci. (I remind myself that it's the hand behind it that counts.) It is a box about 12″ × 14″ × 5″ with cables connecting it to my printer, screen, keyboard, and mouse. Inside this box are circuits and chips that activate the microprocessor, registers, buses, RAM, ROM, and whatnot. With good reason, most users do not fiddle around with hardware. This operational level is hard-wired into the machine's "functional architecture." It is used for storing, retrieving, and altering data.

Cognitive scientists use similar terminology, sometimes in a literal sense. In the early days of artificial intelligence (AI), information-processing systems abounded. Most of them probed human techniques for solving logical problems, sometimes pointing to improved methods. Allen Newell, considered to be the dean of information processing, was moved to declare that we are on the verge of a unified theory of cognition. Although such optimism has not yet come to fruition, some spinoffs of AI are of great value. One of them is that the mind is an information-processing system whose formal structure is referred to as the "cognitive functional architecture." This is comprised of hard-wired, or innate, structures that provide the basic information-processing mechanisms that contain data structures and processes to act on representations in order to retrieve information. Wired-in structures cannot be altered by beliefs or goals or tacit knowledge. Exploring the cognitive functional architecture of the human mind is a major research effort of cognitive science and AI.

Cognitive scientists assume that we possess two sorts of representations, descriptive and depictive. We have met these terms before. In this chapter we expand on their meanings. Descriptive representations are given by propositions, which is a notion taken from logic and linguistics. By definition, a proposition is the smallest unit, or atomic symbol, about which we can make the judgment "true" or "false." Propositions obey the laws of a logic system and can be used to represent the meaning of sentences—and, some cognitive scientists claim, pictures too.

As an example of descriptive and depictive accounts, consider the scene of a ball resting on a box. This scene can be represented with either logical propositions (description) or a picture (depiction). Most likely a picture of a ball resting on a box has already come to mind. A propositional account or representation of a ball resting on a box can be made with the proposition ON(BALL, BOX), where ON plays the role of a predicate connecting BALL and BOX. A propositional representation need not be in a natural language like English. In fact, ON(BALL, BOX) is similar to the way statements are represented in computer language. Whereas neither BALL nor BOX is a complete proposition, ON(BALL, BOX) is and can be either true or false. Propositions combine according to rules of syntax, that is, of logic.* If reasoning took place only with propositions, we would never make any mistakes, because we would function like calculators. But we do make mistakes. To explore this point, let us turn to the problem of how information is stored in the mind's functional architecture.

Cognitive scientists assume that we possess two sorts of knowledge, declarative and procedural. Declarative knowledge refers to static, fact-like representations that serve as innate structures and can be propositional or imagistic. Declarative knowledge is about facts and things; it is operated on by processes, which is the problem of procedural knowledge. The distinction between declarative and procedural knowledge is akin to the difference between knowing a fact and knowing a procedure. A simple case is the difference between knowing someone's telephone number (declarative knowledge) and knowing how to reach that person by phone (procedural knowledge). Often there is an ambiguity between declarative and procedural knowledge. For example, suppose you are about to drive from Boston to New York City. You read directions from a road map, a declarative data base. On the other hand, you may have acquired a written set of directions, or procedural knowledge, that read like this: take the Massachusetts Turnpike west to Route 91 and then proceed south to Route 95, and so on. But putting this scenario into action ultimately requires driving a car, which entails looking for road signs, observing distances, and so on. Consequently, the written set of directions seems now to be a knowledge structure, that is, a piece of declarative knowledge.

Throughout this book we have been investigating these two sorts of knowledge. We have studied mainly the imagistic mode of declarative knowledge and its use in problem solving, which is an aspect of procedural knowledge as is the use of metaphor. This chapter probes deeper into how declarative knowledge is represented.

*Syntax is the grammatical structure of uninterpreted symbols. Semantics, or theory of meaning, is how meaning is assigned to these symbols.

Later in this chapter, we will see that the mental processing of information involves a mix of propositional and imagistic representations. Since this is the case, what gives cognitive scientists the right to make the distinction between kinds of knowledge? They do it for the same reason physicists carve the world up into four basic forces—gravitational, weak, electromagnetic, and strong—even though they all operate simultaneously: Nature kindly allows them to do so. After a while, scientists move toward unifying these compartments of nature. We saw this with Galileo and Newton with astronomical and terrestrial phenomena, Maxwell with light and electromagnetism, and Weinberg and Salam with electroweak theory. Many cognitive scientists seek similar unifications in their subject.

In 1968 the AI researcher M. Ross Quillian asked, "What constitutes a reasonable view of how semantic information is organized within a person's memory?" Quillian addressed the deep problem of what "language" declarative knowledge is represented in. By "memory," Quillian was referring to long-term memory, in which information is saved for days and sometimes years. Long-term memory preserves the meaning of a term or event rather than its exact structure. Consequently, rather than attempt to encode the meaning of a sequence of English words (or any other language, for that matter), Quillian sought a more general notation. He proposed one based on a network of propositions, because propositions can capture relations and still be true or false. Subsequent developments employed special notations for propositions, which allowed certain theories of meaning to be specified. In Chapter 5 we touched on some of the complex problems of meaning, which surface here as well.

As an example of a propositional network, consider the following sentence:

1. Mary gave a beautiful dog to Bill Clinton, who is President of the United States.

This sentence can be decomposed into the following simpler ones:

2. Mary gave a dog to Bill Clinton.

3. The dog was beautiful.

4. Bill Clinton is President of the United States.

If any of sentences 2 through 4 are false, then so is sentence 1. Each of the simpler sentences is a unit of meaning.

Is there any experimental basis to this technique? The answer is yes. Experimenters have found that although subjects did not remember the exact wording of any sentences of the sort in 1 through 4, they recalled the gist of them. For example, instead of reporting hearing what appears in sentence 2 above, the subjects might report that they heard the following:

5. Bill Clinton was given a dog by Mary.

Sentences 2 through 4 can be written in a propositional notation as a *relation* followed by an ordered list of *arguments*. The relations are verbs, adjectives, and relational terms such as President-of, while the arguments are nouns. In this notation, sentences 2 through 4 become lists:

6. (gave, Mary, dog, Bill Clinton)

7. (beautiful, dog)

8. (President-of, United States)

Experimenters developed this propositional notation after finding that although subjects usually could not recall hearing sentences 1 through 8, they recalled the *meaning* of a message contained in sentences 6 through 8. Figure 3 represents how the information in sentences 6 through 8 is stored in a propositional network in long-term memory. Each proposition is represented by a circle with a number inside, which merely distinguishes one proposition from another. Each proposition is connected to its arguments and relations by arrows called links. The propositions, relations, and arguments are called nodes.

An immediate problem arises in the propositional network in Figure 3 with the term "dog." A proposition is useful only if we have a general definition of it. But the term "dog" is not a specific fact because it concerns extremely general knowledge, and so it is denoted as a *concept*.

(a)

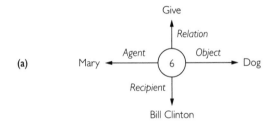

(b)

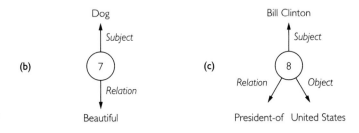

(c)

(d)

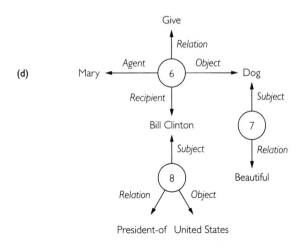

FIGURE 3.

The propositional networks in (a)–(c) represent propositions 6 through 8; (d) represents the combined networks (a)–(c).

Most people cannot precisely define what a dog is. This problem necessitates expanding the propositional network in Figure 3 into networks for concepts, which Quillian called semantic networks. Such a representation for the concept "dog" would link the term with complex biological and anatomical properties, and with other, less scientific, properties such as makes sounds like bow-wow, has four legs (although losing a leg doesn't mean it's no longer a dog), walks in a certain way, has a certain smell, is a living thing, and so on. The latter group includes intuitive knowledge of the thing called a dog, such as how to recognize dangerous dogs. We are combining propositional representations with intuitive and sensual information. This code is not propositional or language-like and so is referred to as an analog code or representation. Sometimes this cut is not unambiguous. For example, a map is both an analog and a propositional representation because it contains not only topographical features but names as well. For example, on a map of the United States, New York City is represented with words and not another picture. Experiments strongly indicate that geometric objects are stored in long-term memory spatially, while words are stored linearly in a propositional representation. Figure 4 represents what a semantic network for the concept of a dog can be. Such a network is an abbreviated version of the propositional representation used in Figure 3 and should be imagined as appended onto Figure 3 wherever the concept "dog" appears.

In Figure 4, the word "DOG" gives a name to the concept node "dog"; this operation is denoted by the thick single-headed arrow connecting "DOG" with "dog." The thinner single-headed arrow stands for "is a." So, a dog "is an" animal and an animal "is a" living thing. Double-headed arrows link the concept node to analog information, while the dashed single-headed arrow links it to propositional information. Figure 4 is an example of the simplest kind of semantic net, appropriately called an "is-a" net.

The integration of propositional and analog information into a network representation is essential for our thought processes. It permits

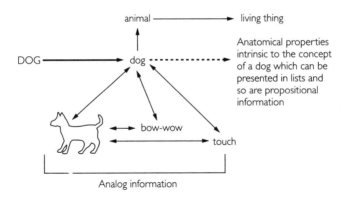

FIGURE 4.

A conceptual network illustrating how analog and propositional information are combined to define a concept.

us to think of the word "dog" when we see or hear a dog. Interactions between propositional and analog representations have been established experimentally. The most-quoted experiment was performed in 1932 by L. Carmichael, H. P. Hogan, and A. A. Walters, who showed subjects line drawings such as the one in Figure 5(b) and described it either as a dumbbell or as eyeglasses. Later the subjects were requested to draw the presented figure from memory. The subjects who were told that the presented figure was eyeglasses produced drawings similar to that in Figure 5(a), while those told it was a dumbbell produced figures like that in Figure 5(c). This means that the customary appearance of a concept in memory influenced the remembered visual appearance of a drawing. Consequently, visual images can be altered by general knowledge and so are more malleable than pictures. Generally, visual images are more abstract than pictures because their representations are tied to other modalities of thought, as we noted in Figure 4.

Granting that the functional architecture of long-term memory is an extensive or associated network of propositional and analog information, how is all of this processed? Consider the following dire situation. A very big dog is running toward us, barking and snarling. We quickly react. How do we know how to react? This is an aspect of procedural knowledge.

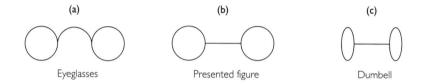

(a) Eyeglasses (b) Presented figure (c) Dumbell

FIGURE 5.

The manner in which the experimenter describes a visual pattern influences the way in which subjects later draw it from memory.

The architecture of long-term memory is assumed to be a complex semantic network. Clearly, all of this information cannot be processed simultaneously, or else we would be in a state of constant confusion. Only certain information is activated and processed in what is defined as short-term memory. Information processed in short-term memory eventually recedes back into long-term memory. From the case of the threatening dog, we know that processing occurs extremely quickly and in parallel—that is, a lot of information is activated and then processed at once. Activation spreading has been demonstrated in laboratory experiments in which narratives are read to subjects. Certain words purposely are set up to prime others by properly setting up links, which, we recall, are the arrows in Figure 3.

How procedural knowledge is represented is a problem on which there is presently no consensus. Its representation depends on methods for problem solving which involve behavior directed toward achieving a goal and often how the problem itself is represented. We have touched on this situation with regard to Einstein, Galileo, Heisenberg, Schrödinger, and Newton, among others. In this chapter and the next we continue this exploration, with focus on the scientific creativity of Einstein and Poincaré.

So far we have omitted any neurophysiological knowledge of how the brain operates. We will explore that subject later in this chapter. For the most part, researchers in AI and cognitive science look to neurophysiological experiments to set constraints on the form of cognitive theories. We will see an example of this when we discuss the computer model of vision proposed by Kosslyn.

What we have seen is that reasoning cannot occur only with propositions. As mentioned earlier, if it did then we would never make any mistakes, because we would function like calculators. But we do make mistakes. This leads us to conclude that thinking is comprised of a collection of processes or functions, some that possess truth and falsity and some that do not, perhaps like visual images. Psychological experiments have shown that the mental representations with which we reason combine descriptive and depictive components. We have already noted this to be the case in the creativity of Galileo, Einstein, and Heisenberg, and we will pursue this point further in Chapter 9. The cognitive scientist Philip Johnson-Laird is among many researchers who have demonstrated this mixed mode of reasoning. He refers to such representations as "mental models." Johnson-Laird summarizes these findings as indications that "there can be reasoning without logic."

A properly set-up scheme for long-term memory should contain mental images, particularly visual ones. Visual imagery at its most basic level resembles what we actually see.

If there are analog representations such as visual images—and if these representations are separate from propositional ones—then there must be a medium to support them, just as there is a part of the cognitive architecture in which propositional networks flourish. This strong claim requires experimental verification. We can now turn to the imagery debate.

THE IMAGERY DEBATE

In the pristine world of debates, distinct and often extreme sides are taken. It usually turns out that more moderate positions are preferable. The position closer to what "actually" happens in image generation and processing is more complex and revealing of the issues at stake. This debate offers another case study of how science is actually done and how scientists vie with one another—at times in the face of conflicting data—to support their own research efforts. The imagery debate is also illuminating because it concerns disciplines other than physics, which we have so far focused on almost exclusively. The protagonists in some of the other debates we have discussed are Kaufmann and Abraham versus Planck, and Heisenberg versus Schrödinger. We will see that the protagonists in the imagery debate hold to their positions just as tenaciously. Often they gather the same data, while interpreting them differently within their own theoretical frameworks, amid cycles of critical exchanges. Although the names, nomenclature, and fields of study differ from what we have discussed in earlier chapters, these hallmarks of scientific research persist.

The sides in this debate, which has been raging since the 1970s, are referred to as the anti- and pro-imagists. Both sides agree that there are visual images in the mind. But the anti-imagists, led by Zenon Pylyshyn, claim that visual images are merely a particular aspect of a processing system, akin to a computer's external lights set flashing by its running a program. Smashing these lights would not affect the computer's internal workings. Thus, visual images have no causal role in thought processes; they are epiphenomena. Pro-imagists such as Kosslyn, on the other hand, consider visual images to play a fundamental role in thinking, and so visual imagery is a part of the cognitive functional architecture with a dedicated representational medium.

The debate centers on the mind's cognitive functional architecture, which provides structures for processing information stored in semantic networks. Since propositions are discrete "atomic" entities, the anti-imagists argue that reasoning ought to be a discontinuous procedure. Yet Kosslyn and Shepard infer continuity in mental processes such as rotation and scanning. Consequently, they claim there ought to be another part of the cognitive functional architecture to support continuous, or "analog," mental processes in which information processing need not be only with propositions. Moreover, certain processes in this special-purpose component of the cognitive architecture are distinct from those that support propositional representations. Kosslyn calls this analog medium the "visual buffer." This is analogous to the computer's bit-mapped memory in which each cell is a pixel (short for "picture element").

The core of the imagery debate is that if visual images play a fundamental role in thinking, then the capacity for thinking with images is a distinct hard-wired component of the mind's cognitive functional architecture. Pylyshyn is willing to accept a dedicated representational medium in the mind to support analog processes, if empirical data warrant it. Thus far he maintains that no such data exist.

Pylyshyn is adamant about a propositional or descriptive representation for visual imagery in which images play no role in thought processes. This is contrary to experimental data providing evidence for a dual representation, such as the experiment of Carmichael et al. Pylyshyn is forthright about this extreme stance, as he writes in *Computation and Cognition*:

> The notion of a discrete atomic symbol is the basis of all formal understanding. . . . Small wonder, then, that many of us are reluctant to dispense with this foundation in cognitive psychology. . . . Unless [analog properties] can be reduced to either atomic symbol foundations or to physical foundations, they remain intellectual orphans. . . . To state the matter more precisely, when we refer to [analog properties or visual imagery], there is an important sense in which we do not understand what we are talking about!

While Kosslyn agrees that propositional representations are part of visual imagery, he claims that once generated, visual images need not drop back into a propositional representation but remain encoded in a web, or associated network, that contains spatial relations stored in the visual buffer. The visual buffer is capable of processing propositional and analog representations. Kosslyn believes that information from certain areas of the brain that are responsible for visual perception is the source of propositional and spatial representations that generate visual images. His computer model of visual imagery is a mix of propositional and depictive representations. It is a reply to Pylyshyn's accusation that, concerning such analog properties as visual images, "we do not understand what we are talking about!" Kosslyn's computer model is a valuable step toward operationally defining the notion of a visual image.

We have found that visual imagery is often essential to creative scientific thinking and that scientists strongly prefer the visual mode of thought in their research. This emerged in our discussions of Bohr, Einstein, Heisenberg, Maxwell, Fermi, Salam, and Weinberg. Consequently, we are pro-imagery and so are initially attracted to Kosslyn's program. Yet despite his and our high expectations, it will turn out that Kosslyn has not yet found his hypothesized visual buffer.

A MODEL OF MENTAL PROCESSING AND ITS CRITICS

To open up this debate further, we turn to a computer model of visual imagery Kosslyn offered in the 1980s based on the image–perception link. According to this hypothesis, visual imagery and perception share the same internal processes. For example, the visual buffer is shared by imagery and perception. So, for example, the fact that images fade quickly can be explained in analogy with perceptions vanishing after attention is turned elsewhere. Assuming a concrete image–perception link is a reasonable way to begin because Kosslyn could utilize Shepard's data on rotation and his own on scanning, in addition to his numerous other experiments. Kosslyn structures the model in a "top-down flow" of hypotheses in "high-level" vision processing. In order to define high-level vision processing, let us begin at the other end of the spectrum, with the theory of "low-level" vision, which focuses on the initial extraction of information from the pattern of light on the retina. By definition this is a "bottom-up" process because it is the first step in constructing the object being imaged.

A widely used theory of low-level vision is due to David Marr.*
Marr writes in his 1982 book *Vision* that the goal of such a theory is to
make "explicit important information primarily about the intensity
changes over a scene and their geometrical distribution and organization." Marr refers to this bare representation as the "primal sketch" (Figure 6).

The breakdown of vision processing into three levels—low, intermediate, and high—is essential to get a start on the problem of vision. It
is, of course, somewhat artificial because, as we will learn, these levels
overlap. The next step is to fill in the primal sketch in a manner that is
computable, that is, programmable on a computer. This stage is called
intermediate-level processing. There is a rich variety of means for developing the primal sketch. One obvious source of information is visual
motion, which is often the way we distinguish objects from their background.

Then there is high-level vision, the central problem of which is
object recognition. This entails how abstract propositional representations of objects stored in our memory can be rapidly retrieved by the in-
termediate- and low-level representations. Consequently, this is referred
to as the top-down flow of hypotheses. The end result of high-level vision is that we recognize objects.

The general problem of vision leads us to conclude that knowledge,
be it in the form of words or visual images, is not stored somewhere in the
mind like file folders in a filing cabinet. Imagine the time it would take to
identify, retrieve, and collate information from such a system before acting in response to an automobile speeding toward us. The quick reaction
times in such a situation suggest knowledge storage and representation
systems that can be accessed in a massively parallel manner.

---

*David Marr was the first researcher in vision to construct computer models that take ac-
count of results in psychology and physiology. He accomplished this highly important work at the
Massachusetts Institute of Technology during the last five years of his tragically short life. Marr's
results, and more importantly the philosophy of his approach, appear in his book *Vision*, published
posthumously in 1982, two years after his death at age 35.

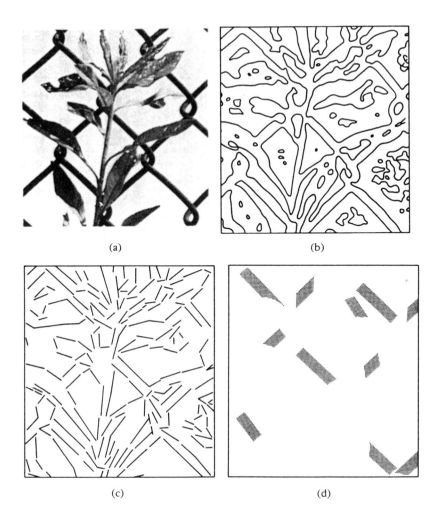

FIGURE 6.

The original image is shown in (a). Figure (b) contains only sudden intensity changes, called "zero-level crossings." Figure (c) is the result of analysis on (b) by edge detectors and (d) by bar detectors. Edges are isolated lines and bars are formed out of two nearby parallel segments. Figures (c) and (d) are then encoded in terms of the positions, orientations, lengths and widths of edges and bars. Figures (c) and (d) are primal sketches of the scene in (a).

After this brief excursion into vision theory, to which we will return in the next section, we can understand why Kosslyn piggybacks his own model of imagery on the fundamentals of high-level vision— for which formulations exist, however tentative. At this point in time, Kosslyn claims that he is interested only in architecture. So scenes are "suggested" to a computer, which it then constructs in the way Kosslyn claims his subjects scan maps and other figures for details—hence, the importance to Kosslyn of the image–perception link. A typical request is that the computer draw the picture of an automobile. How does it do this?

The skeletal outline of an automobile is plotted on a "visual buffer," which is the hypothesized dedicated representational medium for visual imagery. In practice, the visual buffer is a dedicated graphics processor hard-wired into the machine (the "brain"). The visual buffer receives input from two underlying "deep representations," which are encoded in propositions as well as in graphical coordinates. "Processing components," such as FIND (scans the skeletal image for foundation parts), IMAGE (accesses graphical coordinates of parts), PUT (checks propositional file for proper location of parts), and PICTURE (prints the skeletal image), are brought into play. In addition, there are SCAN, PAN, and ROTATE, among others.

Kosslyn contends that the scanning and rotation results are analog processes in the following sense: They "unfold" through the same intermediary steps as the actual physical process, and without reasoning or any tacit knowledge accrued from having actually witnessed rotations and translations in the world of perceptions. This means that there are deterministic constraints in the filling of cells in the visual buffer as the process unfolds continuously and through the same intermediary states as the actual physical process.

Since the late 1980s, Kosslyn has adventurously upped the ante in his model because he has begun to delve into neurophysiology in order to further substantiate his hypothesized processing components. What about the visual buffer? Kosslyn claims neurophysiological evidence for its existence, on which I should like to defer comment until we discuss the present status of the imagery debate.

Pylyshyn, on the other hand, interprets scanning and rotation experiments quite differently. He claims that Kosslyn's data could equally well be explained by assuming that the mind stores and processes information with a propositional encoding comprised of lists of symbols. For example, consider a list of ordered pairs of symbols such as "to the left of" and "to the right of." One can add operations that correspond to rotations and translations. Pylyshyn was able to set up a propositional account of Kosslyn's scanning experiment in Figure 2 in which distances between pairs of objects are represented in lists of propositions. Objects farther away from one another have longer lists. In Pylyshyn's propositional model, response times increase with increasing distance because they depend on the length of a list representing relative distances between pairs of points on a map. Similarly, a propositional account can be given to Shepard's data.

Pylyshyn goes on to claim that Kosslyn's experiments are basically flawed because the experimenter inadvertently instructed subjects to use visual mental imagery, a result the researchers were trying to prove. This danger always lurks in experiments of this sort. Consequently, claims Pylyshyn, Kosslyn has no grounds to infer from his data that the subject's response is an example of a visual image set into a dedicated representational medium, because this medium is not hard-wired into the brain: The subjects are asked to introspect and then reason toward their visual imagery aided with suggestions made by the experimenter and/or by their tacit knowledge in responding to instructions worded as "look for" and "compare" distances between points on a map.

Additional evidence by Pylyshyn that the rotation operation is not part of the cognitive functional architecture is in the experiment in Figure 7. This experiment indicates that judgment of whether the test pattern was a "good" or "poor" subfigure influenced rotation rate. Consequently, the rotation operation is not hard-wired into the mind. On the other hand, can we not interpret Pylyshyn's data as demonstrating the

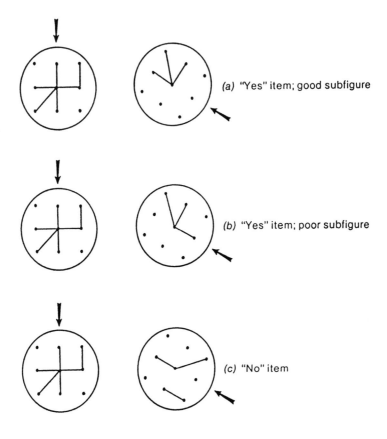

(a) "Yes" item; good subfigure

(b) "Yes" item; poor subfigure

(c) "No" item

FIGURE 7.

Samples are shown from Pylyshyn's 1979 experiment. The subject rotates mentally the figure on the left until its top is coincident with the arrow in the figure on the right. In this way the subject judges whether the right-hand figure is a subpart of the left-hand one. The correct replies are "Yes" in (a) and (b), and "No" in (c). The mental rotation rate is faster when the test item is a good subfigure, as in (a), than when it is a poor subfigure, as in (b). Good subfigures on the right contain a reasonable configuration of lines for triggering recognition of the ones on the left.

importance of Gestalt psychological principles of organized wholes? These principles are proximity, similarity, closure, good continuation, and symmetry.* A propositional representation for how a scientist or artist chooses to call a configuration symmetrical is particularly difficult to imagine.

A key criticism of Pylyshyn is that while propositions have truth value, visual images do not. While this is true for logicians, because there is no mathematical logic of pictures, it is not true for physiologists, nor for us either. And luckily so, because just imagine if we couldn't identify places and people we know. I can tell with incredible accuracy whether or not the woman I see walking across the street is my wife. Cognitive scientists who are computationally oriented to the extreme, like Pylyshyn, push the criticism of the lack of truth or falsity of images to the conclusion that images are not useful for reasoning. We have already seen examples to the contrary, and others given later in this chapter are among more evidence that refute this claim.

Pylyshyn and other anti-imagists contend that the assumed link between imagery and perception means that visual imagery merely mimics perception and so adds nothing new to analyses of thinking. If this were not bad enough, the image–perception link means that we cannot imagine anything other than what we have already witnessed in the world of sense perceptions. Therefore, visual imagery is a constrained description of the world. Although the image–perception link is essential to his model at this point in time, Kosslyn is willing to concede that images and perception are not exactly equivalent, in that imagery is a more transient sort of representation than perception.

As the reader has undoubtedly surmised, ever lurking in the background of these viewpoints is the underdetermination thesis. For the sets of data from the experiments of Shepard and Kosslyn, different hypotheses can be proposed for the link between stimulus and re-

---

*These principles are discussed in some detail later in this chapter, in the section "Gestalt Psychology and Vision."

sponse. We have studied the propositional viewpoint of Pylyshyn based on the premise that visual images are epiphenoma, and we have examined Kosslyn's, in which visual images play a role in thought. In 1978 the cognitive scientist John R. Anderson provided a simple example to show that both viewpoints can be transformed into the other. One person's propositional representation is another's analog one.* Consequently, despite Pylyshyn's philosophical arguments and Kosslyn's counterarguments and refined experiments, in the 1980s the imagery debate became a standoff. Pylyshyn writes in his 1984 book *Computation and Cognition* that while he congratulates Kosslyn on his "fascinating data," he "finds less laudable . . . the interpretations made of them."

The standoff was a curious one. Kosslyn, with the biggest stake in experimental data, dismissed Anderson with short shrift, claiming that as a "statement of faith . . . it will be possible to distinguish empirically between two views." Kosslyn's argument in favor of his computer model was to claim that what Pylyshyn proposes are "Rube Goldberg" models.

Pylyshyn, the theorist, was more receptive to Anderson's result and emphasized the important role of the underdetermination thesis in psychological research. But Pylyshyn took the interesting view that "one should not appeal solely to data." Rather, what ought to distinguish one model from the others is "how general it is, and so on." Can we assume that "and so on" alludes to an operationalizable notion of aesthetics or intuition?

---

*Consider the following example for which I am indebted to Professor J. M. Brady. We can represent any image on an (x,y) coordinate system with some function I of x and y, which we write as $I = I(x,y)$. For a circle, the function is $I = x^2 + y^2$. As a first approximation to making sense of an object, the brain is "uninterested" in smoothly shaded regions but is interested in edges, in order to construct a primal sketch. So, the brain deals with some function of sharply varying changes in the light entering the eye. Assume that we look at a wire ring and call the function containing information on light intensity reflected from the ring $E(x_i,y_i)$, where there are far fewer points (x_i,y_i) than on the "actual" object being viewed whose spatial coordinates are the set of points (x,y). But the set of points (x_i,y_i) can be represented in any one of a number of data structures and so can be as propositional as you like and can approach as closely as you want the function I of the "real" wire loop with all its details.

Other anti-imagists believe it is still too early to know in advance what sort of experimental data will be decisive in supporting or rejecting ontological claims for visual images. For support, they turn to a historical episode in physics in which the ontological status of atoms was settled in a most unexpected manner. We recall from Chapter 4 that Jean Perrin's measurements on the *macroscopic* quantities involved in Brownian motion provided the data that tipped the scales decisively in favor of atomism.

In the mid-1980s, a new experimental tool that displays images of the functioning brain appeared: positron emission tomography (PET). Kosslyn immediately turned to PET as the means for deciding issues decisively in his favor. He writes with emphasis in his 1994 book *Image and Brain* that these data *"arise* and are *used"* within the context of the processing system as a whole." In order to pursue Kosslyn's interesting proposal, we turn to the subject of vision itself.

VISION

In 1967, Marvin Minsky and Seymour Papert at M.I.T. made computer vision a summer project, entitled "Summer Vision Project," to be solved by a recently graduated student. This assignment shows how naive people were then about the mind's structure and function. By the early 1970s, everyone realized that the "problem of visual imagery" interfaces with just about every fundamental problem in cognitive science, including issues in philosophy of mind and neurophysiology.

The faculty of vision is closely connected with knowledge stored in one's memory. This is reasonable because "seeing" and "looking for" involve interpretation. We have already come across this in the difference between "seeing as" and "seeing that." Some of these nuances emerge in such AI areas as robot vision. Imagine asking a robot to go into the kitchen and fetch a coffee cup. How can the robot distinguish this object from everything it comes across? How much information must be programmed into the robot's memory for it to make necessary inferences? Every time we look for something, different parts of the mind's architecture work in parallel. Vision is massively parallel.

What do we see? We have discussed previously that what is in front of the eye is not what we "see." Consider Figure 8. To a young child, it is an odd-shaped bottle with metal fragments inside. To a scientist, it is a cathode ray tube. What do you see in the arrangement of black and white flecks in Figure 9? After some inspection, we see a dalmatian hound. Someone who had never seen a dalmatian would never have figured it out. Subsequently, every time we look at Figure 9, we immediately see a dalmatian hound. The black and white flecks have "snapped" into place, and a Gestalt switch has occurred.

The psychologist Richard Gregory's research on illusions has revealed further complexities in what we mean by "seeing." His main point concerns the image–perception link: Perception is not completely "*driven* by knowledge [because if it were, then] it would be almost impossible to perceive anything highly unusual." What Gregory means is that if we were completely driven by "knowledge" = logic, then how can we be fooled by illusions such as Kanizsa's triangle, in Figure 10, which elicits illusory contours bounding an illusory surface?

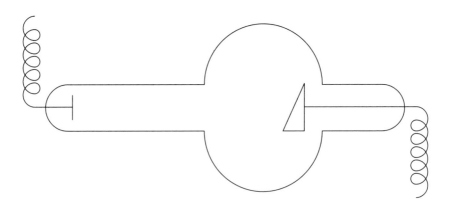

FIGURE 8.

A schematic of an x-ray tube. Any physicist would recognize it, but a nonphysicist probably would not, and a child definitely would not.

FIGURE 9.

An arrangement of white and black blotches in which a dalmatian hound can be discerned against a white background.

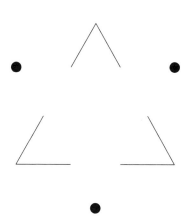

FIGURE 10.

A version of Kanizsa's triangle.

That there is an image–perception link in vision is no surprise. It is manifestly clear from the very act of seeing and understanding what we see. Light reflected from objects passes through the eyes' lenses and then propagates through the vitreous humor, finally striking the light-sensitive cells, or photoreceptors, that comprise the retina's surface (Figure 11). The retina's photoreceptors convert incident light into neural energy, which they transmit to the optic nerve. After a complicated journey, this encoding of the two-dimensional retinal image reaches other parts of the brain. The brain's task in visual perception is to decode this message and arrive at an image of a three-dimensional world.

The eye is far from perfect as an optical instrument. What is transmitted to the retina is not an exact version of what is happening before the eye. For example, there is a complex scattering of light in the aqueous and vitreous humors whose indices of refraction differ, and there

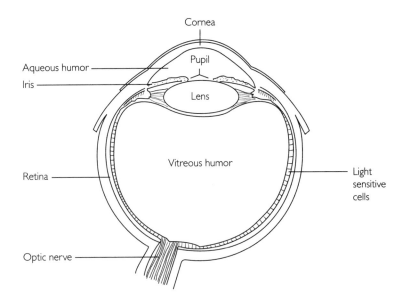

FIGURE 11.

Light enters the eye through the cornea and then proceeds through the aqueous humor, lens, vitreous humor, and finally strikes the retina with its light-sensitive cells.

could be problems with the eyes' lenses that must be corrected optically. As Hermann von Helmholtz remarked, any optician who designed an optical instrument of such poor quality as the eye would go bankrupt.

Figure 12 is a schematic diagram of the visual system, which is comprised of the eyes, optic pathways to the brain, and the brain. A great deal is known about the neurophysiology of the visual system. Via complex pathways, electrical and chemical impulses are transmitted from the retina to the primary visual cortex, referred to as V1 and located in the rear of the brain. This is carried out by two main routes, one for light rays entering the right visual field of each eye, and the other for the left visual field.

We must be careful at this point not to say that the brain in some way directly "sees" a retinal image. If this were the case, then there would have to be an additional eye in the brain to see this image, and so on ad infinitum. This is the "homunculus fallacy" first written about by Descartes in the seventeenth century. Rather, what the brain records is the cumulative result of many processes, which include electrical and chemical phenomena, as well as rules of organization of forms such as those in Gestalt psychology. Nor need the disposition of the object viewed by the retina be the "actual" shape of the object viewed. For example, a square viewed other than head-on will be imaged as a trapezoidal figure. Similarly, a circle can be viewed as an ellipse, or even as a straight line. In the light of such phenomena, we sometimes read statements to the effect that the brain is truer to external objects than the eye. How is this the case?

Presently the consensus reply to this complex question is that processing of visual information in the retina and primary visual cortex is parallel. The reason is that processing visual information requires such input as perceptual knowledge, perhaps even scientific knowledge; aspects of color perception; in addition to visual motion and binocular disparity, which result in depth perception, that is, folding together the two two-dimensional retinal images into a single three-dimensional image. Let us explore the parallel nature of vision processing.

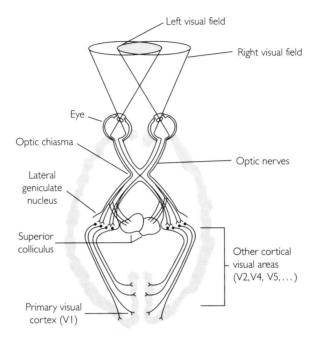

FIGURE 12.

Pathways from the eye to the brain. The gray fuzzy outline indicates the brain's surface. Optic nerves emanating from each eye meet at the optic chiasma. Nerves from the nasal side of the retina cross over and proceed to the opposite side of the brain, while nerves from the outside of the retina continue to the same side of the brain as the eye in question. Consequently, light from the right halves of both eyes goes to the right brain. Note that light from the left side of the visual field falls on the right side of each eye and so is transmitted to the left brain. Conversely, information from the left side of the visual field is passed to the right brain. Optic nerve fibers connect up with cells in either the lateral geniculate nucleus or the superior colliculus. Both of these areas are connected to the primary visual cortex, area V1. It is hypothesized that the lateral geniculate nucleus plays a role in perceiving details and in object recognition, while the geniculate nucleus is involved in the localization of objects in space. Later in this chapter we discuss other cortical areas such as V2, V4, and V5.

In the 1960s David Hubel and Torston Wiesel discovered that neurons in the primary visual cortex are grouped into columns according to whether they are sensitive to lines of a particular orientation. These vertical groupings are called "hypercolumns" and are on the order of 1 mm thick (Figure 13). Hubel and Wiesel made this discovery through studying the response of laboratory cats to different patterns of

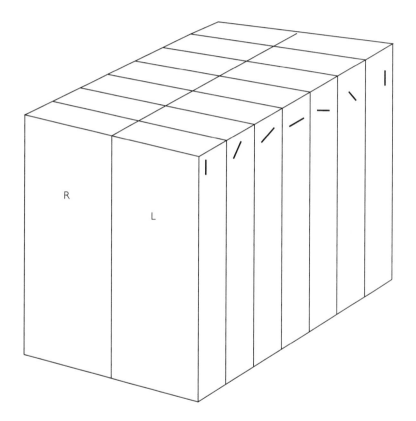

FIGURE 13.

Hubel and Wiesel found that the primary visual cortex is organized into vertical slabs, or "hypercolumns," which are the basic computational units. There are hypercolumns that are responsive to the right (R) and left (L) eyes. The slabs are sensitive to edge orientations in steps of about 10 degrees. This R and L pattern of hypercolumns is repeated throughout the primary visual cortex.

visual stimulation. They inserted recording electrodes into the cat's brain. Electrodes inserted at right angles registered only for a single line orientation, while those inserted tangentially registered for a multiline orientation. This locus of neural activity is called a "topographic map" onto cortical "space." Topographic maps indicate an area in which neurons are highly organized. For their astonishing research, Hubel and Wiesel were awarded a Nobel prize in 1981.

Let us consider how Hubel and Wiesel's discovery fits into image processing. Figure 14 is a schematic for incorporating results on how retinal images are turned into neural images, which are then analyzed toward reassembly of the object that is before the eyes. The figure is divided into the three hypothesized levels of image processing: low, intermediate, and high. As I mentioned earlier, this categorization is mainly heuristic, that is, it is theoretically expedient for formulating computational models of the brain. Said simply, one has to start somewhere, and luckily this split into three levels, however artificial, turns out to be useful.* Neurophysiologists rarely use it. Even in computational models, this categorization has to be handled with care because there is a great deal of overlap between intermediate- and high-level processing. With these caveats in mind, I shall take the liberty of continuing to use this nomenclature because the necessarily synoptic discussion of vision in this chapter tries to cut a fine line between the Charybdis of computational psychology and the Scylla of neurophysiology.

The low-level portion relates to information obtained from stimulation of the retina's light receptors. Figure 14 is highly simplified, and the reader should not infer that there is a line drawing of an arrow on the retina. The hypothesized primal sketch results from a combination of low- and intermediate-level processing and is then filled in by the intermediate- and high-level parts.

*As we noted earlier in this chapter, there are precedents for hypotheses of this sort, for example, in research in elementary particle physics. We recall as well that Einstein's light quantum hypothesis was heuristic. He proposed it as a starting point toward understanding phenomena in which light is converted into matter, like the photoelectric effect.

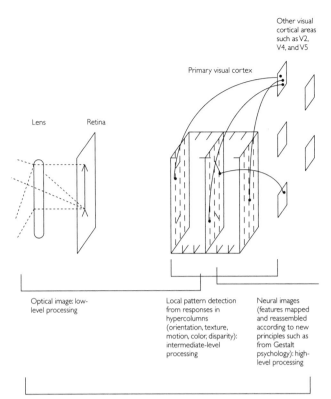

Other visual
cortical areas
such as V2,
V4, and V5

Primary visual cortex

Lens   Retina

Optical image: low-
level processing

Local pattern detection
from responses in
hypercolumns
(orientation, texture,
motion, color, disparity):
intermediate-level
processing

Neural images
(features mapped
and reassembled
according to new
principles such as
from Gestalt
psychology): high-
level processing

In practice, there is a complex interplay between levels.

FIGURE 14.

A schematic of the formation of neural images that models the processes occurring in Figure 13. The retinal image of an arrow is formed by light rays (dotted lines) that enter the eye (lens). Low-level vision processing result in the outline (primal sketch) of the arrow. The "image" is transmitted as energy via neurons to the primary visual cortex, V1, whose different sections are feature detectors sensitive to various features of the arrow. The front and middle panels or hypercolumns of V1 are comprised of columns of neurons that are sensitive to the arrow's oblique lines, and the third one to vertical lines. There are no actual lines on these panels, of course. Rather, the actual features of the arrow are drawn to indicate which panels "fire," that is, are activated by lines with these orientations. This is the realm of intermediate-level image processing, where various faculties begin filling in the primal sketch. Other visual cortical areas such as V2 (responsible for shape processing), V4 (responds to color), and V5 (contains neurons that respond selectively to motion) assist in "reassembling" the image with the help of Gestalt laws of good organization. This is high-level processing, which is a top-down process.

297

High- and intermediate-level processing involves, for example, motion analysis, which is important toward distinguishing objects from their background. Considerable evidence has been gathered to support the assumption of dedicated neurophysiological circuits for motion processing in area V5 of the visual cortex, which is exterior to V1. In V5, for example, some cortical cells respond most efficiently only to edges that move across the visual field. There are other specialized cortical areas external to V1 such as V4, in which Semir Zeki of University College London has found evidence for color processing of images. Zeki's research has been instrumental toward the realization that certain areas of the brain have specific functions.

"Reassembly" of the image is the result of parallel outputs from V1 to other areas, which, in turn, produce parallel input replete with a top-down flow of hypotheses. At this point the property of "arrowness" is assembled. Of importance to this process are lines defining edges. The reassembly operation is most sensitive to groups of lines that underlie Gestalt psychological principles of organization.

GESTALT PSYCHOLOGY AND VISION

The key work in Gestalt psychology was begun in the 1920s by the psychologist Max Wertheimer, about whom we shall have more to say in Chapter 9. The point of Wertheimer's research is that there must be something more in the visual system than cells that respond to variations in light intensities, that is, to what is figure and what is ground. We don't comprehend objects as disconnected parts; rather, we see wholes and, more precisely, we identify patterns. Some patterns are more easily discernible than others because their parts can be easily connected. Wertheimer called good patterns *gestalts*, or organized wholes. The subject became known as Gestalt psychology. The notion of Gestalt had been discussed some years before by Christian von Ehrenfels in 1890, who emphasized that form quality (*Gestaltqualität*) was the essence of vision, rather than individual parts. Interestingly, von Ehrenfels' realization was in response to the following problem raised by Mach. Why is it that we can recognize a melody in any key? Mach hypothesized that in addition to hearing individual notes, we hear relations as well. In 1890 von Ehrenfels proposed a new term for this relation—Gestalt—in order to emphasize that the individual notes of a melody had form quality. Figure 15 displays some Gestalt principles of organization for visual perception.

These principles are the dynamics behind the *Prägnanz* principle, which is the tendency toward an organization of perceptual elements that possesses proximity, similarity, good continuation, closure, and symmetry. Experiments indicate that these laws are hard-wired into the visual system. If they were not, we would be overwhelmed by the constantly changing patterns of incoming perceptual information. Researchers have found evidence across cultures for Gestalt principles of organization.

Gestalt principles of organization can also have a profound effect on how we view paintings. Fritz Heider, a colleague of Wertheimer in Berlin during the 1920s, has an interesting view on how Picasso achieved some of the eye-catching effects in certain paintings. In a reminiscence of 1973, Heider writes that, for example, Picasso's *Three Musicians*, 1921, is so eye-catching because Picasso intentionally destroyed the "natural units of familiar objects by opposing one unit forming factor to another" (Figure 16). Heider does not claim that Picasso should in any way be considered as a founder of Gestalt psychology but that he surmised some of its basic principles as essential to the Cubist aesthetic.

THE PRESENT STATUS OF THE IMAGERY DEBATE

The philosophical and scientific merit in the imagery debate notwithstanding, after surveying visual research we are led to ask whether the debate addresses visual imagery in the depth it claims. The consensus is no. On the issue of encoding, the opposing views can be transformed

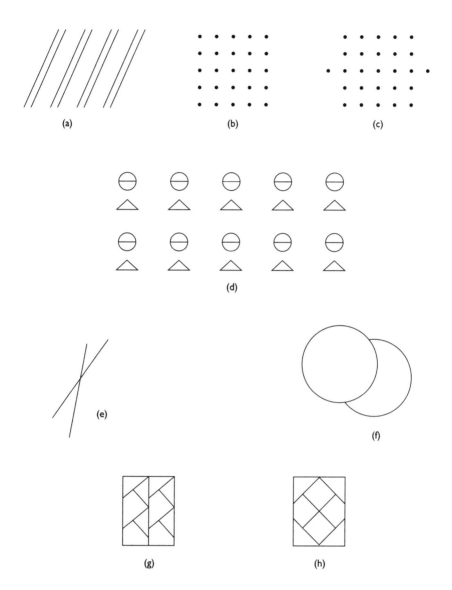

FIGURE 15.

The patterns in (a) through (g) are representations of Gestalt principles of perception. Patterns in Figures (a) to (c) illustrate proximity; (d) illustrates similarity; (e) illustrates good continuation; (f) illustrates closure; and (h) illustrates symmetry relative to the pattern in (g). Adapted from A. L. Glass, K. J. Holyoak, and J. L. Santa, *Cognition* (Reading, Mass.: Addison-Wesley, 1979).

FIGURE 16.

Pablo Picasso's two versions of *Three Musicians*, 1921. Top: Philadelphia Museum of Art: A.E. Gallatin Collection. Bottom: The Museum of Modern Art, New York. Mrs. Simon Guggenheim Fund. Photograph © 1996 The Museum of Modern Art, New York.

one into the other. While the descriptivists such as Pylyshyn neglect developments in the neurophysiology of the visual system and ignore experimental data on a dual representation of knowledge, Kosslyn pushes neurophysiological data further than is merited at present.

Despite the unsurprising formal translatability between the pro- and anti-imagery positions, the discovery of a dedicated representative medium would be a coup in favor of the pro-imagists. They claim such evidence exists in data from neurophysiology. The principal research tool is positron emission tomography (PET) to demonstrate that V1 is topographically mapped in response to stimuli that can be external *or* internal. By "external or internal stimuli," I mean the following. In recent experiments performed by Kosslyn and coworkers, as well as by Hanna Damasio, Thomas Grabowski, and their colleagues at the University of Iowa, subjects are given an object to inspect visually. Then they are asked to close their eyes and image the object. PET scans indicate that V1 fires in *both* situations. Does this prove the case for the existence of a dedicated representational medium for visual imagery? In order to reply, let us make a foray into images of the brain.

PET measures increased blood flow to a localized area of the brain that occurs when the subject undertakes a particular task. The increase is detected with a radioactive substance injected into the patient's bloodstream. Radioactive oxygen is used because of its short half-life, which results in positrons (positively charged electrons) among its decay products. After traveling a short distance in brain tissue, a positron collides with an electron to produce annihilation light quanta, which are detected with an array of radiation detectors around the patient's head. The greater the brain's activity in a localized area, the greater will be the blood flow, and so a large number of annihilation light quanta are traceable to that area. After about three hours, a topographical map like the one in Figure 17 is obtained.

There are problems with PET. The main drawback is the poor spatial resolution, which is evident in parts (b), (c), and (d) of Figure 17. PET's spatial resolution is presently 8 mm, a region that includes billions

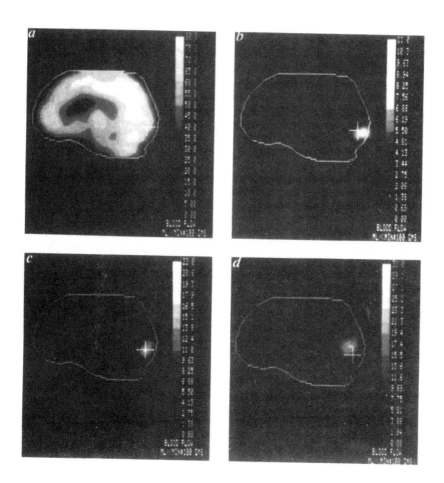

FIGURE 17.

Positron emission tomography (PET) images of local blood flow during stimulation of the brain's visual centers. Figure (a) is the image of blood in the brain "at rest." The scale on the right is calibrated in units of blood flow. Figure (b) is an image of the difference in blood flow between resting and stimulated states obtained by subtracting Figure (a) from the image obtained when the center of the visual field is stimulated. The area of increased activity is in the back of the brain, that is, in the primary visual cortex V1. Figures (c) and (d) were made similarly to (b) and are images of differences in blood flow when areas away from the center of the visual field are stimulated. The areas of activation move accordingly. These figures are topographical maps of visual space in V1. [From P.T. Fox, M.A. Mintun, M.A. Raichle, M.E. Meizen, J.M. Allman and D.C. Van Essen, "Mapping the human visual cortex with positron emission tomography," *Nature* (1986)] of neurons. Neurophysiologists such as Hubel and Wiesel could not have discovered hypercolumns with PET. Instead they used electrical probes inserted at various angles for examining individual layers of the cortex. Physiologists study individual cells. Poor spatial resolution is, in part, due to the subtracting methods necessary in PET. The PET image of the subject's "resting brain" is subtracted from the brain image obtained when a particular part of the visual field is stimulated. The problem here is what is meant by a person's brain being "at rest." We are always thinking of something—even if it's that we ought to think about nothing—and, of course, in PET scans the brain has been sensitized by radioactivity so that even low-level excitement is magnified. Since there is no standard baseline for each experimental run, experimenters are never certain of exactly what they are subtracting, and so experimental runs for an individual subject are averaged. Another factor in spatial resolution is the radioactive dosage. The higher the dose, the better the spatial resolution. There are safety levels.

Besides PET's poor spatial resolution, its temporal resolution is on the order of seconds, while activations in the brain can occur in milliseconds or less. Consequently, although in a given experimental situation we can ascertain that V1, V2, and V4 have been excited, we don't know the sequence of activation.

In summary, PET's poor spatial resolution is compensated for by its value in displaying active regions of the cerebral cortex, thereby also differentiating which parts are active under different conditions. PET is a powerful tool for mapping the primary visual cortex. PET reduces brain functions to a topographic problem because it is informative about where things happen but not how they happen. Semir Zeki writes in his 1993 book *A Vision of the Brain* of

> the enormous promise that [PET] holds for the future if, through adequate means of stimulation, one can demonstrate entire active systems of the cerebral cortex and infer their connectivity.

In this way, results from neurophysiology can be used as constraints on models of the cognitive functional architecture proposed by researchers in AI and cognitive science. Presently, with few exceptions, researchers in cognitive science and AI work pretty much independently of those in neurophysiology.

Although Kosslyn is forthright about PET's problems, he often makes lavish claims about his visual imaging experiments, the gist of which is as follows. Among the experiments Kosslyn performed is a series in which he had subjects figure out through visual imagery how to complete a variety of figures that they had actually viewed and that Kosslyn calls "canonical" (standard) and "noncanonical" (nonstandard). Canonical pictures are easy to identify even though parts are missing. Noncanonical ones are more difficult to decipher because they have fuzzy contours, disrupted parts, peculiar relations among parts, or are shown from a peculiar perspective. Kosslyn claims areas of the brain utilized in high-level processing of perceptions come into play in order to try to make sense of the noncanonical pictures by moving around their parts by visual imagery.

PET scans of the subjects were taken. Some of the results appear in Figure 18. The details of the experiments that led to the data in Figure 18 are unimportant for what follows. I leave it to the reader to peruse Kosslyn's lucid description of his extremely clever set of experiments. The main point is Kosslyn's claims that through subtraction procedures he has identified parts of the brain responsible for making sense of visual images *and* that these are the parts that also figure into high-level vision. Kosslyn himself hypothesized some of the regions for use in his model. The essential point is that the precise areas of the brain involved in high-level vision are as yet unclear and highly hypothetical.

Kosslyn puts great weight on the fact that PET scans indicate V1 is stimulated in the course of visual imagery just as in visual perception. He claims this is important because it indicates that we actually experience an "image." Yet most vision researchers do not consider the excita-

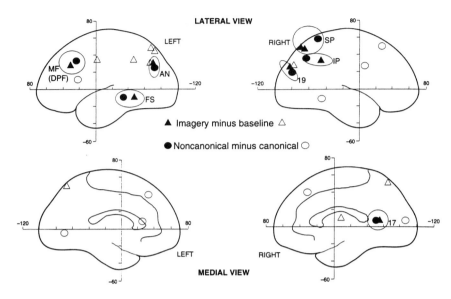

FIGURE 18.

Results from PET scans performed by Kosslyn and co-workers in which they showed subjects canonical and noncanonical pictures. The circles indicate foci of activity which showed up when blood flow in canonical pictures were subtracted from blood flow in non-canonical situations. The triangles in both outline and filled, in addition to the filled in circles, are data from visual imagery experiments whose details are unimportant. The key point is Kosslyn's claim that his data points (triangles and circles) pinpoint regions in the brain which participate in high-level vision processing. From this he claims to have established an image-perception link in visual imagery, which opens up the opportunity for him to piggyback a theory of visual imagery on results from high-level vision.

tion of V1 during visual imaging to be at all surprising. After all, isn't the visual cortex the principal site for visual imagery? So, contrary to Kosslyn, this does not settle the question of whether visual images are epiphenomena. Kosslyn offers interesting neurological evidence to the effect that visual images are not epiphenomena: If they were, we would expect that damaging V1 would not affect a person's capacity for visual imagery. The opposite result was obtained with patients whose occipital lobe (which contains V1) had been removed.

Yet in 1992, data began to appear that indicate that patients with impaired object perception due to head injuries have normal mental imagery. For example, such patients can make detailed drawings of objects from memory but cannot identify visually presented ones. How can this be if there is a strong image–perception link, as Kosslyn hypothesizes? Here emerges a strength that can also be a defect in Kosslyn's theory. The structure of his theory of visual imagery is not unlike those of other theories in the history of science proposed at a time when there is a plethora of interesting but difficult-to-explain experimental data.

This was the case, for example, with Lorentz's electron theory in 1904. Since the theory's formulation in 1892, Lorentz had dealt with sometimes bewildering data by proposing an ever-growing web of interlinked hypotheses. Step by step Lorentz was constructing a theory of matter in motion. Then, in 1905, along came Einstein, who offered a theory based on two axioms from which he was able to *deduce* just about all of Lorentz's hypotheses. Although Einstein's special theory of relativity superseded Lorentz's electron theory, Lorentz's work was essential to Einstein's thinking. Theories such as Lorentz's, which are heavy with hypotheses, are often the way to proceed. In 1904 Lorentz explained this necessity to Henri Poincaré, who had often criticized Lorentz for making too many hypotheses. Lorentz wrote that he "should be permitted the excuse, that we are entering into [the unknown domain of electron physics] cautiously groping forward."

Kosslyn is also "groping forward," in a way similar to Lorentz. Whenever new data that seem detrimental to his theory are published, Kosslyn offers another hypothesis. So, Kosslyn forthrightly notes in *Image and Brain* that "at first blush [the new data on the robust visual imagery of perceptual impaired patients] appears to be an embarrassment for the present theory." To explain away the embarrassment, Kosslyn offers the hypothesis that although imagery and perception rely on the same internal processes, "those mechanisms are not used identically in the two cases." While visual imagery depends on already processed information, this is not the case for perceptions, which must be organized from scratch. This serves to explain these 1992 data within his theoretical framework while preserving a basic part of the theory, the image–perception link.

In *Image and Brain*, Kosslyn claims to have resolved the imagery debate. As evidence that visual images are not epiphenomena, he offers PET scans like that in Figure 17 in which the stimulated area moves in accordance with a shift in external stimulus with respect to the visual field. As to what "looks" at visual images, he offers PET scans like those in Figure 18, which he assumes with great confidence show that visual imagery is processed by the parts of the brain active in high-level vision. He concludes that the "mechanisms underlying imagery seem no more mysterious than those underlying visual perception." QED.

Has Kosslyn really resolved the imagery debate in his favor? To claim PET scans offer proof that certain topographically mapped areas of the cortex are the visual buffer needs further substantiation as well as clarification and sharpening of terminology regarding his hypothesis that topographic maps of imagined situations are visual "images." His hypothesis on a strong link between high-level vision and visual imagery requires higher temporal resolution in order to ascertain the sequence of excitations. Finally, present theories of visual perception are so hypothetical that to piggyback a theory of visual imagery on them and then to make far-reaching claims must be taken with a great deal of caution, however adventurous and meritorious the effort may be.

In contrast to Kosslyn's direct and nonequivocal claims, other vision researchers consider the "degree to which the same neural representations are involved in both visual imagery and perception is unclear," write Alumit Ishai and Dov Sagi of the Weizmann Institute of Science in Israel. Such new results have appeared to suggest, or indicate, that, in fact, cortical cells that process incoming stimuli serve also as memory cells for visual imagery in the sense that for short time intervals they can be activated by stimuli to reproduce a visual image. Consequently, these cells may serve as the interface between visual images and visual perceptions, providing evidence for such a link, as Ishai and Sagi infer cautiously.

In summary, the nature of visual perception is still not completely understood, particularly as to how it is generated in various parts of the brain. But among what is known about visual imagery is that it's a massively parallel process in which theories of perception play some role. Semir Zeki summarizes the situation in vision research today as follows, "The scientific study of the brain is still in its infancy." From our excursion into research on visual perception, we realize that the imagery debate is much like setting up two straw men. The vision researcher J. Michael Brady at Oxford University summed up the situation pungently in the course of a recent discussion, "In the chasm between resides our general ignorance about visual perception. That is, how we as human beings see images." There is a great deal left to do before anything resembling an imagery debate is settled.

THE MISUSE OF LANGUAGE IN VISUAL IMAGERY

We have used terms like neural "images" and topographic "maps" in cortical "space." Does this mean that we have actual maps imprinted on the surface of our brain? If we reply affirmatively, do we fall into the "in-my-head" abyss criticized so strongly by Gilbert Ryle? At this point we have to be very careful about the source of an image. Is it one that we are imagining, like imagining our grandmother? Or is it one *caused* by a visual stimulus? If it is caused by a visual stimulus, then it turns out that under certain laboratory conditions a topographic map on the cortical surface can be exhibited. In this case the notion of "in-my-head" has some limited credibility. The conditions are, in fact, extreme, and we have already discussed one: Hubel and Wiesel's 1978 experiment that substantiated the hypercolumn structure of the primary visual cortex.

Not long after Hubel and Wiesel's initial experiments, in 1982 R. B. H. Tootell and associates produced what is regarded as the first image that shows a remarkable structural similarity between a stimulus presented to the retina and its topographic map in cortical space (Figure 19). The

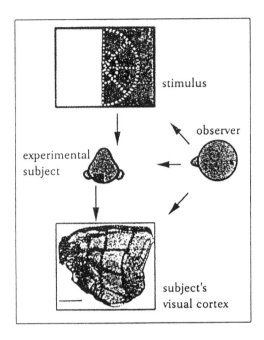

FIGURE 19.

A human "observer" compares a stimulus comprised of a blinking pattern with the topographical map on the brain from an "experimental subject" which is a monkey. The observer is amazed by the approximate correspondences between the stimulus and the topographic map on the brain's visual cortex. The data are from R.B.H. Tootell, M.S. Silverman, E. Switkes, and R.L. De Valois, "Deoxyglucose analysis of retinoptic organization in primate striate cortex," Science, 218, 901–904 (1982) and the figure is from A.R. Damasio, Descartes' Error (1994).

following question arises: Can we reconstruct objects from such a topographic map? Clearly, the answer is no. What Tootell's monkey witnessed was a strongly flickering, well-lit pattern of dots (Figure 19(a)). (In fact, it was the last thing the monkey ever saw!) Since the visual stimulus was static in space and properly enhanced, some approximation to it was "imprinted" on the monkey's brain because of neuronal firings in the proper hypercolumns.

Less clearly imprinted on everyone's brains for minute instants of time are topographic maps from our daily life. They constantly change and are, of course, nowhere near as intense as the one on the monkey's brain in Figure 19. The reason is that the light reflected off the face of someone with whom you are speaking is not nearly intense enough to produce a map of the sort on the monkey's brain in Figure 19. Moreover, there is no reason to assume that topographical maps are evidence for a dedicated representational medium for images constructed from visual stimuli. These brain "imprints" come and go, having been assembled by many parallel processes.

What about visual imagery? This is the problem of key interest to us. When, for example, we close our eyes and think of our grandmother, is her image imprinted on our mind in a topographic organization of neurons that corresponds roughly to her face? I think not. Rather, what we image has the form only metaphorically of a topographic map. Its stimulus is internal and emanates from long-term memory, which causes the firing of many different parts of the brain. So that what we "image" is an assemblage of an immense amount of information about our grandmother, some of which is not of visual modality: the odors in her house, the sound of her voice, the touch of her hand.

For visual imagery, terms like "image," "map," and "space" are fraught with dangers, such as the homunculus fallacy mentioned earlier. Perhaps the field of visual imagery research needs a Heisenberg uncertainty principle paper to provide restrictions on words like "image" and "map" so that they are used unambiguously.

For the most part, the everyday world comes in twos: yes or no; light or darkness; wave or particle. Consequently, perhaps it is not coincidental that as a start toward investigating the mind's functional architecture, cognitive scientists assume that there are two types of declarative knowledge representations, propositional and pictorial. We have found that in practice it is difficult to draw a distinct line. That thinking involves suitable combinations of both modes of representation will be further discussed in this chapter and in Chapters 9 and 10. And just as was the case in quantum physics, we must be prepared for surprises in future developments in vision and in exploring visual imagery.

We have noted the importance of visual imagery in scientific research again and again throughout this book in examples from the history of scientific thought. This subject offers the opportunity to explore the problem of visual imagery from a fresh viewpoint and to provide evidence that they are not epiphenomena. In what follows we discuss situations in art and science in which creative thinking is carried out in visual imagery. As Antonio Damasio writes in his book *Descartes' Error,*

> Images are probably the main content of our thoughts, regardless of the sensory modality in which they are generated and regardless of whether they are about a thing or a process involving things.

In earlier chapters, we explored the importance of visual imagery in scientific research for Galileo, Einstein, Bohr, and Heisenberg, among others. Let us discuss thought experiments in more generality.

THOUGHT EXPERIMENTS

In their initial stages, all experiments are thought experiments. Then what sets aside the thought experiments of a Galileo, an Einstein, a Maxwell, a Bohr, or a Heisenberg? We begin our reply by returning to Ernst Mach, about whose philosophy we have been extremely critical. Nevertheless, Mach was a superb scientist. As Einstein said of Mach in 1922, "Mach was a good mechanician, however he was a deplorable philosopher."* In 1905 Mach wrote an article entitled "On thought experiments," which we can take as indicative of what most scientists at the time considered to be the role of this type of thinking. Mach wrote, "Our ideas are more readily at hand than physical fact: thought experiments cost less, as it were." A wonderful example of this cost policy is

---

*The occasion for this uncharacteristically sharp public comment was that, while in Paris, Einstein was informed of the posthumous publication of Mach's *Theory of Optics.* In the book's preface, Mach repudiated relativity theory in no uncertain terms.

Nikola Tesla, the electrical engineering genius from the early twentieth century. Tesla described how he played mentally with designs for complex electrical motors. He claimed to be able to dissemble visual images of motors and, from time to time, check their parts for wear.

Mach succinctly described what to most everyone then and now is a thought experiment, namely, an "idealization or abstraction of existing physical conditions." So, for example, Mach suggested we imagine situations in which there is no wind resistance for projectiles. We can also think of the mathematical limiting process of letting an infinite number of infinitesimal line elements pass to zero while their sum is finite; this is the process of mathematical integration. These were the sorts of thought experiments that also appealed to Galileo, who, as Mach put it well, "is a master of this kind of experiment." After all, Galileo imagined what it was like for bodies to fall through vacuums. An important consequence of Galileo's thought experiment is that all bodies fall through vacuums with the same acceleration regardless of their weight. But there were no laboratory vacuums available for him to demonstrate this point. So, to argue for his rather astounding law of fall, Galileo proposed a thought experiment in the *Dialog on the Two Chief World Systems*, which we discussed in Chapter 1.

Galileo's thought experiments, as well as those of later scientists, were used to present arguments for hypotheses that had already been proposed. We recall Carnot's engine and Maxwell's demon in Chapter 4. But Einstein's thought experiments were different: His were unplanned for and were insights that resulted in discoveries. I have in mind his two great thought experiments of 1895 and 1907. In Chapter 1 we discussed the 1895 thought experiment, which led a decade later to the special relativity theory. On the basis of his readings in electromagnetic theory, the 16-year-old boy studying in a Swiss preparatory school that emphasized the power of visual imagery conceived in his mind's eye what it would be like to catch up to a point on a light wave.*

*Details of Einstein's life are discussed in Chapter 9.

Another key thought experiment occurred to Einstein in 1907, while he was working in the Swiss Federal Patent Office in Bern. Twelve years later, Einstein recalled it as the "happiest thought of my life." We can imagine Einstein at the Patent Office sitting on his high stool at a large, tilted drawing board. Sunlight streamed in through the floor-to-ceiling windows. Every now and then he would look around to see if the director, Friedrich Haller, was observing his "boys." If not, then Einstein would carefully slip open his top desk drawer and sneak a look at some of his own calculations. While all other physicists were abuzz over Kaufmann's data on the mass of high-speed electrons, Einstein's concern was an asymmetry in the view of space and time he offered in 1905: the preferred status of inertial reference systems. While special relativity can deal with accelerated motions, it takes account only of *measurements* made in inertial reference systems. Einstein considered the preferred status of inertial reference systems as an asymmetry in the theory, a blemish to be removed. But how? The seeds of the solution lay in the "happiest thought" of his life, shown in Figure 20. Two years earlier, Henri Poincaré had tried his hand at extending Lorentz's theory of the electron to include gravity. But no one before Einstein thought of relating gravity and acceleration. This masterstroke led to one of the most beautiful theories ever conceived of—the general theory of relativity, which Einstein completed in 1915.

Since thought experiments occur to the prepared mind, what could have prepared Einstein in 1907? Clues abound in the 1905 special relativity paper. The paper opens with Einstein inviting the reader to imagine riding on a magnet in uniform motion relative to a wire loop. He then asks the reader to imagine this experiment from the alternative viewpoint of someone on the wire loop. This situation is much like the one in which Galileo stands on a bridge overlooking the river Arno and observes a boat passing from under the bridge, which we discussed in Chapter 1.

Einstein imagines how these two sets of observers interpret their measurements of the current arising in the wire as a result of the relative

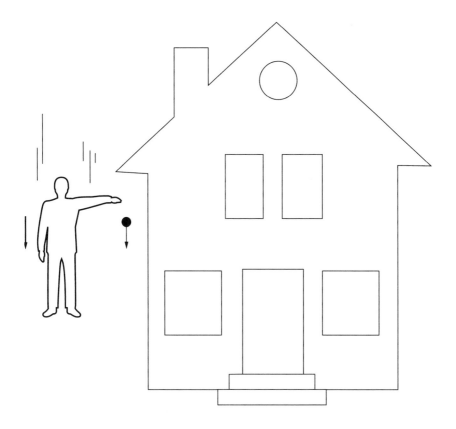

Figure 20.

In 1907 Einstein conceived of a thought experiment that led to a basic part of the general theory of relativity, the equivalence principle. The thought experimenter jumps off the roof of a house and simultaneously drops a stone. He realizes that the stone falls at relative rest with respect to him, while they both fall under the influence of gravity. It seems, therefore, as if in his vicinity there is no gravity. Einstein's great realization here is that the thought experimenter can consider himself and the stone to be at relative rest by replacing the Earth's gravitational field with an acceleration equal in magnitude but oppositely directed to the Earth's gravitation. Eureka! Gravity and acceleration are relative quantities.

motion between wire and magnet (Figure 21). They measure the *same* quantity: the electrical current generated in the wire due to the relative motion between magnet and conductor. But according to electromagnetic theory in 1905, the two observers have totally different interpretations for the current's origin, which has a single cause (relative motion between magnet and conductor). Einstein wondered why there need be two different interpretations for a phenomenon that depends on only a single quantity. He interpreted this redundancy in explanation as one of the asymmetries that beset recent physics. He considered this situation so serious that he presented it in the very first sentence of the 1905 relativity paper:

> That Maxwell's electrodynamics—the way in which it is usually understood—when applied to moving bodies leads to asymmetries that do not appear to be inherent in the phenomena is well known.

Recollecting how he formulated special relativity in 1919, Einstein recalled that he considered this redundancy in explanation to be "unbearable" because the "difference between these cases could not be a real difference, but rather, in my conviction, only a difference in the choice of reference point."

Whereas Galileo's thought experiment with falling bodies was used to demonstrate an irreconcilable internal inconsistency in Aristotelian physics, Einstein's was meant to demonstrate the need to overhaul the physics of 1905. Other thought experiments in the 1905 paper reveal further redundancies in explanation, also referred to by Einstein as asymmetries. Getting rid of these redundancies required a new view of space and time. The 1907 thought experiment led to generalizing special relativity to measurements made in accelerating reference systems. This, in turn, required a folding together of geometry and physics with such astounding predications as the bending of light near massive objects (see Chapter 3).

Let us look back over Einstein's thought experiments of 1895 and 1907 from the viewpoint of intermediate- and low-level processing. Vi-

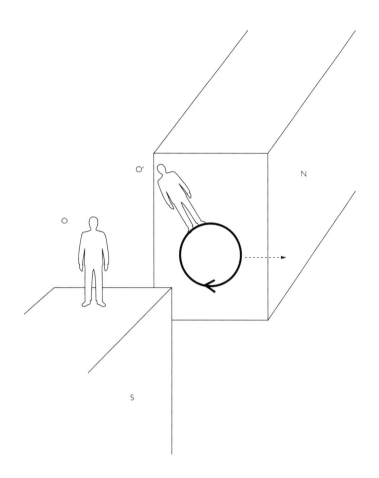

FIGURE 21.

N and S designate pole faces of a permanent magnet between which a wire loop travels to the right as indicated by the dashed arrow. Observer O' rides on the conducting loop and O stands on the magnet's south pole. Both observers measure the same quantity—the current generated in the wire loop as a result of its motion relative to the magnet. The current's direction is clockwise, as indicated by the arrow on the loop. Although the measured quantity (electric current) depends only on the relative motion between loop and magnet, the two observers explained the current as arising in two fundamentally different ways. To observer O, the current arose from the force on positive charges in the wire due to their motion through the magnetic field. On the other hand, O' interprets the current as arising due to an electric field in the positive charge's vicinity. The origin of this electric field is the magnet, which O' sees moving by it with an equal and opposite velocity to the one O measures. Einstein found this duality of explanations to be "unbearable."

sual imagery in these experiments is abstracted from phenomena we have actually experienced. These images are, of course, knowledge-loaded with the twist that Einstein reassembled them in a unique way. Einstein "noticed" that the symmetry of relative motion in Figure 21 is broken by asymmetry in explanation. So reassembly of the image's parts in the thought experimenter's imagination is incomplete according to, for example, the Gestalt laws of good organization. This factor may well have led to Einstein's emotional reaction that the situation in electromagnetic induction is "unbearable." The element of completeness, which restores continuity and symmetry, is his realization that any differences between the two observers resides only in the "choice of viewpoint."

On the other hand, what Bohr and Heisenberg were up against in 1927 was visual imagery—or the lack thereof—of a totally different sort, due to the peculiar "particle" nature of light quanta. The quotation marks indicate that we ought not understand the particulate nature of the light quantum to be in any way like particles from the daily world, which are localized entities. Having in mind the prequantum meaning of the term "particle," Bohr and Heisenberg struggled with what it means for a light quantum to be polarized. How can a particle be polarized? The visual imagery of visualization will not do here. The term "polarization" had always been interpreted according to the wave theory of light, which had become intuitive even though we do not "see" light waves. Basically, the polarization of a light wave is defined as the direction of its plane of vibration. But the light quantum is a particle and so has no plane of vibration. (Recall in Chapter 2 Bohr's attempt at a resolution to this intuitive impasse with the complementarity principle.) In this case, any attempts at visual imagery failed. Their resolution was a formal one, dictated by their interpretation of the mathematical formalism of quantum mechanics.*

*Quantum mechanics interprets the polarization of a light quantum in a statistical sense. In classical physics, if a vertically polarized light beam is passed through a polarizer at, for example, 45° to the vertical, then we say that part of the beam is passed while the remainder is absorbed. But since light quanta are indivisible, then in a similar experiment quantum mechanics says that there is a certain probability that some photons will be passed through the polarizer, while others will not.

We may inquire why Einstein, one of the great masters of the thought experiment, conceived of no more seminal ones after 1907. The reply could well be that research in atomic physics, as well as any moves beyond general relativity, required thought in a direction to which he was not disposed, namely, toward visualizability (visual imagery generated by scientific theories) and not visualization (visual imagery abstracted from phenomena we have witnessed in the world of perceptions). The theories produced by Bohr, and particularly Heisenberg, required permitting the theory's mathematics to generate proper visual imagery. This mode of thought invites an adventurousness and play with visual imagery far beyond how this is carried out in classical physics, of which the relativity theories are essentially a part. In short, Einstein was unwilling to alter his mode of visual imagery.

In 1963, Heisenberg described as best he could his daring play with visual images. While quoting Heisenberg *in extenso*, I will take the opportunity to provide annotations in brackets and italics to indicate how the conceptual problems he mentions touch on those in vision theory:

What frequently happens in physics is that, from seeing some part of the experimental situation, you get a feeling of how the general experimental situation is. That is, you get some kind of [*This is the situation in low-level processing.*] picture. Well there should be quotation marks around the word 'picture.' This [*Note his sensitivity to the term "picture" in quantum mechanics, just as is the case for the term "image" in visual imagery.*] 'picture' allows you to guess how other experiments might come out. And, of course, then you try to give this picture some definite form in words or in mathematical formula. Then what frequently [*Again, close contact with a propositional-based visual theory.*] happens later on is that the mathematical formulation of the 'picture' or the formulation of the 'picture' in words, turns

[*Difference between what the eye sees and what the mind "sees."*]

[*Importance of nonverbal thought to scientific creativity.*]

[*Difference between high- and low-level vision processing.*]

[*Difference between what one learns from analysis of the retinal images and the image constructed by high-level vision; that is, between the primal sketch and the end-product of high-level vision.*]

out to be rather wrong. Still the experimental guesses are rather right, that is, the actual 'picture' which you had in mind was much better than the rationalization which you tried to put down in the publication. That is, of course, a quite normal situation, because the rationalization, as everyone knows, is always a later stage and not the first stage. So first one has what one may call an impression of how things are connected, and from this impression you may guess, and you have a good chance to guess the correct things. But then you say 'Well why do you guess this and not that?' then you try to give rationalizations, to use words and say, 'Well, because I described such and such.' The picture changes over and over again and it's so nice to see how such pictures change.

VISUAL IMAGERY, COGNITIVE SCIENCE, AND THE HISTORY OF SCIENTIFIC THOUGHT

We have explored the importance of visual thinking to Bohr, Einstein, Feynman, Heisenberg, Schrödinger, Salam, and Weinberg. To these we can add instances such as James Watson and Francis Cricks' molecular models made from tinker toy-like sets that were essential for their discovering the structure of DNA. These cases contain conclusions about visual imagery in creative scientific thinking:

1. Visual imagery plays a causal role in scientific creativity (Einstein's thought experiments).

2. Visual imagery is usually essential for scientific advance (Bohr, Einstein, Feynman, Heisenberg, Schrödinger, Salam, and Weinberg).

3. Visual imagery generated by scientific theories can carry truth value (Feynman diagrams).

These conclusions go far to substantiate that visual images are not epiphenomena and are essential to scientific research.

The development of quantum physics is an especially interesting case because it displays the dramatic transformations in visual imagery due to advances in science. The basic reason is the transition from classical to nonclassical concepts. The solar system imagery for Bohr's original theory was imposed on its foundation in classical mechanics. This phase of development of Bohr's theory concerned the *content* of a visual representation, which is what is being represented.

In 1923 the visual imagery of the solar system atom was discarded in favor of permitting the available mathematical framework itself to *represent* the atom. This phase focuses on the *format* of a representation, or the representation's encoding. Mathematics was the guide and led to Heisenberg's breakthrough in 1925. Soon after, in 1927, the quest began toward a new representation of the atomic world that culminated in Feynman diagrams. This transition in visual imagery is depicted in Figure 22.

Whereas imagery and meaning were imposed on physical theories prior to quantum physics, the reverse occurred after 1925. Quantum theory presented to scientists a new way for "seeing" nature. It is similar to the image resulting from high-level processing of retinal data. Heisenberg began to clarify the new mode of "seeing" with the uncertainty principle, while Bohr's complementarity principle approached the problem from a wider viewpoint that included an analysis of perceptions. The principal issue turned out to be the wave/particle duality, which rendered terms such as "position," "momentum," "particle," and "wave" ambiguous. In Chapters 1 and 7, we began to trace the research culminating in Feynman diagrams. A principal upshot of the content–format–content shift in Figure 22(c) is the ontological status

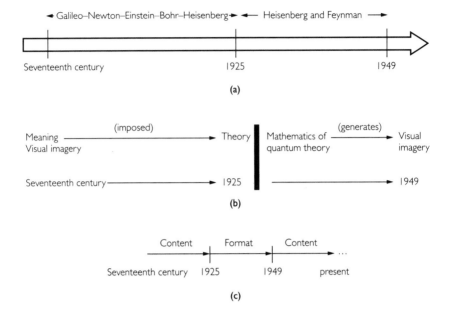

FIGURE 22.

(a) The major figures in the conceptual transition in theorizing, from the seventeenth century until 1925, and then from 1925 to 1949. (b) The major change from visual imagery and its meaning being imposed on physical theories (seventeenth century through 1925) to the mathematics of quantum physics generating the relevant physical imagery with its meaning. (c) This is the transition from content to format to content.

accorded to Feynman diagrams, which are the new visual imagery. It is apropos to complete this argument in Chapter 10.

THE HISTORY OF SCIENCE AND THE IMAGERY DEBATE

Approaching developments in quantum mechanics with notions from cognitive science leads us to the following results. Einstein's imagery concerned the knowledge content of visual images abstracted from our world. Symmetry was essential for proper reassembly of scenes such as electromagnetic induction. Einstein focused mainly on the content of a representation, or what is represented.

As we saw in Chapter 2, Heisenberg pursued a line of scientific research that permitted the only partially understood mathematics of quantum mechanics to generate its own meaning, part of which involved restrictions on words such as position, momentum, and electron. We can express this as follows: The mathematics (syntax) of quantum mechanics to some extent generates its own meaning (semantics), part of which are the uncertainty relations. The thought experiments of Bohr and Heisenberg necessarily used figures from the world of perceptions whose operations are restricted, however, by the laws of quantum physics. Since the Feynman diagrams of quantum mechanics are generated by its mathematics, they have truth value in the sense of logic. This point negates the criticism that visual images never have truth value.* Feynman diagrams also reply to the criticism that visual imagery is constrained by the image–perception link. If this were the case, we could not visualize things we have never perceived *and* visual imagery would have no predictive value whatsoever because it merely mimics what we would expect from perceptions. On the other hand, we would not have been able to draw Feynman diagrams without the proper mathematics.†

Can the history of scientific thought, when studied with concepts from cognitive science, yield new insights into visual imagery in general? In replying positively to this question, we will combine insights from this chapter with those from Chapter 7 on the assimilation/accommodation process and metaphors. We proceed as follows:

1. Figure 22 indicates that there is an analogy between examples of visual imagery and the lack thereof in quantum mechanics from 1913

---

*A straightforward counterexample to this anti-imagist claim are Venn diagrams for analyzing syllogisms in logic.

†In Chapter 10 we will come across the physicist Gregor Wentzel's work in which he drew Feynman-like diagrams in 1943 as purely didactic aids in the early quantum theory of elementary particles. He came across them in a purely fanciful way and offered no mathematics to back them up, nor were they used by physicists.

to 1948, with the problem of how we can generate visual images in the cognitive functional architecture that are not abstractions from phenomena we have actually witnessed.

2. The mathematical structure of advanced scientific theories contains partially interpreted mathematical symbols based on visual imagery related to visual perceptions. The next step is for this partially interpreted symbolism to generate results like the uncertainty relations.

3. The partially interpreted symbols in scientific theories can then attain meaning of a descriptive and depictive sort. The accompanying visual imagery is effective in problem solving and can have truth value.

4. Let us speculate that steps (1)–(3) can occur in the cognitive functional architecture as follows. Our semantic network reaches a point of development where it conflicts with our understanding of the world. In order to accommodate this situation, we create metaphors that enable the generation of new knowledge structures that contain requisite analog information, particularly visual images.

This conjecture includes how scientific research enlarges the facility for visual imagery, as we have come to expect from examples considered throughout this book. As we have found again and again, great scientific discoveries are not made by sweeping deductions but rather by some dazzling new combination of imagery and metaphor. To further pursue this theme, we turn next to exploring creative scientific thinking in its different modes of mental imagery.

9
SCIENTIFIC
CREATIVITY

C reative experiences like those of Archimedes (stepping up into his bathtub), Poincaré (stepping up into a coach), Charles Darwin (reading Malthus), Marcel Proust (biting into a piece of toast), Einstein, Pablo Picasso, Joan Miró, and Mark Rothko (describing their most creative ideas as appearing with no conscious forethought), and so many others, seem to avoid discrete rational reasoning. Their irrational insights led to startlingly new representations of nature. The seminal ideas emerge not in any real-time sequence but in an explosion of thought. Analog processes do just this because they seem to produce their output in a single explosive step without any resemblance to discrete and purposeful rational reasoning.

Cognitive scientists are willing to accept analog representations in the cognitive functional architecture of visual imagery, if empirical data indicate this to be the case. They claim that thus far no such data exist for visual imagery. On the basis of our discussion of vision in Chapter 8, we find ourselves in agreement, even though we are pro-imagery. This chapter explores why certain modes of creative thinking in science strongly resemble analog thought, for which we present a model. We focus on procedural knowledge, that is, problem solving. In Chapter 10 we enlarge our exploration of creativity to include art. This is natural because art and science at their most fundamental are adventures into the unknown, in which artists and scientists seek aesthetic representations of worlds beyond appearances.

CREATIVITY AND DIGITAL THOUGHT

As we saw in Chapter 8, digital processes are carried out with propositions that can be executed in a massively parallel manner. If thinking can be reduced to digital processes executed at the symbolic level, then there ought to be some sort of mental algorithms* for the cognitive architecture to process. How this can be? How, in any "rational" way, can we obtain a result that goes beyond the logic system used for calculations? This is what Einstein accomplished in going beyond the axioms of Newtonian science.

What about simulating creative thought on a computer? After all, computers are finite, closed mathematical structures with far fewer transistors than the brain has neurons.† We can speak of "brain plasticity," which is the brain's ability to alter itself continuously either by growing

---

*An algorithm is a well-defined sequence of operations.

†The human brain has on the order of 10^{11} neurons, while the largest computers presently have about 10^9 transistors.

new connections between neurons or by altering existing ones. How can a computer have any plasticity if its architecture is fixed?

Zenon Pylyshyn defines "extreme plasticity" for a computer as the many different ways one can specify the same "effectively computable functions," which are mathematical functions on which computers can produce numerical results in a finite time. But this suggestion actually pushes the problem of creative thought back one notch because someone has to formulate the effectively computable function. Pylyshyn claims that the mind's architecture will very likely turn out to be unlike any computer architecture presently known. So, he warns us, be prepared for surprises!

Whereas Pylyshyn writes in a purely theoretical vein, one of the doyens of artificial intelligence (AI), Herbert A. Simon, and collaborators at Carnegie Mellon University, have constructed "discovery" software (to be discussed in a moment) for which they have tried to toe the historical line as best they can. Simon was the first cognitive scientist to directly confront the logical positivist insistence on a split between the context of discovery and that of justification. Logical positivists took justification to be the proper domain for the philosopher, who is supposed to be interested only in rigorizing already established theories. Utilizing some of his important results on problem solving, in the 1960s Simon first proposed that creativity can be explored by quantitative means. While I agree with this proposal, I disagree with Simon's definition of a "logic of scientific discovery."

Simon's basic assumption is that there are no differences in anyone's thought processes. It is just that certain people like Poincaré and Einstein have better heuristics, that is, problem-solving ability. This assumption is essential for constructing discovery software for major scientific discoveries. Such programs utilize results from research on problem solving by subjects of ordinary intelligence. While this is a useful heuristic for constructing discovery software, it completely misses the mark in real life. After all, just about every scientist, artist, and writer can practice his or her craft 10 hours a day every day and not produce the astonishing work of a Darwin, Einstein, Poincaré, Picasso, or Shakespeare.

According to Simon, the scientist commences a search along the branches and nodes of a decision tree. A decision tree, data, and theory constitute what Simon calls the "problem space." This is not a real space with extension, but a metaphorical way of expressing available options that can be implemented in a computer. An important prerequisite in highly creative problem solving is to choose the right problem. Scientists like Poincaré and Einstein excelled at this. Certain people have better heuristics and can reach a solution node with a minimum of trial and error.

Simon's logic of discovery is embodied in computer programs capable of detecting patterns in data. But it is difficult to escape the nagging feeling that his discovery programs contain the result to be "discovered." Take, for example, Simon's program KEPLER, which "discovers" Johannes Kepler's third law for planetary motion.* In addition to omitting Kepler's reliance on such extra-scientific notions as Neoplatonism, Simon's KEPLER deals with better data than Kepler himself ever had. KEPLER is essentially a number-crunching program taking powers of available data, such as cubes and squares, and then comparing ratios, seeking a constancy. In the end, Kepler's third law emerges, of course. At this stage, to claim any general results concerning scientific creativity is premature, to say the least.†

---

*The square of the time for one complete revolution of a planet about the sun is proportional to the cube of its average distance from the sun.

†Recently, Simon and Deepak Kulkarni produced a computer simulation, which they call KEKADA, of the biochemist Hans Krebs' theory of the ornithine cycle, which he formulated in the 1930s, in which such organic substances as urea are synthesized. Simon and Kulkarni's data are taken from the historian of science Frederick L. Holmes' extensive interviews with Krebs, as well as Holmes' detailed study of Krebs' laboratory books. Knowing what the end result is, and imposing certain heuristic guides selected after studying Holmes' interviews, not unsurprisingly KEKADA produces Krebs' theory. But the path chosen by KEKADA differed slightly from Krebs' actual one. As Holmes points out insightfully, this might be a useful aspect for discovery software. The point is that historians often ask questions such as "what if events had been otherwise. . . ." Discovery software can offer replies that might also have scientific significance.

With the present computer architectures, we cannot expect a great deal from the algorithms of discovery software. Nevertheless, just for the sake of argument, suppose that 1,000 years from now some fantastic mathematics is invented that can be instantiated into some incredible computer architecture, which can run extraordinary computer programs, which can write sublime music like Mozart and formulate astonishing scientific theories like Einstein.* This computer architecture will also be able to deal with mental imagery on par with propositional representations. One mathematical argument against this sort of program's success is a theorem proven in 1931 by the Austrian logician Kurt Gödel, which says, among other things, that there is no general algorithm for use in proving all mathematical theorems. But what if our futuristic algorithms run on an inconceivably fantastic architecture of such a kind for analog representations, for which Gödel's theorem doesn't apply? Moreover, the mind is a finite machine and so Gödel's theorem does not apply to it anyway.

In general, arguments against computers doing anything creative are essentially moralistic and wrongheaded. So what if this sort of situation comes to pass? Even if it does, the computer's creativity will still not be reducible to *individual physical occurrences*, but only to the machine's functional organization. This is not the view that the philosopher John R. Seale referred to as "strong AI," which claims that mental processes can be run on computers using algorithms and that computers can also have mental qualities. Consequently, thought and consciousness are the results of algorithms that are simply more vastly complex than those used for thermostats, for instance. First, circa 2000, this view is difficult to maintain. Second, it is not necessary to attribute mental qualities such as consciousness to a creative computer. This aspect of the brain can be entirely different and perhaps nonalgorithmic.

Recently the eminent English mathematician and physicist Roger Penrose has written on points similar to ones discussed in this chapter.

*To which I add that I am rather pleased with the sequence of adjectives in this sentence which I found in my word processing software's Thesaurus.

Penrose's work has created somewhat of a stir, particularly his book, *The Emperor's New Mind*. Penrose attacks strong AI, claiming that there is something in thought that escapes computation. His arguments are based on the restrictions that Gödel's theorem impose on algorithms. Penrose invokes consciousness as the key to creative thought because he claims it is a means to overcome the restrictiveness of algorithms. Although he suggests that unconscious thought is algorithmic, he claims that this is not the case for judgments such as "understanding" and "artistic appraisal," for which consciousness is needed. To support his argument, Penrose offers Poincaré's introspection (discussed in this chapter). In it he substitutes the term "consciousness" for what Poincaré called "intuition." Among other evidence, Penrose presents his own creative episodes, one of which he claims suddenly occurred after an encounter with a problem situation very similar to Poincaré's. Quoting from Einstein and Mozart, he links their "inspirational thought" to consciousness, which is nonalgorithmic. In a sequel, *Shadows of the Mind*, Penrose conjectures a scientific basis for consciousness by linking it with assumed quantum properties of cytoskeletons and microtubules, which are the substructure of neurons. He considers this combination of biology and physics to be a new physics. As yet there have been no concrete results.

An interesting discussion about machine intelligence has recently emerged from the first regulation chess tournament between the current world champion, Garry Kasparov, and a chess-playing computer, in this case IBM's Deep Blue. Chess-playing computers, such as Deep Blue and its notable predecessor Deep Thought, excel at chess problems as well as games in which moves have to be made quickly. Deep Thought beat several grandmasters before losing to Kasparov, who declared afterward that he had saved the human race.

Chess-playing computers can evaluate millions of moves per second with certain weightings for certain preferred moves, while exploring ahead in some depth. Deep Blue can process more than 100 million moves per second, while evaluating as much as six moves in advance. Preference is given to chess positions yielding maximal material advantage. "Understanding" the game, however, is still another story. For example, like most chess computers, Deep Blue has not been programmed to offer a draw, which can lead to absurd solutions to chess problems and actual chess games. Chess masters, on the other hand, have the intuition to cull out from the vast number of possible moves pretty much the proper one, often with as much as the next six moves worked out in advance. Some years ago I remember attending a lecture given by Bobby Fisher, one of the greatest of the grandmasters. Fisher took us through a game he had played against another grandmaster, Mikhail Tal. After one of his middle-game moves Fisher commented that at this point Tal retired because he knew that Fisher had him in no more than six moves. Fisher then went through several replies that Tal could have made. They were all to no avail.

Presently, and perhaps for a long time to come, a major difference between humans and machines is that humans are able to operate in vague situations. On the other hand, in order to operate effectively, chess-playing computers must be in a situation where they are responding to a clear-cut attack, say on a king or another key piece. This insight was Kasparov's path to victory. But he didn't realize it immediately, and everyone was shocked when he lost his first game with Deep Blue. But he learned from his mistakes. One mistake Kasparov made was to fall into the psychological trap that he was playing a human opponent whom he could intimidate with, for example, a thrilling sacrifice or a dazzling, slashing attack. The computer was unmoved. Kasparov realized that if you don't play solid positional chess the machine will punish you severely. Part of Kasparov's incredible chess-playing ability is to learn from past mistakes. The strategy he adopted was one in which the computer has trouble: seek out positions that seem to have no imminent strategic objective such as the unfurling of a threat against a king. As Kasparov conceded after his 4–2 win, "This was a serious opponent. . . . You have to limit its unlimited potential."

We need not be surprised if some time in the future a descendent of Deep Blue will defeat all comers. Does this mean that computers exhibit intelligence? An immediate negative response is based on the fact that the computer is not demonstrating anything like intuition or judgment. Rather, it is operating as an extremely sophisticated number cruncher. An interesting experiment with the future world's champion chess computer would be to set up a chess match in which the machine's human opponent is not told beforehand that he or she is playing a computer. There would have to be some arrangement in which each "side" agrees to play in separate rooms on separate chess boards and their moves are transmitted to each other and then displayed on each others' boards. The problem is whether the real human can tell that he or she has been playing a computer. If not, then would everyone have to concede that the computer has developed a very special sort of artificial intelligence? This argument is often referred to as the "Turing test," after the pioneer of computer science Alan Turing, who proposed it in 1950. Surely the computer's artificial intelligence would be extremely specialized since further interrogation of the "other side"—for example, asking the computer how it feels after either losing or winning—would reveal that one is dealing with a machine. As Kasparov put it: "I think I did see some signs of intelligence, it's a weird kind, an inefficient, inflexible kind."

CREATIVITY AND ANALOG THOUGHT

Psychological tests indicate that people do not experience sudden illuminations without previous conscious or unconscious reasoning. This agrees with observations and statements by scientists such as Einstein, Poincaré, and von Helmholtz and musicians such as Mozart, which are indicative of unconscious thought. All of these episodes can be taken as evidence supporting the hypothesis of unconscious streams of thought that can hold problems in an active but unconscious state over long periods of time. Before proceeding, let's consider whether there is any evidence of unconscious streams of thought that surface into consciousness.

This concept appeared for the first time forcefully in a widely read book by William James, *Principles of Psychology*, published in 1890. James writes,

> Consciousness does not appear to itself chopped up in bits. . . . It is nothing jointed; it flows. A "river" or a "stream" are the metaphors by which it is most naturally described. In talking of it hereafter, let us call it the stream of thought, of consciousness.

Behaviorist psychologists or associationists (aka positivists) will, of course, reply in the negative to all conjectures about internal mental processes. In their view, only what emerges in speech and action counts, not any theorizing on what goes on in the head. Gestalt psychologists will also answer negatively. In their opinion, the moment of realization is catalyzed by the injection into the field of knowledge of a fact that suddenly snaps existing and heretofore patternless data into a highly symmetric pattern. Metaphorically speaking, this is analogous with systems in physics, which tend toward patterns of highest symmetry because they are the lowest energy states and so the most stable.*

Unconscious parallel processing of information is fully consistent with information being stored and processed in long-term memory in a semantic network of the sort we discussed in Chapter 8. There we saw experimental evidence supporting this hypothesis. For the most part, we access this information unconsciously.

---

*The famous triumverate of Gestalt psychology—Wertheimer, Karl Koffka, and Wolfgang Köhler—had training in physics. In fact, during World War I Wertheimer did research for the Austrian-Hungarian army on devices to detect hidden artillery batteries by a triangulation technique based upon muzzle blasts. This effect has to do with perception because Wertheimer discovered that sound travels to the brain at different speeds from the left and right ears. Based on this result, Wertheimer designed a helmet with protruding tubes at the positions of the observer's ears.

Psychological tests indicate propositional and analog representations, with crossovers. Examples abound of situations whose outcome depends on a division of attention between information in the two representations as well as input from senses other than visual. Consider driving a car, or moving a soccer ball down a field, or dribbling a basketball down a court. A great many unconscious decisions are involved.

George Mandler, a leading researcher in memory and general problems of consciousness, has extensively studied processes in which problem solutions seem to appear suddenly after a period of incubation, a phenomenon he calls "mind popping." He has found that incubations and subsequent "mind popping" "draw primarily on activation and spread of activation" in a network of information. Very possibly, continues Mandler, the "demand for the solution, the question posed," somehow keeps relevant information activated in the subconscious. This is the problem of how procedural knowledge is represented, that is, of how best to solve a problem. Researchers in this field conjecture that while consciousness plays the important role in our daily lives of restricting the bounds of our actions, in the unconscious we can activate complexes of information without boundary. He concludes that

> Much of our daily action and thought is occasioned without
> deliberation or intent, and the study of memory in particular
> has started to move away from the study of deliberate
> searches and retrievals.

So far, we have seen that psychological data lend support to the assumption that information held in long-term memory can be processed in parallel in the unconscious and find its way into conscious thought. An innovative idea emerges not in any real-time sequence but in an explosion of thoughts. This is characteristic of analog processes. Can we propose a model for the emergence of unconscious parallel processing—that is, of network thought?

A MODEL FOR NETWORK THINKING

Let's assume that the general problem of scientific creativity can be divided into two parts: (1) network thinking, which leads to (2) the nascent moment of creativity. In network thinking, concepts from apparently disparate disciplines are combined by proper choice of mental image or metaphor to catalyze the nascent moment of creativity. This necessarily nonlinear thought process can occur unconsciously, and not necessarily in real time. Concepts combine like light rays focusing at a point. We are reminded here of Poincaré's description of scientific creativity as the "process in which the human mind seems to borrow least from the exterior world, in which it acts, or appears to act, only by itself and on itself."

Let's further assume that network thinking requires an analog medium because mental imagery is an integral part of creative scientific thought. Suppose that the analog or dedicated representational medium consists of:

1. lists of information that concern the problem at hand;
2. lists of information that concern other problems on which the investigator is currently working.

Suppose that the items in lists (1) and (2) are activated from long-term memory. Can the items in lists (1) and (2) be the nodes in some sort of connectionist machine? Maybe. But given the primitive state of connectionist machines, this hypothesis does not explain much, despite its suggestiveness. So let's assume that in the dedicated part of the cognitive architecture for analog thought, connections can form among the items in (1) and (2). The connections form by means of certain processes that are either unique to it or shared with a propositional representation. Connections are tried and rejected. This is where selection mechanisms or guidelines such as intuition and aesthetics come into play to weed out certain combinations of connections. Where do these guidelines come from?

In a flight of fancy, Einstein wrote that scientific theories are "free creations of the human intellect." Similarly, Poincaré described, as best he could, the choice of the proper path for generalizing from experimental data as "free" and guided by "our instinct for simplicity." Did both men really mean complete freedom? Poetically, perhaps. Scientifically, they meant freedom "modulo constraints placed by nature." There are guidelines on the form scientific theories must take in order to serve as probes into worlds beyond sense perceptions. (Notice that because I am writing as a scientific realist, I said constraints placed *by* nature and not *on* nature.)

Ever since Thales of Miletus, scientific research has revealed new guidelines for scientific creativity. Among the guidelines we have discussed thus far, some of which Einstein and Poincaré kept in mind—most can be instantiated only in *combined* depictive and descriptive representations. Of course, certain guidelines in the following list can be written in a strictly mathematical (propositional) manner (e.g., the principles of energy conservation and relativity). New scientific theories must meet the following criteria:

• continuity with previous theories (This is rather straightforward because we must begin somewhere.);

• metaphysical assumptions such as conservation of energy, momentum, electric charge, and other more esoteric quantities, and the principle of relativity, too;

• symmetries of nature such as invariance of equations under certain transformations, or how we intuitively expect nature to behave (Sometimes scientists invent new symmetries that have only an intuitive basis; for example, Einstein's notion of minimalism in scientific representations which led him to propose the light quantum, as we shall discuss later in this chapter and in Chapter 10.);

• metaphors and analogies that permit probing unknown or little-understood domains with more familiar quantities;

- assumptions indigenous to a particular era such as Neoplatonism, *Nature Philosophy, Anschauung,* and *Anschaulichkeit;*

- the importance of mental imagery as a guide in creative scientific thought;

- aesthetics (not unconnected with symmetry) used as a means for discovery and for decision between theories (Nothing mystical is meant here regarding aesthetics, because to mathematicians like Poincaré the term "aesthetics" has well-understood connotations like invariance and symmetries.);

- intuition, which scientists define as best they can and which sometimes can include visual imagery;

- unification of known forces, which has roots going back to the beginning of science and is emphasized in *Nature Philosophy.*

Figure 1 is a model for how network thinking can be represented as it concerns creative scientific thought; it incorporates the guidelines listed above. Figure 1 is a representation for procedural knowledge. Already existing theories, data (declarative knowledge), and operators for reasoning are "mapped" onto the dedicated representational medium, which can support reasoning with propositions and analog entities such as mental images. All of these quantities are assumed to be in long-term memory. This medium may be metaphorically like a telephone switchboard, with connections yet to be made. Activation is maintained in the unconscious as the result of a previous intense conscious desire to solve the problem at hand. This activation can spread in the unconscious in ways that might not have been possible within the confines of conscious thought. The representational medium is divided into columns that contain unconscious lines of thought working in parallel while interacting among themselves. In order to further constrain and guide the new combinations that can emerge, guidelines for creativity come into play. Among them are aesthetics, conservation laws, and metaphors. As we discussed in previous chapters, many of these guidelines are them-

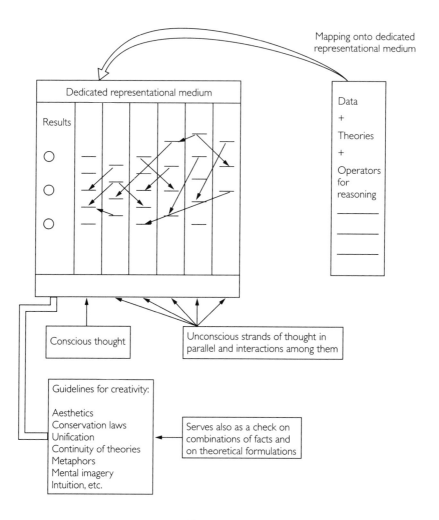

FIGURE 1.

A model for network thinking.

selves the result of scientific discoveries and turn out to be universal laws of nature. The interplay among unconscious parallel lines of thought eventually emerges into the conscious thought as the problem's solution.

It is unenlightening to argue the pros and cons of propositional versus analog thought on the basis of hypothetical tales of scientific creativity. Instead, let us focus on an actual comparative case study of two titans of the history of scientific thought, Henry Poincaré and Albert Einstein.

WHY POINCARÉ AND EINSTEIN?

We can learn a great deal about scientific creativity by taking certain introspections seriously and seeing what follows. Poincaré and Einstein are particularly interesting to study because they produced introspections as a direct result of their interest in cognitive psychology as well as the very foundations of their subjects. When they looked into their minds, they took care with the essays they produced. In Poincaré's case, we have an independent check through interviews some years earlier with a psychologist. In both cases, we find no inconsistencies with later or earlier writings, and their history is precise.

Poincaré's and Einstein's lives overlapped, and at one time so did their research. They are the key players in a classic episode in the history of ideas: In 1905 they both possessed the same data and mathematical formalism that Poincaré interpreted as a step toward an electromagnetic basis for physics but Einstein used to formulate the special theory of relativity. Later in this chapter we will look at why this happened.

The two men knew of each other through their work and met at least once. Sometime between 1902 and 1904, in a private study group, Einstein read Poincaré's reprint volume *Science and Hypothesis*. Another member of the group, Maurice Solovine, recalled that Poincaré's book "profoundly impressed us and kept us breathless." The sheer sweep of the material presented in Poincaré's lucid style is dazzling. Essays address the foundations of mathematics and physics, taking into account problems on the very frontiers of these subjects.

Of their one meeting at the Solvay Conference in the fall of 1911 in Brussels, Einstein wrote to his friend Heinrich Zangger in November 1911 that "Poincaré was in general simply antagonistic (against the theory of relativity) and, for all his acuity showed little understanding of the situation." Despite the outcome of their conversations, Poincaré was sufficiently impressed with Einstein to write a letter supporting him for a position at the Swiss Federal Institute of Technology in Zurich:

Einstein is one of the most original thinkers that I have ever met. . . . Since he seeks in all directions, one must expect the majority of the paths on which he embarks to be blind alleys.

No doubt Poincaré considered relativity to be one of those blind alleys. Nevertheless, Einstein got the job.

Both men were interested in the psychological origins of scientific concepts—the psychology of creativity—and consequently believed that, as Einstein put it, "scientific thought is a development of prescientific thought." Both men left introspections that too often are used to argue for the mystery that must forever surround the creative act.

POINCARÉ AND EDOUARD TOULOUSE

Although a scientist's introspection can help to unravel creativity, it must be handled carefully, within a web of mutually confirming historical data. This is the situation with Poincaré's widely cited and much analyzed introspection, "Mathematical invention," presented as a lecture on May 23, 1908, at *l'Institute Général Psychologique* in Paris. As George Mandler described, Poincaré's introspection "has become the obligatory prolegomenon for any discussion of incubation in the literature." It is cited in just about every discussion of scientific creativity. Yet can we trust Poincaré's self-analysis, which focuses on a mathematical discovery he made some 27 years earlier, in 1881? After all, Poincaré's "mind-popping" mathematical solution had not been monitored. Had Poincaré ever actually probed his *own* thought processes before 1908? Is it possible that on the spur of the moment he thought up the scenario presented in "Mathematical invention"?* It turns out that Poincaré's 1908

---

*This was, in fact, what August Kekulé did in 1890 to amuse readers of the *Proceedings* of a conference in Berlin celebrating his discovery some 30 years before of the cyclical structure of benzene. Kekulé intensely disliked speaking in public. Afterwards, he usually convinced himself that he had been an embarrassment, and often these self-recriminations resulted in severe headaches. So, when asked to submit a written version of his Berlin lecture, Kekulé was particularly cross at the introspection agrees with conversations and psychological tests performed on him in 1897 by the French psychologist Edouard Toulouse and published in 1910 with Poincaré's imprimatur as *Henri Poincaré*.

Toulouse was Chief of Medicine at the asylum of Villejuif and Director of the Laboratory of Psychology, *l'Ecole des Hautes Etudes*, Paris. In the preface to *Henri Poincaré*, published in 1910, Toulouse writes that since 1895 he had been examining with the "methods of clinical medicine and of the psychological laboratories" men who had demonstrated "superior minds" through their work. Attesting to Toulouse's reputation, and to the seriousness of his intent, among those who agreed to be interviewed by him, besides Poincaré, were the writer Emile Zola, the sculptors Jules Dalou and Auguste Rodin, and the composer Camille Saint-Saëns. The best observations turned out to be for Zola and Poincaré, who was interviewed last.

Although Toulouse completed the book with knowledge of Poincaré's introspection, this did not taint Toulouse's analysis of his 1897

---

organizer for not having provided a stenographer to record his speech, which he considered to have been rather stupid. Archival evidence found by two chemists, John H. Wotiz and Susanna Rudofsky, indicates that in retribution Kekulé decided to give everyone something really sensational. He concocted his "snake dream," which he claimed to have had sometime during the winter of 1861–1862 while on the faculty of the University of Ghent in Belgium. Dozing in front of his fireplace in a comfortable chair (on display in the university's museum), Kekulé writes that he had a vision of a snake swallowing its tail. Kekulé described this dream in the preface to his remarks made in Berlin, which is why no one at the Kekulé festival heard about it. We know this from newspaper accounts of Kekulé's lecture, which say nothing about the dream, which would have caused a sensation. In the preface, Kekulé "recalled" a similar dream on a London bus sometime during 1854–1855 which led to his earlier carbon-chain theories.

It is noteworthy that Kekulé described neither of these dreams until the preface to his 1890 lecture, more than 30-odd years later. Wotiz and Rudofsky have traced the general idea of Kekulé's dreams to a text published in 1881 and which he had definitely read sometime before 1890. This text describes molecules dancing. Then there is a gentle spoof of Kekulé published in 1886 with the cyclic benzene structure illustrated with six monkeys holding each others' feet. Wotiz and Rudofsky speculate that Kekulé constructed his two dream sequences primarily from these two sources, and without proper attribution either. They conclude that "although we would like to correct the record, we have our doubts that the truth will get in the way of a good story." This has been the case, because while Wotiz and Rudofsky's paper was published in 1954, Kekulé's dream remains a much-used example in the creativity literature.

data. Toulouse mentions Poincaré's 1908 lecture almost at the very end of his book, and then only in support of his own findings regarding the emphasis Poincaré placed on unconscious processing.

To the best of my knowledge, Toulouse's book contains the most complete psychological profile ever undertaken face to face with a major scientist. Archival material of Poincaré, which I had the good fortune to discover in 1976, independently supports Toulouse's observations and Poincaré's introspection. Toulouse's psychological analysis indicates that not only had Poincaré introspected before 1908, but he used the results of introspections to set his highly successful method of research in mathematics: Poincaré depended on unconscious thought for many of his key discoveries.

We now turn to Toulouse's data obtained from Poincaré in interviews and psychological tests as well to Poincaré's 1908 introspection. At present there are only informative biographical sketches of Poincaré, and so Toulouse's interviews add much to what little we know, in addition to unpublished material from the Poincaré archives.

PORTRAIT OF THE MATHEMATICIAN

Poincaré was born on April 29, 1854, in Nancy, France, where his father, Léon, age 26, was a physician and professor of medicine at the university; his mother, Eugénie (née Launois), was 24. Henri spoke at nine months and read at six years. At the lycée in Nancy, from 1862 to 1873, since renamed the Lycée Henri Poincaré, Poincaré was one of the top students in every subject, exhibiting unusual ability in mathematics and receiving high praise from all his teachers. In 1873 Poincaré entered l'Ecole Polytechnique, from which he graduated in 1875. Already there were numerous good-natured stories of his distracted demeanor. Studies in problem solving in which there is a dependence on incubation, or unconscious processing, indicate that distraction and daydreaming permit a wider ranging of activation. Poincaré liked to read science popularizations, gradually moving into more detailed works, enjoyed listening to music, and tried unsuccessfully to learn the piano. He recalled no important influence on his education at l'Ecole Polytechnique from the lectures, milieu, or professors.

From 1875 to 1879 he was a student at l'Ecole des Mines. Poincaré's meticulous notes taken on field trips while a student there exhibit a deep knowledge of the scientific and commercial methods of the mining industry, a subject that interested him throughout his life. Poincaré's school career at both *grandes écoles* was exemplary. In 1879 he was awarded a doctorate in mathematics from the University of Paris.

In a confidential report, dated August 12, 1879, to the Ministry of Education, the Faculty of Sciences of the University of Paris reported their assessment of Poincaré's doctoral thesis: The thesis contains some interesting results at the very beginning, but the

> remainder of the thesis is a little confused and shows that the author was still unable to express his ideas in a clear and simple manner. Nevertheless, considering the great difficulty of the subject and the talent demonstrated, the faculty recommends that M. Poincaré be granted the degree of Doctor with all privileges.

Poincaré's doctoral thesis was intended to improve a method for solving differential equations that had been suggested some years earlier by the French mathematician Augustin-Louis Cauchy. Confidential assessments of his teaching at the University of Caen, where Poincaré taught from 1879 to 1881, mention a sometimes disorganized lecture style.

In 1881 Poincaré was called to Paris, where he spent the rest of his illustrious career, eventually assuming chairs at the Sorbonne as well as at l'Ecole Polytechnique. He was elected to l'Académie des Sciences in 1887 and to its presidency in 1906. Poincaré was the only member of l'Académie des Sciences whose research merited election to all five of its sections: geometry, mechanics, physics, geography, and navigation.

In 1908 Poincaré was elected to l'Académie Française and to its directorship in 1912. He had strikingly recouped the earlier somewhat negative assessments of his writing style.

Poincaré's curriculum vitae is a 110-page book. He published about 500 papers and 30 books and received numerous honorary degrees and every scientific prize except the Nobel prize—for which a great deal of lobbying was done on his behalf. Poincaré took all philosophical, scientific, and mathematical knowledge to be his province. Besides being one of the greatest mathematicians ever, he made significant contributions to every branch of physics and astronomy, and he formulated a unique viewpoint of the philosophy of science called conventionalism (see Chapter 6). He did all of this in addition to occupying the apex of the pyramid that constituted the structure of French science.

THE SCIENTIFIC CREATIVITY OF HENRI POINCARÉ

Toulouse reports that Poincaré worked at his desk specifically on mathematics research for no more than four hours a day: 10 A.M.–noon and 5 P.M.–7 P.M. Evenings were reserved for journal reading. Poincaré's rigid schedule is not uncommon for mathematicians. For example, the British mathematician G. H. Hardy worked only from 9:00 A.M. to 1 P.M., whereupon he retired to the cricket field. A working regimen is essential for productivity in any profession. From time to time, highly creative artists attempt to hide from the public their preliminary work, instead consciously trying to weave a story of sudden inspiration of a single canvas. This is not the case. A great deal of preliminary work is done. Picasso often dated his preliminary sketches for posterity. Laplace left parties before midnight in order to be able to arise early the next morning and work. No matter what time the binge ended the evening before, in his heyday Ernest Hemingway always arose at six in the morning and then appeared at noon on his balcony in Havana with the first drink of the day in hand.

Although Poincaré was a remarkably clear observer, Toulouse found that his memory of visual images was not good. Whereas Toulouse found that the "interior language" of most mathematicians is visual, he repeatedly expressed surprise that Poincaré "neglected visual imagery altogether." For example, when shown a table of letters or numbers, Poincaré repeated the setup to himself, rather than trying to recall its visual image.

Generally, Poincaré found memorization difficult and was poor at rote learning, instead seeking patterns. By rote learning I mean the following. In every subject a certain amount of material must be learned by memory. Some people find no problem in memorizing large amounts of material like poetry, vocabulary and grammar for a foreign language, names and dates in history, and genus and species in biology. Others, such as Poincaré and Einstein, have trouble memorizing material of this sort for which there are no clear patterns or which they cannot figure out from fundamental principles taken as being true. In the sciences, there are also varying degrees of memorization. For example, when given a complex problem, many scientists attempt to remember certain previously derived results. Others work from what scientists call "first principles," that is, they delve down into the most fundamental assumptions pertaining to the problem at hand and work their way through to a solution. Working from first principles requires deep understanding.

Several of Poincaré's published books are notes from his classroom lectures. They exhibit his style of working from first principles. The books are self-contained, with everything developed from the basic equations of that discipline. An excellent example is Poincaré's published lecture notes *Electricité et Optique*, comprised of his lectures from 1888, 1890, and 1899, given at the Sorbonne. These lecture notes were highly influential in the development of electromagnetic theory in Europe.

Toulouse concluded that Poincaré's problem-solving methods were "intuitive, rapid and spontaneous." When writing a paper, Poincaré did not use notes, nor did he have any definite overall plan or goal in mind, nor even any idea of whether the problem at hand was solvable. This is research at its most fundamental, as Toulouse reports from an interview with Poincaré:

> [Poincaré] does not make a grand plan when he writes a paper. Ordinarily he starts it without knowing where he will conclude. . . . Starting is generally easy. Then he seems led by his work and has not the impression of any willful effort. At that moment it is difficult to distract him. When he searches, he often writes a formula automatically in order to awaken some association of ideas. If the starting is painful, M. H. Poincaré does not persist and abandons the started work, in contrast to Zola who persists. This is why he has no patience. In certain work, M. H. Poincaré proceeds by sudden blows, taking and abandoning a subject. During intervals he assumes . . . that his unconscious continues the work of reflection. Cessation of work is difficult if there is not a sufficiently strong distraction, especially in the case where the work is not judged complete. . . . It is for this reason that M. H. Poincaré never does any important work in the evening in order not to trouble his sleep. . . . It is a method of work uncommon in scientific matters and it constitutes a character well suited to the mental activity of M. H. Poincaré.

Note Poincaré's complete confidence in his power of unconscious work, which comes about through a "play of associations" in the manner of network thinking. In fact, Toulouse implies that Poincaré knew just how to activate areas of long-term memory: "[He] writes a formula automatically in order to awaken some association of ideas." Toulouse continues with an observation about the depth and breadth of Poincaré's knowledge in long-term memory:

> [Poincaré's] memory and faculty of assimilation and comprehension of the most varied things—mathematics, physical

sciences, philosophy, literature, languages—permitted him to examine problems in diverse ways.

An examination of Poincaré's unpublished manuscripts reveals the results of concentrated efforts of four hours of work per day by someone of his caliber. Incredibly, he could work through page after page of detailed calculations, be it of the most abstract mathematical sort or pure number calculations, as he often did in physics, hardly ever crossing anything out (Figure 2).

As Toulouse recognizes, this mode of work explains Poincaré's incredible production rate of discoveries. It also reflects his repugnance for administrative or practical tasks, which he performed effortlessly. Poincaré's doodling on memos for faculty meetings are ample proof that he worked everywhere (Figures 3 and 4).

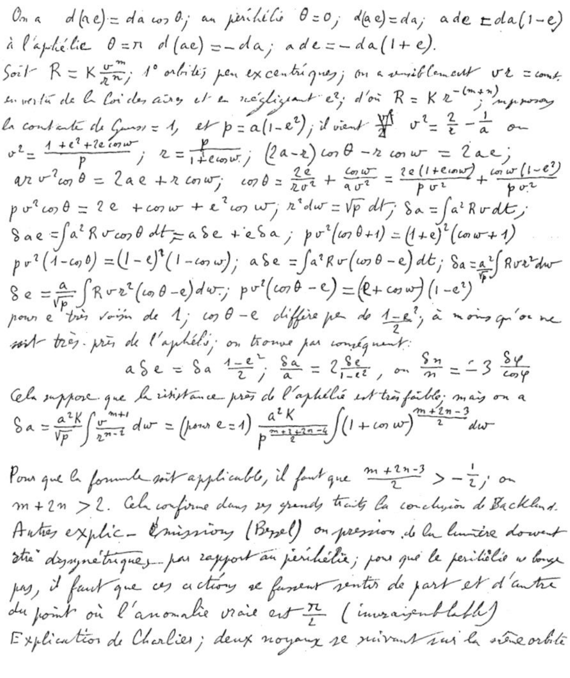
On a $d(ae) = da \cos\theta$; au périhélie $\theta = 0$; $d(ae) = da$; $a\,de = da(1-e)$
à l'aphélie $\theta = \pi$ $d(ae) = -da$; $a\,de = -da(1+e)$.
Soit $R = K\frac{v^m}{r^n}$; 1° orbites peu excentriques; on a sensiblement vr = const. en vertu de la loi des aires et en négligeant e^2; d'où $R = K r^{-(m+n)}$; supposons la constante de Gauss = 1, et $p = a(1-e^2)$; il vient $v^2 = \frac{2}{r} - \frac{1}{a}$ ou
$v^2 = \frac{1+e^2+2e\cos w}{p}$; $r = \frac{p}{1+e\cos w}$; $(2a-r)\cos\theta - r\cos w = 2ae$;
$ar v^2 \cos\theta = 2ae + r\cos w$; $\cos\theta = \frac{2e}{rv^2} + \frac{\cos w}{av^2} = \frac{2e(1+e\cos w)}{pv^2} + \frac{\cos w(1-e^2)}{pv^2}$
$pv^2\cos\theta = 2e + \cos w + e^2\cos w$; $r^2 dw = \sqrt{p}\,dt$; $\delta a = \int a^2 R v\,dt$;
$\delta ae = \int a^2 R v \cos\theta\,dt = a\delta e + e\delta a$; $pv^2(\cos\theta+1) = (1+e)^2(\cos w+1)$
$pv^2(1-\cos\theta) = (1-e)^2(1-\cos w)$; $a\delta e = \int a^2 R v(\cos\theta - e)\,dt$; $\delta a = \frac{a^2}{\sqrt{p}}\int R v r^2\,dw$
$\delta e = \frac{a}{\sqrt{p}}\int R v r^2(\cos\theta - e)\,dw$; $pv^2(\cos\theta - e) = (e+\cos w)(1-e^2)$
pour e très voisin de 1; $\cos\theta - e$ diffère peu de $\frac{1-e^2}{2}$; à moins qu'on ne soit très près de l'aphélie; on trouve par conséquent:
$$a\delta e = \delta a\,\frac{1-e^2}{2};\quad \frac{\delta a}{a} = 2\frac{\delta e}{1-e^2},\ \text{ou}\ \frac{\delta n}{n} = -3\,\frac{\delta\varphi}{\cos\varphi}$$
Cela suppose que la résistance près de l'aphélie est très faible; mais on a
$$\delta a = \frac{a^2K}{\sqrt{p}}\int \frac{v^{m+1}}{r^{n-2}}\,dw = (\text{pour } e=1)\ \frac{a^2K}{p^{\frac{m+1+2n-4}{2}}}\int (1+\cos w)^{\frac{m+2n-3}{2}}\,dw$$
Pour que la formule soit applicable, il faut que $\frac{m+2n-3}{2} > -\frac{1}{2}$; ou $m+2n > 2$. Cela confirme dans ses grands traits la conclusion de Backlund.
Autres explic.— Émissions (Bessel) ou pression de la lumière doivent être dissymétriques par rapport au périhélie; pour que le périhélie ne bouge pas, il faut que ces actions se fassent sentir de part et d'autre du point où l'anomalie vraie est $\frac{\pi}{2}$ (invraisemblable)
Explication de Charlier; deux noyaux se suivant sur la même orbite

FIGURE 2.

A page from one of Poincaré's workbooks. Not a line crossed out, not a diagram drawn.

FIGURE 3.

A sample of Poincaré's doodlings from his student days at l'Ecole des Mines.

Poincaré literally skimmed the cream off the top of research areas, leaving to others the task of providing rigor. On December 30, 1890, Hertz wrote to Poincaré with a plea to have pity on mere physicists and work out some special examples, with reference to Poincaré's recent groundbreaking research on the three-body problem, which contains the seeds of modern chaos theory.

Toulouse reports that his results on Zola and Poincaré are poles apart and entirely unexpected:

> The one [Zola's] was an intelligence that was willful, conscious, methodical, and seemingly made for mathematical deduction: it gave birth entirely to a romantic world. The other [Poincaré's] was spontaneous, little conscious, more taken to dream than for the rational approach and seemingly throughout apt for works of pure imagination, without subordination to reality: it triumphed in mathematical research. And this is one of the surprises, which calls for direct studies touching on the deepest mechanisms. . . .

(a)

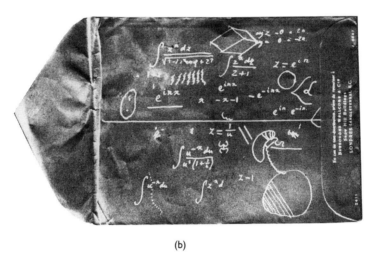

(b)

FIGURE 4.

The marginalia on a memo in (a) indicate how Poincaré sometimes occupied himself during a thesis defense. In (b) is literally a "back of the envelope calculation" by Poincaré.

Zola worked like a scientist and Poincaré like a writer or poet. Toulouse fell into the typical stereotyping between poets and scientists. Toulouse's report of Poincaré's dreamlike approach to his mathematical research is important because this state of mind opens up the boundaries of thought beyond the restrictions imposed by conscious deliberations.

Toulouse's conclusion on *"le problème de génie"* ["the problem of genius"] was that there is a tendency for the "primary play of associations" among elements in the field of knowledge to achieve the proper synthesis that is an "illumination." Fittingly, Toulouse recalled that Poincaré referred to this inarticulable process as *"le travail de l'inconscient"* ["the work of the unconscious"].

POINCARÉ AND CREATIVITY RESEARCH CIRCA 1900

Before moving on to Poincaré's 1908 introspection, it is apropos to inquire what could have prompted him to agree to a lengthy self-analysis with Toulouse. After all, Poincaré had an overwhelming schedule of research, teaching, and academic duties. In large part there was, of course, his interest in the psychology of creative thinking. This may have been the whole reason. But it is tempting to conjecture that Poincaré's interest in Toulouse's project to study creativity was also connected with the rise of the psychoanalytic movement. Although many famous scientists wrote introspections during this period (e.g., Hermann von Helmholtz in 1891), unique to Poincaré's was his elaboration on the workings of the unconscious. This interest may have resulted from Poincaré's familiarity with medical researchers at the University of Paris like Jean-Martin Charcot and Pierre Janet.

Then there was the philosopher Emile Boutroux, who was Poincaré's brother-in-law and a friend of the American psychologist–philosopher William James. Like the Danish philosopher Søren Kierkegaard, James discussed at some length streams of unconscious thought, a notion crucial to Poincaré's view of creativity. Poincaré cited Boutroux in the 1908 psychology essay. From correspondence in the Poincaré archives, we also know that he often attended soirées of a future fervent supporter of Freudian psychoanalysis and an intimate of Freud himself, Marie Bonaparte.

What directly prompted Poincaré to write the 1908 introspection most likely was the psychological questionnaires on scientific creativity distributed between 1902 and 1908 by the French psychologists E. Claparède, T. Flournoy, and H. Fehr, which they published in *L'Enseignement Mathématique.* The returns were disappointing. One of the most interesting questions—"To what sort of internal imagery, of what form of 'internal dialogue', do you avail yourself?"—was left to the end, and no one of any note replied. The psychologists concluded that

> The question of an interior language did not seem to very much captivate our correspondents, who responded only very laconically. This is extremely excusable. They had the right to be tired from so long a questionnaire.

Thirty questions were sent out from 1900 to 1908. There was some good-natured playfulness in some of the responses. Boltzmann, the only well-known physicist to reply, wrote, "I have only one piece of advice for young mathematicians, 'Be a genius'!" In a later reply, he added, " . . . the rest is unimportant." *Tant pis!*

INTUITION AND LOGIC

Poincaré's interest in the psychology of concept formation is amply demonstrated in his classic writings on the foundations of geometry that he began to publish in 1887 (see Chapter 6). Some of these essays were published in 1902 in *Science and Hypothesis.* In an essay of 1904, "Mathematical definitions in education," he wrote, "It is by logic we prove, it is by intuition we invent." In 1908 Poincaré concluded that "Logic, therefore, remains barren unless fertilized by intuition." I think Poincaré's descriptions of logicians as those who merely cleaned up proofs divined by intuitionists was brought about by his displeasure at the direction mathematics was taking in the first decade of the twentieth century. This was a time when Giuseppe Peano, Bertrand Russell, and Alfred North Whitehead spearheaded a group of mathematicians whose goal Poincaré interpreted to be that of reducing all mathematics to a cookbook procedure in which creativity, with its intuitive aspects, would be excluded. Next would come the sciences.

Poincaré severely criticized logicians for neglecting the psychological origins of mathematical concepts. For example, Poincaré wrote in his book review of the mathematician David Hilbert's 1902 *Foundations of Geometry* that

> The logical point of view alone appears to interest him. Being given a sequence of propositions, he finds that all follow logically from the first. With the foundations of this first proposition, with its psychological origin, he does not concern himself.

For Poincaré, the fact that the psychological origin of concepts is necessarily not rational, as witnessed in the origins of geometry (see Chapter 6)—then a purely propositional description of mathematics—cannot capture its creative dimension. Poincaré believed that any propositional description omits intuition, aesthetics, and mental imagery. So much for digital thinking, in Poincaré view.

POINCARÉ ON INTUITION, AESTHETICS, AND MENTAL IMAGERY

Just as with science, with philosophy Poincaré was as meticulous as possible when defining terms. When working as a physicist, Poincaré defined intuition as something abstracted from our senses. But when working as a mathematician, he adhered to his cautionary remark that "intuition is not necessarily founded on the evidence of the senses; the senses would soon become powerless" because, for example, we cannot image multidimensional spaces or figures. Rather, in this case he took recourse to a *process* definition of intuition, which, in Poincaré's view, was more appropriate to creativity in general:

> [T]o make geometry, or to make any science, something else
> than pure logic is necessary. To designate this something else
> we have no word other than *intuition*.

What is not logic in a demonstration is intuition. This is a process definition because it describes intuition as a catalyst for thought.

Poincaré described one class of intuitive mathematician as thinking in visual images that are images of the senses. But, regarding creativity, Poincaré never included himself in this group of intuitionists. This
agrees with Toulouse's observation that Poincaré "seemed to neglect visual imagery altogether." I am not claiming that Poincaré considered visual imagery to be ancillary to scientific research. Quite the contrary.
For example, Poincaré's pioneering research on dynamical systems, only
recently having been understood, was ingeniously carried out using
topological concepts in two and three dimensions in lieu of rigorous
and far too difficult investigations of the exact solutions of partial differential equations. Such rigorous investigations, concluded Poincaré,
would have obscured essential points. He was right.

Poincaré referred to his mental imagery in mathematics as "sensible intuition," which is the ingredient in a mathematical proof that, as
he writes in his 1908 introspection, is not of a "simple juxtaposition of
syllogisms," but of their "order . . . the feeling, so to speak, the intuition
of this order, [the] ability to perceive the whole of the argument at a
glance." While we have no visual imagery of the steps in a mathematical
proof, we do have some sense *beyond logic* of what form a mathematical
proof ought to take or of what approach or tactic is best for an overall
strategy of attack: "intuition is this faculty." The *form* of the new creation
is of the essence; the details follow.

Emphasis on form over content, culling away inessentials, and
proper problem selection are hallmarks of high scientific creativity.
Basic to these descriptions is that the scientist, musician, artist, or writer
has completely mastered the technicalities of his or her field. Scientists
such as Poincaré and Einstein soared over technical difficulties and ze-

roed in on essential problems. In this vein we recall Galileo's realization that the key problem in motion was not how bodies fall through viscous media or are blown around by air currents, but the "simpler" case of free fall through vacuum. The other impediments to motion were just that—impediments, to be included later as merely complicated arithmetic. Yet this simpler case required such far-reaching assumptions as the independence of weight on free fall through vacuum.

In "Mathematical invention" Poincaré recalls that in his important discovery in 1881 of what became known as automorphic functions, there were stages or cycles of thought: conscious work, unconscious work, illumination (hopefully!), conscious work, and then verification. Conscious work is followed by a period of rest that actually involves "unconscious work": "What strikes us immediately are these appearances of sudden illumination, obvious indications of a long period of previous unconscious work." Data from introspections are strong indicators of lines of thought operating in parallel, and they signal caution about the veracity of "Aha" experiences.

Because "invention is selection" to Poincaré, how does the unconscious select and assemble the appropriate combination of mathematical facts? In reply, Poincaré defined intuition as the ingredient of creativity in addition to logic:

> The rules that guide choices are extremely subtle and delicate, and it is practically impossible to state them in precise language; they must be felt rather than formulated. Under these conditions how can we imagine a sieve capable of applying them mechanically?

Some of these rules, such as aesthetics and intuition, appear in Figure 1. Poincaré believed that the choice of the appropriate combination of facts cannot be articulated: "This too is most mysterious [because the] useful combinations are the most beautiful, I mean those that can charm the special sensibility that all mathematicians know." And, not unexpectedly, wrote Poincaré: "Among the combinations we choose, the most fruitful are often those which are formed of elements borrowed from widely separated domains." Network thought is of the essence.

The mathematician's "special aesthetic sensibility" or sensible intuition plays the role of the "delicate sieve" that filters out all but the few combinations that are "harmonious" and "beautiful." Liberation of thought in the unconscious "permits unexpected combinations of mathematical facts." This is the "play of associations" observed by Toulouse that leads to scientific invention. This, too, is a point of modern psychological research that indicates the freedom in the unconscious for ranging over and activating information in long-term memory. Poincaré concludes, "In a word, is not the subliminal ego superior to the conscious ego?" We have seen all of this in Toulouse's psychological profile.

In terms of aesthetics, Poincaré is among the most quotable of scientists, because he actually used it as a guideline for his scientific research:

> The scientist does not study nature because it is useful; he studies it because he delights in it, and he delights in it because it is beautiful.

He continues by linking beauty to the credo of the scientist and to life, itself:

> If nature were not beautiful, it would not be worth knowing, and if nature were not worth knowing, life would not be worth living.

Poincaré's notion of stages of thinking was not unique; for example, the great German philosopher–scientist Hermann von Helmholtz's description in 1891 of his creative episodes was similar. Other scientists were introspecting as well. von Helmholtz recalled a period of hard work culminating in intellectual fatigue:

> Then after the fatigue of the work had passed away, an hour of perfect bodily repose and rest was necessary before the

fruitful ideas came. Often they came in the morning upon awakening. . . .

On the phenomenon of illuminations occurring on awakening in the morning, von Helmholtz quoted the great German mathematician Karl Friedrich Gauss: "The law of induction was discovered January 1835 at 7 A.M., before rising." On the workings of the unconscious, von Helmholtz recalled Goethe's words:

What man does not know
Or has not thought of
Wanders in the night
Through the labyrinth of the mind.

While the type of problems on which Poincaré did creative work often yielded a solution in relatively short periods of time, those pursued by von Helmholtz sometimes took "weeks or months," resulting in a "sharp attack of migraine." Besides the basic difference in the areas on which Poincaré and von Helmholtz pursued their basic research, so differed their approaches or styles. Poincaré almost invariably found the royal road to invention, whereas von Helmholtz described himself as a mountain climber who

ascends slowly and toilsomely [and] who, finally, when he has reached his goal, discovers to his annoyance a royal road on which he might have ridden up if he had been clever enough.

The royal road, continued von Helmholtz, is the one displayed in scientific papers.

POINCARÉ INTROSPECTS

Most analyses of Poincaré's 1908 introspection focus on the cycle of conscious thought–unconscious thought–illumination–verification. In the light of Toulouse's observations, however, a hitherto unnoticed point emerges from Poincaré's introspection: Poincaré's critical and most far-reaching illuminations occurred in periods that involved unconscious work with *no* predefined path and/or *no* set goal. Since Poincaré reported to Toulouse that this is the creative mode on which he explicitly relied for his research, I will focus on this aspect of Poincaré's introspection.

The background to Poincaré's mathematical research in 1881 is as follows. Poincaré's doctoral dissertation dealt with a restricted class of differential equations. But he believed that his results were generalizable to much more complex situations. He was not sure, however, whether this would actually work out.

First Poincaré tried to prove that the solutions he sought did not exist. He was not completely successful, and this bothered him. After a sleepless night, he awoke with the realization of how to establish one class of these hybrid functions, which he quickly generalized and named Fuchsian functions after the German mathematician Lazarus Fuchs whose work on differential equations had been of some importance to Poincaré. In Germany these functions were soon referred to as automorphic functions, which is their modern name. The key part of the scenario of Poincaré's scientific invention follows.

Poincaré journeyed from Caen, where he was on the faculty, to a geological conference at nearby Coutances. The "vicissitudes of the journey made me forget my mathematical work." That is, he *consciously* forgot the work. About to embark on a sightseeing drive at Coutances, he stepped up into the carriage when the "idea came to me, though nothing in my former thoughts seemed to have prepared me for it." The idea came from network thinking on the unconscious level: The means to generate Fuchsian functions were from an apparently unrelated branch of mathematics, non-Euclidean geometry. Immediate written verification was unnecessary and so he resumed conversing with the other passengers. Back at Caen, Poincaré verified the result *"à tête reposée"* ["at his leisure"]. As Mozart wrote to his father of work on *Idomeneo*, "Everything is composed, just not copied out yet."

A second episode in this research immediately followed with unconscious work and with no conscious goal. Poincaré turned to certain arithmetical research "without suspecting that they could have connection with my previous researches." Disgusted with his lack of success, he went on vacation and "thought of entirely different things." While walking on a cliff beside the sea, he experienced a flash of illumination like the one at Coutances, "with the same characteristics of conciseness, suddenness and immediate certainty"; namely, that application of arithmetical transformations to his work on Fuchsian functions permitted further generalizations. Verification followed back at Caen.

Throughout his invention of Fuchsian functions, Poincaré focused on groups of transformations that left these functions unchanged (invariant). A key step in Poincaré's invention of Fuchsian functions occurred when he suddenly realized—illumination after unconscious thought with no predetermined path—that these groups could be obtained from considerations based on an apparently unrelated discipline, non-Euclidean geometry. Groups are mathematical quantities that play a key aesthetic role in mathematics and physics. So Poincaré actually followed what he advocated concerning intuition and aesthetics: His "special aesthetic sensibility" selected the most aesthetic solution, which involved bringing into play concepts from a discipline that heretofore was considered to have no relation to analysis of functions that solve differential equations. As he wrote in "Mathematical invention": "Among the terms which have exercised the most happy influence, I would point out group and invariable."

FOUR-DIMENSIONAL AESTHETICS

Of all Poincaré's physics papers, the two that epitomize his notion of aesthetics he wrote in 1905 with the same title, "Sur la dynamique de l'électron." A short version was published in June of 1905 in the *Comptes Rendus* and a longer version in March of 1906 in the *Rendiconti del Circolo*

Matematico di Palermo. In these two papers Poincaré reformulated Lorentz's 1904 theory of the electron into a form that, unbeknownst to him at the time, turned out to be mathematically identical to special relativity. Poincaré had realized that for mathematical analysis of electromagnetic theory, the correct manner of writing distances is in terms of four-dimensional space, where time is the fourth dimension. In three-dimensional space, the distance from the origin of coordinates to a point (x,y,z) is given by the function $F = x^2 + y^2 + z^2$. In four dimensions, the distance from the origin to a point designated by (x,y,z,t), where t is time, is given by the function $G = x^2 + y^2 + z^2 - c^2t^2$, where c designates the velocity of light. Poincaré discovered that the distance G is left unchanged (invariant) when the spatial coordinates and the time are replaced by expressions that also constitute a group of transformations, which he named the Lorentz transformations.

Aesthetics of a quasi-visual sort enter here through Poincaré's comparison of F and G. Poincaré designated the group of transformations in four dimensions that leave G invariant as the Lorentz group, which he interpreted as the rotation group in four dimensions. We can see why he did this by comparing G with the function $F = x^2 + y^2 + z^2$ that we discussed earlier. If we take F to be the constant R^2, the equation $R^2 = x^2 + y^2 + z^2$ describes a sphere of radius R in three dimensions. A sphere in three dimensions exhibits rotational symmetry, because a sphere remains a sphere despite which diameter we rotate it about. But the minus sign in front of c^2t^2 in G breaks any straightforward analogy with F, and so too with spheres. Being a master of group theory and symmetry considerations, Poincaré made contact with the mathematical theory of invariant quantities by boldly substituting in G for the "distance" ct the quantity ict, where $i = \sqrt{-1}$, and so $(ict)^2 = -c^2t^2$. Then Poincaré replaced G with the new function $G' = x^2 + y^2 + z^2 + (ict)^2$. Taking G' to be a constant, say, Q^2, means that the equation $Q^2 = x^2 + y^2 + z^2 + (ict)^2$ represents a sphere of radius Q in four-dimensional space, with coordinates (x, y, z, ict), and so exhibits rotational symmetry in that space.

This brilliant move permitted Poincaré to propose a means to seek other quantities that remain unchanged in inertial reference systems: Seek quantities that can be written like the function G' because their Lorentz invariance is automatic. Basically, the physicist should seek quantities that transform under the Lorentz group like the space and time coordinates. The methods that Poincaré introduced into physics for constructing invariant quantities in four dimensions were elaborated in 1907 by the German mathematician Hermann Minkowski and are still essential today (see Chapter 6). A study of Poincaré's notebooks reveals that he began this work with aesthetics in mind, that is, to immediately seek Lorentz-invariant quantities.

POINCARÉ AND RELATIVITY

The invariance of the quantity $G = x^2 + y^2 + z^2 - c^2 t^2$ has far-reaching consequences, among them are that the velocity of light c is the same in every inertial reference system and that time is a relative quantity, on which Poincaré never agreed. Poincaré became fully aware that Einstein's special relativity and Lorentz's electron theory are mathematically identical. Consequently, in agreement with the underdetermination thesis, Poincaré contended that they are also observationally equivalent regarding existing data. Since the relativity of time is beyond our sense perceptions—after all, the motion of a clock does not seem to affect the time it registers (unless, of course, you drop it)—then this prediction of special relativity need not necessarily be what is actually the case. Poincaré was wrong on this point and should have known better, because by 1912 data had accrued in support of the relativity of time.

What Albert Einstein and Henri Poincaré accomplished in 1905 continues to fascinate historians and philosophers of science. Some scholars interpret the situation to the effect that both men arrived at the special theory of relativity and, consequently, Poincaré ought to share the accolades with Einstein. Existing archival and primary sources render the co-discovery claim insupportable. But this is not merely a debate about priority. Key issues are involved. Among them are the effect of Poincaré's writings on Einstein's thoughts toward the special theory of relativity, the question of when scientists should elevate hypotheses to the lofty heights of being untestable by experiment and unquestionable by theory, the notions of underdetermination and conventionalism, differing opinions on the weight of empirical data, and the problem of scientific creativity. Interestingly, from 1905 through 1912 (the year of Poincaré death), co-discovery was never an issue for Einstein and Poincaré. But it served a darker purpose in Nazi Germany, where relativity was permitted to be taught under the proviso that Poincaré be considered its discoverer.

In physics the notions of intuition and aesthetics can be still more subtle than in mathematics, as will become clear in our discussion of Albert Einstein.

EINSTEIN AND MAX WERTHEIMER

We are fortunate that Toulouse published his extensive psychological data on Poincaré; no psychologist ever did this for Einstein. Throughout his life Einstein maintained discussions with colleagues in virtually every discipline. But he was generally reticent about speaking with psychologists seeking to analyze him. Einstein's interactions with Sigmund Freud on scientific matters are known not to have been satisfactory. They were at loggerheads over whether psychoanalysis is a science. For example, in a draft reply written in 1927 to someone who suggested Einstein permit himself to be psychoanalyzed, Einstein wrote, "I regret that I cannot accede to your request, because I should like very much to remain in the darkness of not having been analyzed."

Max Wertheimer, the pioneer of Gestalt psychology, was an exception, most likely because he and Einstein were kindred souls politically and he had a firm physics background, which gave him some depth of understanding of Einstein's scientific views. They began discussing creativity in 1916 when they were colleagues at the University of Berlin. From Wertheimer's manuscripts we know that in 1943 he began to write up these discussions for publication as a chapter in his posthumously published book of 1945, *Productive Thinking*. The chapter on Einstein is called "Einstein: The thinking that led to the theory of relativity." What an astounding title! Can we believe it? Is Max Wertheimer, the eminent psychologist, really going to let us in on how Einstein discovered the theory of relativity? Alas, no. Instead Wertheimer gives us a highly interesting Gestalt psychological reconstruction of Einstein's discovery of relativity. Wertheimer's scenario de-emphasizes the role of thought experiments and aesthetics, while focusing instead on experimental data that were supposed to be of crucial importance to reorganizing Einstein's field of thought. Wertheimer's scenario conflicts with historical data.

A PORTRAIT OF THE PHYSICIST AS A YOUNG MAN

Presently there is no biography of Einstein that gives a balanced account of his science, philosophy, public and personal life. This is a daunting task in light of the rich archival material available and the complexity of the subject.

Biographical studies of Einstein and Poincaré indicate dramatic contrasts in their school careers, personal lives, and early research efforts, in addition to those attributable to the intellectual cultures in which they lived and worked. Were I to write a biography of Poincaré, I would alternate chapters between his scientific work and other parts of his life. The scientific life was dominant, which is not unusual for scientists, and as far as we know there never was any reflection of one on the other. With Einstein it was altogether different, at least through 1909. As organized as Poincaré's life was, Einstein's was the opposite. Einstein's life is the stuff of which movies are made. The scientist whose early life Einstein's parallels closest is Galileo, whose biography is sketched in Chapter 1.

Albert Einstein was born on March 14, 1879, in Ulm, Germany, where his father, Hermann, age 32, owned a featherbed business; his mother, Pauline (née Koch), was 21. As a child Albert entertained himself, doubtlessly due to difficulties learning language. While Poincaré spoke at 9 months, Einstein did not speak well until 2½ years and showed no early precocity. About age 13 or 14, Einstein demonstrated the ability to solve difficult mathematical problems posed to him by his paternal uncle, Jakob Einstein. These diversions, in addition to insightful recognition of young Einstein's talents by a medical student who boarded in the Einstein household, were necessary counterbalances to the disastrous situation he faced at school. Some years later Einstein described his entire student career as a "comedy." The humor was found only in retrospect.

Einstein's rebellious attitude against the authoritarian teaching methods at the Gymnasium he attended in Munich led eventually to a situation so intolerable that he left in 1894 without a diploma. He essentially was a high school dropout. (The Gymnasium he attended was destroyed during World War II. Ironically, it was rebuilt and renamed the Albert Einstein Gymnasium.)

Until the fall of 1895, Einstein traveled through northern Italy. As had been the case for Goethe some hundred years before, the Italian sunshine and landscape impressed the young man, freeing him of the *Sturm und Drang* of the Munich years. During this period, however, Einstein did not neglect his love of science. By this time he knew integral and differential calculus, self-taught at about age 13. In the summer of 1895, Einstein wrote his first scientific essay, which he sent to his uncle, Caesar Koch. The essay demonstrates that Einstein was conversant with such advanced topics in theoretical physics as Heinrich Hertz's version of Maxwell's electromagnetic theory. Even so, there are no signs of genius in the essay. Yet in retrospect, the perseverance and self-discipline needed to learn difficult subjects were an indication of things to come. Einstein was an autodidact.

As Einstein had promised his parents prior to withdrawing from the Gymnasium, he was also preparing himself for the entrance examination to the Eidgenössische Technische Hochschule (ETH), Zurich.* Whereas Poincaré went from a stellar stay at the lycée in Nancy directly to l'Ecole Polytechnique, Einstein failed the entrance examination to the ETH because of deficiencies in foreign languages, biology, and history, subjects that require rote learning. Because of Einstein's excellent grades in the mathematics and physics portions of the entrance examination, one of the school's most eminent professors, Heinrich Friedrich Weber, encouraged Einstein to attend his lectures if Einstein stayed in Zurich. Instead Einstein decided to take the advice of another professor to spend a year at a preparatory school in the Swiss canton of Aarau, in order to correct the deficiencies that had caused him to fail the entrance examination.

The cantonal school made a strong impression on Einstein due to its unpretentiousness and its seriousness, which in no way depended on a teacher's authority. The school also emphasized the power of visual thinking, a mode of thought to which Einstein found himself disposed. Sometime during his sojourn in Aarau during 1895 and 1896, Einstein realized a thought experiment in highly visual terms over which he would ponder tenaciously for 10 years. This ended when he realized that it contained the "germ of the special theory of relativity," as he related to Max Wertheimer sometime in 1916, and then described in print in his "Autobiographical notes."† He flourished in Aarau, finishing with the highest grade average in his class, and gained admission to the ETH.

Einstein's educational experience at the ETH (from 1896 to 1900) was bittersweet. Difficulties arose almost immediately. The role of visual thinking was de-emphasized, and the outdated physics curriculum focused

---

*Referred to also as the Swiss Federal Institute of Technology and the Swiss Polytechnic Institute.
†See Chapter 1, Figure 9.

on applications. Einstein liked neither the subject matter nor being co-erced to memorize large quantities of what to him was unessential material. So at home in the evenings and during cut classes, he studied the masters of theoretical physics like Ludwig Boltzmann and von Helmholtz. From them he learned the kind of physics not taught at the ETH, as well as the importance of visual thinking in the making of a scientific theory. As Boltzmann wrote in a book that Einstein undoubtedly read as a student,

> Unclarities in the principles of mechanics [derive from] not starting at once with hypothetical mental pictures but trying to link up with experience at the outset.

Yet when Einstein attended classes, he was attentive. Contrary to Poincaré's, Einstein's notebooks display well-taken notes with a lean style that concentrates on concise statements.

Contrary to Poincaré's lack of success in learning to play the piano, Einstein had more than moderate success with the violin. Unlike Poincaré, Einstein detested Wagner: "His musical personality is inde-scribably offensive," wrote Einstein in 1939 with quite apparent reasons. Bach and Mozart were Einstein's favorite composers. Like Poincaré, Einstein read widely outside of science. Whereas Poincaré's somewhat frail constitution restricted his physical exercise to long walks, Einstein's ro-bustness is reflected in his attraction to hiking and in particular sailing, which he pursued into the 1930s.

EINSTEIN IN LOVE

Poincaré's correspondence during his years at l'Ecole Polytechnique reveal a bon vivant student life. Since most of our picture of Poincaré as student is gleaned from letters to his mother, it is not surprising that there are no recorded female involvements. Einstein, too, was an active member of student café society. We do know that he had a number of correspondences with female friends. Then there was the tumultuous love affair with the only female member of his class, Mileva Marić. They met in October 1896 and until August 1899 were just close friends. From 1899 to 1902 their love affair had all the trappings of Romeo and Juliet, with a considerable amount of *La Bohème* too.

At least from 1899 through 1902, the dynamics of Einstein's personal life drove his scientific work, as in the lives of artists, musicians, and writers. The love letters were recently published. For the most part they read similarly to ones you or I may have written. They are important in retrospect because they were written by Albert Einstein. Can you recall ever having seen a book of love letters written by a scientist?

In August 1899 Einstein wrote Mileva (they were separated much of the time due to opposition from both sets of parents) of his concerns over the present state of the "electrodynamics of moving bodies," which would be the title of the 1905 relativity paper. Despite suggestive phrases such as this one in their correspondence, it is straightforward to conclude that Mileva acted as at most a sounding board for Albert's ideas. There is no evidence that Mileva should have shared with Albert the accolades for the special theory of relativity. The couple who could not live without one another married in 1903 and found they could not coexist in peace. Albert and Mileva separated in 1914 and divorced in February 1919. Later that year he married a distant cousin, Elsa, whom he had known since 1912.

Einstein recalled that his independence of thought was not appreciated by the professors at the ETH, particularly not by the eminent Professor Weber, with whom he had many intense personality conflicts. As a result, Weber, the man who had encouraged Einstein to pursue studies at the ETH, tried to prevent him from graduating. Having failed, Weber refused Einstein any letter of recommendation upon graduation. Einstein was the only one of four students in his class who passed the final examination to be refused a position as *Assistent* to a professor at the ETH. (Mileva failed twice and so never received a university diploma.) To the contrary, Poincaré was accepted for advanced studies at the prestigious l'Ecole des Mines, while also pursuing advanced mathematical study at the University of Paris with the famous mathematician Charles Hermite. Whereas Poincaré eagerly pursued mathematical problems after graduation, Einstein recalled in 1946 that it took him a year to recover from the ETH and to reacquire his taste for scientific research.

From 1900 to 1902, Einstein had only intermittent employment and was denied positions as assistant to several major physicists. He was convinced that somehow Weber was behind this situation. In this period Einstein developed a chronic stomach ailment, caused by malnutrition, that plagued him throughout his life. As Einstein wrote to Mileva on April 4, 1901, "Soon I will have honored all physicists from the North Sea to the southern tip of Italy with my offer." He persevered: In 1901 Einstein submitted a doctoral thesis to the University of Zurich, which rejected it, but he succeeded in publishing his first paper in the prestigious German physics journal, the *Annalen der Physik.*

Finally, through intercession of the father of a college friend, Marcel Grossmann, Einstein obtained a position as technical expert third class (provisional) at the Swiss Federal Patent Office, Bern. In reply to someone's comment that he might be bored in this position, Einstein wrote to Mileva in February 1902, "Certain people find everything boring—I am sure that I will find it very nice and I will be grateful to Haller [the director of the Patent Office] as long as I live." And he was.

THE CREATIVITY OF ALBERT EINSTEIN

Einstein's Bern period (1902–1909) was the most creative of his life. While working at the Patent Office eight hours a day, six days a week, he published about 50 papers. Although he had published five papers in the *Annalen* during 1901 to 1904, nothing prepared the world for what would happen in 1905. After all, in 1905 Albert Einstein was a middle-level junior civil servant with an academic record that in retrospect was distinctive only by its lack of distinction. His score on the cumulative final exam at the ETH was 4.91 out of 6.0, good but not superlative; he had failed once to obtain a Ph.D., and he was denied letters of reference from his undergraduate school. Yet at eight-week intervals, starting in March 1905, Einstein submitted three papers to the *Annalen* that changed the course of physics in the twentieth century, not to mention life itself on our planet. I think we can say that Einstein, along with Freud, "created" the twentieth century.

In the first of the 1905 trio of papers, Einstein proposed that light has a particle nature. The second paper solved the problem of why dust particles in the air perform an erratic dance, known as Brownian motion. An offshoot of this paper was a means for demonstrating the existence of atoms. The third one was the special relativity paper. These three papers appeared in Volume 17 of the *Annalen*, which is so valuable that it is customarily removed from library shelves and placed in a safe. Einstein published one more paper in 1905, which contained a result he had overlooked in the relativity paper. He couldn't believe it himself—energy and mass are equivalent, $E = mc^2$, a result that is not merely an equation any longer. It has become the signature of the twentieth century. Its most spectacular experimental affirmation occurred 40 years later at Hiroshima.

Contrary to Poincaré's discovery of Fuchsian functions, Einstein's early research results were at first appreciated, mostly for the wrong reasons, if at all, including the 1905 paper on special relativity. That Einstein had an *Annus Mirabilis* in 1905 became clear only in retrospect from the 1920s when *all* of his contributions from that year were duly acknowledged. Special relativity was not recognized as an achievement until 1911, when the frontier of physics moved from the discredited electromagnetic world-picture to the nature of light. Whereas Poincaré clearly exhibited a certain genius in mathematics by age 17, there was no forewarning of Einstein's creative outburst in 1905.

In order to gain further insight in Einstein's creative thought, we must come to grips with his various definitions of intuition and aesthetics.

EINSTEIN INTROSPECTS
ON INTUITION AND AESTHETICS

One of Einstein's definitions of intuition was a feel for nature:

> There is no logical path leading to these laws [of nature], but
> only intuition, supported by sympathetic understanding of
> experience.

According to Einstein, there is no logical path between necessarily imprecise experimental data and exact statements of a scientific theory. The scientist's only guide is intuition "resting on" a particular understanding of what are good data. This resembles Poincaré's process definition of intuition because it enables something to happen. Yet it differs from Poincaré's with regard to the facts or data to which it pertains because facts or data in mathematics and physics are fundamentally different.

On this point, in his 1946 "Autobiographical notes," Einstein recalled that while at the ETH he had fleetingly considered mathematics as a career. But due to the "following strange experience" he decided otherwise. The "strange experience" was the illumination that he had no nose for what is a good problem in mathematics and saw himself as a mathematician in the position of "Buridan's ass." Einstein continues in a quite pungent manner:

> This was obviously due to the fact that my intuition was not
> strong enough in the field of mathematics in order to differen-
> tiate clearly the fundamentally important, that which is really
> basic, from the rest of the more or less dispensable erudition.

Thus, another of Einstein's definitions of intuition: the feeling for what is a fundamental problem in physics.

Central to Einstein's creative thinking was his visual imagery, which is based on another definition of intuition as visualization and visualizability (see Chapters 2 and 8). Many scientists in the German-cultural environment believed that creative thinking occurs in visual imagery and words follow; this statement can be found in a revealing letter Einstein wrote on June 17, 1944, to Jacques Hadamard and in Einstein's introspection, "Autobiographical notes."

Poincaré began his 1908 introspection with the problem: "What, in fact, is mathematical invention?"; Einstein began his 1946 introspection by asking, "What, precisely, is 'thinking'?"—he inquired one level deeper than Poincaré. Our first impressions of an external world are "sense impressions" from which "memory-pictures emerge." Certain memory pictures form series. For Einstein this does not yet constitute thinking. Thinking begins when one memory picture occurs a great many times in several different series. The picture serves as an "ordering element" for the different series, an element that could also relate heretofore unconnected series. Einstein referred to this ordering element as a "concept." Thinking is

> operations with concepts, i.e., the creation and use of definite functional relations between concepts and the coordination of sense experiences to these concepts.

Thus far Einstein's testimony on thinking is almost identical to analyses by von Helmholtz on thinking in his 1894 essay, "The origins and correct interpretation of our sense impressions," and to Boltzmann's 1897 definition of visual images as "concepts."

But Einstein departs from von Helmholtz's view in two essential ways. First, to Einstein, thinking is "a free play with concepts," which is similar to Poincaré's view. Second, any coordination between sense experience and concepts can be obtained only by using intuition to leap over the abyss between sense data or the data of the laboratory, on the one hand, and the exact laws of physics, on the other.

That our creative thinking is essentially nonverbal seemed reasonable to Einstein, for how could "we 'wonder' quite spontaneously about some experience"? Einstein was as specific as he could be on the meaning of "wondering," which "seems to occur when an experience comes into conflict with a world of concepts which is already sufficiently fixed within us." For example, Einstein recalled that as a boy of 5 or 6 he saw a compass and "wondered" how its needle stayed fixed on magnetic north as if held by an unseen hand. This image remained in his mind. It became the basis of his preference for field-theoretic formulations of physics, of the sort pioneered by Faraday and Maxwell. This representation of nature is an abstraction of the way phenomena occur in the world: action by contact.

It is in Einstein's use of the term "wonder" that his three definitions of intuition fuse. The first time that Einstein used the term *Anschauung* extensively was in "On the development of our intuition [*Anschauung*] of the existence and constitution of radiation" (1909) (see Chapter 2). As the title indicates, Einstein explored the clash between the *Anschauungen* or intuitions of the time-honored wave mode of light and the particle mode, or light quantum, that he had invented in 1905. Since Einstein considered the principles and concepts of special relativity theory to be extensions of those in Newtonian mechanics, despite special relativity's new notions of space and time, he persisted in later years to emphasize the continuity between the two theories. On the other hand, as Einstein wrote in a letter to his friend Conrad Habicht in the spring of 1905, he considered light quanta to be *"sehr revolutionär"* ["very revolutionary"], and we can now suggest why: Light quanta were what Einstein referred to in 1946 to be a "wonder," since their characteristics conflict with already formed concepts, that is, the visual images of *Anschauung*.

AESTHETICS

Poincaré's notion of aesthetics was of a formal mathematical sort, in which he explored results in mathematics by seeking ways to alter variables in equations in a way that left the equation itself unchanged, or invariant. Einstein, on the other hand, concerned himself with symmetries in how theories ought to *represent* nature.

An example concerns dropping a stone into a still pond of water. Circular waves move outward from the stone's entry point. This is the representation used for how an electron produces spherical light waves. All scientists in 1905 judged the visual representation of particle and wave (discontinuity/continuity) side by side to be aesthetic and to carry no hint of contradiction. It was the contradiction in *terms* that resulted in the physics community's suspicions about light quanta. To Einstein alone, however, there was an asymmetry in this tension between discontinuity/continuity, or particle/wave—redundancy in representation. He saw no reason to burden a theoretical representation with particle and wave side by side. As he stated in his 1905 paper on the constitution of light, this "profound formal distinction" science makes is wrong. This aesthetic discontent was precisely Einstein's reason in 1905 for proposing that light can be *represented* as a particle. By 1909 Einstein found it necessary to relax this guideline and explore the wave/particle duality of light.

Although we can visualize light moving through space as *either* a wave or a particle, we cannot visualize or even imagine light as both wave *and* particle. Consequently, it is no wonder that for more than two decades physicists puzzled over and resisted Einstein's light quanta. Their criticisms were based not on any empirical data, but on the problem of visually representing light, and particularly on how certain optical phenomena that had traditionally been the hallmark of light could be explained with light quanta. For example, no visual representation could be constructed for how light quanta produce interference. One existed for light waves, namely, as we have already discussed in Chapters 2 and 7, comparison with how water waves produce interference. Einstein never came to grips with the riddle of light quanta. From about 1919 on, he focused on a unified field-theoretical description of matter and light based on such *concepts* as continuity and action by direct contact, because they are among the mainstays of prescientific thought, which is our most basic knowledge constructed from our early explorations of the world. Particles were supposed to emerge as knots in space-time.

Then we have Einstein's astounding declaration about the state of electrodynamics, which he announces in the very first sentence of the relativity paper. Despite what every other scientist thinks, Einstein claims that there are "asymmetries" which beset the very foundations of electrodynamics.

One asymmetry was the redundancy in explanation of Maxwell's theory for electromagnetic induction, which Einstein developed with a thought experiment in the 1905 relativity paper (see Chapter 8). Being master of the understatement, Einstein's phrase in this opening sentence "is well known" is far from the truth: To *everyone else*, having two explanations for the generated current in electromagnetic induction was perfectly natural (See Chapter 8).

These notions of aesthetics, combined with his "feel" or intuition for the proper subset of data to use to bridge the gap in one fell swoop (analog thought) to the exact statements of physical theory, enabled Einstein to formulate the special relativity theory. But this required Einstein to permit his scientific speculations to take him beyond the immediate world of sense perceptions where, for example, there seems to be no relativity of time. Einstein, alone, was willing to *redefine* the concept of intuition to a level of abstraction higher than in the mechanics and electrodynamics of 1905.

POINCARÉ'S AND EINSTEIN'S CREATIVE THINKING

Introspections supported by historical data are invaluable for exploring creative scientific thought. We have found this to be the case for the introspections of Poincaré and Einstein. In particular, for Poincaré we have the results of detailed psychological tests in which he participated during 1897 under the direction of the psychologist Edouard Toulouse. Poincaré's and Einstein's introspections provide evidence that network thinking ought to be explored further. For this purpose I have provided a network model based on the interdisciplinary methods of cognitive science. Analysis of introspections from Einstein and Poincaré enable us to cull away some of the mystery that is supposed to surround aesthetics and intuition.

Poincaré considered intuition and aesthetics as critical guidelines for unconscious network thinking. He deemed what is not logic in a mathematical proof to be intuition. This process definition of intuition describes it as a catalyst for creative thought. Poincaré described his *nonvisual* mode of mental imagery as a "sensible intuition," which is consistent with Toulouse's finding that in his creative work Poincaré neglected visual imagery altogether. Sensible intuition is the faculty for perceiving the whole of a mathematical demonstration at a glance. Aesthetics is a guideline in network thinking because it acts as a "delicate sieve" that filters out all but the most "beautiful" combinations of mathematical facts. For a mathematician, the concepts of aesthetics and beauty have definite connotations, such as invariance and symmetry. Poincaré's recollection of his great mathematical invention in 1881, in conjunction with his published papers and unpublished manuscripts, and Toulouse's results, indicate that in his research, Poincaré actually *relied* on aesthetics to guide his network thinking.

Particularly interesting support for the model of network processing offered in Figure 1 is that Poincaré's most far-reaching illuminations (or "mind popping") occurred when he was working on the cutting edge of scientific research with no predefined path or set goal. From what we know about activation of mental networks of information in whatever coding they may be in (analog or propositional), only the freedom of unconscious thought will do in this situation. Einstein worked similarly, being facile enough to incorporate information from diverse areas into physics problems.

Einstein's introspection of 1946 in conjunction with his published papers, correspondence, and archival documents indicates that his creative mode was also network thinking but of a sort emphasizing visual images. Consider the two thought experiments that were essential to his discovering the special and general theories of relativity. In Chapter 1 we discussed the 1895 thought experiment, which concerned what it is like to catch up with a point on a light wave. This experiment led to the special theory of relativity. We know how this experiment emerged only from his 1946 introspection, "Autobiographical notes" and from Wertheimer's Gestalt reconstruction written some years after their conversations. What is interesting is that in his "Autobiographical notes" Einstein describes the thought experiment as something that he "had already hit upon at the age of 16." Perhaps his use of the word "hit" denotes the suddenness of this realization.

We are luckier with the 1907 thought experiment, which we explored in Chapter 8 and which led to the general theory of relativity. While discussing his work in a seminar at Kyoto, Japan, in 1922, Einstein recalled that this experiment emerged suddenly while he was daydreaming at the Patent Office. Daydreaming is akin to unconscious thought in that it knows no boundaries. Consequently, there is a "play of associations," to use Toulouse's terminology.

The differences in mental imagery between Poincaré and Einstein are also among the keys to the intellectual puzzle concerning how they interpreted in dramatically different ways the identical mathematical formalisms that they produced independently of one another in 1905. The rich visual imagery in Einstein's thought experiments permitted him to interpret the mathematical scheme as a theory of space and time, while Poincaré interpreted Einstein's theory as an improved version of Lorentz's electron theory.

Two of Einstein's uses of the concept of intuition overlapped with Poincaré's. Intuition is the scientist's feel for the correct way to bridge the gap between the exact statements of a scientific theory and inexact data. This is a process definition of intuition similar to Poincaré's. Then there is Einstein's definition of intuition as the sense for what is a fundamental problem. Einstein's third use of intuition differed from Poincaré's because it concerned the mode of mental imagery central to Einstein's creative thinking—visual imagery. Here I refer to the important term in

German philosophy and science, *Anschauung*. Einstein used this concept of intuition in his great thought experiments of 1895 and 1907, which were important to his formulating the special and general theories of relativity, as well as in his early work on the nature of light. His notion of aesthetics differed from Poincaré's, which was useful essentially to mathematicians. For Einstein, aesthetics played a key role in deciding how physical theories ought to represent nature.

In terms of declarative knowledge and procedural knowledge we may summarize our findings on Einstein and Poincaré as follows. Their representations of declarative knowledge differed in Einstein's preference for visually representing facts. Regarding procedural knowledge, their goals in problem solving differed in that Einstein realized that what had to be sought was a new theory of space and time, while Poincaré worked toward a unified theory of matter. The differences in their declarative knowledge affected their procedural knowledge, or problem-solving techniques, which is evident in Einstein's adroit use of thought experiments.

Can any computational theory of cognition based purely on propositions or any general problem-solving program capture Poincaré's and Einstein's creative thinking? To be optimistic, this is certainly not possible with any available today and certainly not with today's computer architectures. I say this in view of the fact that creative thinking results from analog and digital processes, which include mental imagery.

We discussed why analog thought better captures creative thinking than digital thought. In order to provide substance for this argument, a model of analog thought in Figure 1 encompasses unconscious parallel lines of thought. If parallel lines of thought are excluded, then we would have to fall back on, for example, Gestalt psychological-oriented schemes in which a previously disequilibrated field of thought is snapped into an orderly one. The facts in this field are not encoded as propositions but are just "there." We recall from Chapter 8 that according to the *Pragnänz* principle, the tendency toward a good Gestalt is irresistible. But for this to occur, Gestalt psychological scenarios must depend on critical experiments resulting in discontinuous Gestalt switches, none of which matches what we know about actual historical situations.

It is an understatement to say that scientific creativity is a difficult subject to study. So Gestalt psychological scenarios are not to be dismissed out of hand, because their emphasis on pattern recognition based on symmetries is an important part of our understanding of vision. Also of importance is Jean Piaget's genetic epistemology, whose fundamental hypothesis is that there is a parallelism between the construction of intelligence and the history of scientific thought. So, for example, Piaget advocates the emergence of abstract scientific theories from those based on more concrete objects, whose existence must be established as a prerequisite to any scientific progress. We saw this occur in Einstein's special relativity theory with the establishment that the velocity of light is unchanging, Bohr's establishment of a stable atom as the prerequisite to any atomic theory, and Poincaré's proposal for the origins of geometry, which, Piaget tells us, was extremely influential in the formation of his own view of psychology. I have incorporated aspects of genetic epistemology into the discussion of scientific progress in Chapter 7, with the assimilation/accommodation process. The severe limitation of Piaget's total scenario is that the end result of intelligence is thinking primarily with logical symbols, so that the faculty of visual imagery ought to play a subsidiary role. Similarly, Piaget predicts that this should be so in science, which we found not to be the case. Nevertheless, there are virtues in Piaget's work, particularly the assimilation/accommodation process that is incorporated into Figure 1 through the use of metaphors as a guideline for creativity.

This chapter has explored scientific creativity and modes of thinking, including digital and analog processes. We found that among the factors playing an important role are aesthetics and visual imagery. For the most part, we focused on Poincaré and Einstein. In order to expand our results, in addition to drawing together themes we have discussed throughout the book, such as intuition, representation, and creativity, it is natural to turn next to the relations between art and science. We will find that at the moment of creativity, boundaries dissolve between artists and scientists, as they have done often in art and science in the twentieth century.

10
ART,
SCIENCE,
AND
THE
HISTORY
OF
IDEAS

The abstract artist Piet Mondrian wrote in 1937

> For there are "made" laws, "discovered" laws, but also laws—a truth for all time. These are more or less hidden in the reality which surrounds us and do not change. Not only science but art also, shows us that reality, at first incomprehensible, gradually reveals itself, by the mutual relations that are inherent in things.

What better way to introduce this final chapter than to widen the theme of scientific realism and its integral role in the history of ideas to include the interplay between art and science.

Particularly in the twentieth century, scientists have struggled to find ways to represent unseen worlds. As we discussed in earlier chapters, this situation is especially acute in atomic physics, where scientists try to "read" nature from heavily theory-dependent data such as bubble chamber photographs. At first scientists assumed they could deal with unseen quantities such as electrons with concepts abstracted from the world of sense perceptions. By degrees they became convinced of their errors and, by 1927, realized the restrictions on visual imagery and language in this enigmatic domain. Mathematics became an essential guideline, leading eventually to a proper visual representation of the unseen atomic world: Feynman diagrams. This required dramatic transformations in the concepts of visual imagery and intuition.

Representing nature has always been a central problem for scientists and artists. We have testimony on this from Albrecht Dürer and Leonardo da Vinci, among many other Renaissance artists who struggled with mathematical and physiological problems concerning linear perspective. In the late nineteenth and early twentieth centuries, artists once again became interested in problems of space and time. This was the subject of Georges Braque, Paul Cézanne, and Picasso. It was also part of the *Zeitgeist* that at about the same time Braque and Picasso in Paris, and Einstein in Bern, brought forth new notions of space. Braque and Picasso discovered Cubism, while Einstein discovered relativity.

We are struck by parallelisms between developments in modern art and science, particularly in the late nineteenth and early twentieth centuries. By this I mean the shift in representation in art from the extremely figurative or naturalistic art—as in the Renaissance—to the increasingly abstract art of the late nineteenth and early twentieth centuries. These shifts in art coincided with increased abstraction in physical theory accompanied by transformations in intuition. Was this merely coincidence? As we noted in Chapters 1, 2, and 8, at the same time there was a change among scientists from *imposing* visual imagery abstracted from the world of sense perceptions onto physical theories *to* creating theories *generating* their own visual imagery (see Chapter 8, Figure 22). Such parallels exemplify what I mean by a relation between art and science. Examples from actual scientific research and works of art are essential to elevate this discourse beyond trivial discussions about the meaning of terms. At the nascent moment of creativity, boundaries dissolve between disciplines.

After exploring at some length the notion of aesthetics in art and science, we turn to how artists and scientists—particularly in the twentieth century—have sought to represent nature. This is the proper point to conclude our discussion in Chapter 7 on transformations in the visual representations in subatomic physics. As a case study of the interplay between art and science, we discuss the rise of Cubism in the hands of Picasso and Braque, with its offshoots in nonfigurative art. This discussion includes similarities and differences between art and science, as well as in artistic and scientific progress. We conclude by inquiring into creativity in art and science.

AESTHETICS IN ART AND SCIENCE

While the term "aesthetics" is difficult for philosophers to pin down, artists and scientists have some firm ideas about it. Let us look at the philosophers first, because they can alert us to general problems.

To some extent, most people can convey more easily what they mean by a painting of Michelangelo or Rembrandt being aesthetic or beautiful than they can with Picasso's *Les Demoiselles d'Avignon*, Marcel Duchamp's *Fountain*, which is a urinal, or Damien Hirst's dead animals preserved in formaldehyde. We might say that Michelangelo and Rembrandt express with clarity and emotion a scene that seems familiar, despite the intervening centuries. Their paintings are naturalistic in the classical sense of depicting as accurately as possible a subject's visual appearance. Can we be more precise about the judgmental criteria we bring to bear for aesthetics?

Aesthetic judgment is a field in which there is no consensus. Opinions range from the subjective to the objective. Subjectivists believe there is no rational way to assess beauty. Some subjectivists consider it fruitless ever to attempt to define beauty, while others contend that it makes no sense to discuss beauty without an observer. Thus, statements such as "there exist beautiful flora and fauna in the unexplored depths of the Amazon rain forest" are meaningless. This view of aesthetics will not do for physics, where speculations are made on the symmetries of matter existing in the universe some 15 billion years ago, as well as about unseen elementary particles.

Objective assessments of aesthetics and beauty attempt to remove subjective elements. Can we really believe this? The reason is that an "objective" viewing of a painting requires that "we need bring with us nothing from life, no knowledge of its ideas and affairs, no familiarity with its emotions," wrote the aesthetician Clive Bell. Think about it for a moment. There is no totally objective set of rules for appreciating a painting, as we would expect. The American philosopher John Dewey summarized this situation in his 1934 book, *Art and Experience:*

> No matter how ardently the artist might desire it, he cannot divest himself, in his new perception, of meanings funded from his past intercourse with his surroundings, nor can he free himself from the influence they exert upon the substance and manner of his present being. If he could and did there would be nothing left in the way of an object for him to see.

Although no quantitative measure of aesthetics can emerge from any theory of beauty, the invocation of the sublime signals the presence of an aesthetic that might be only of a very personal sort. Yet the feeling that results from an aesthetic experience signals us to listen carefully to the artist or scientist.

Let's consider some examples of aesthetic experiences from scientists for whom aesthetics is often an explicit guideline in their research. Throughout this chapter we will find that scientists are often much better than artists at articulating what they mean by aesthetics, although artists such as Braque and Picasso were quite straightforward. Besides the aesthetic of a particular art movement such as Cubism, to artists aesthetics is generally a diffuse notion with different definitions for different areas. Certain aspects of aesthetics can be specified, such as balance, symmetry, and their opposites, as they relate to forms and colors. Just as in art, discoveries in science are made by breaking the rules.

We recall Henri Poincaré's introspection in which he describes the mathematician's "special aesthetic sensibility," which is an intuition that filters out all but the few combinations that are "harmonious" and "beautiful" (Chapter 9). Then we encountered Poincaré's view that the very reason why the scientist studies nature is "because it is beautiful." As Albert Einstein announced in 1905, the fact that the accepted formulation of electricity and magnetism implied "asymmetries that do not appear inherent in the phenomena is well known." Well known, that is, to Einstein (Chapter 9). Like Poincaré, Einstein mused over the beauty of the universe and the idea that the "eternal mystery of the world is its comprehensibility." We listen seriously to these great thinkers because their experiences were among the catalysts for some of their greatest discoveries. Both Poincaré's and Einstein's aesthetic experiences are rooted in their own concepts of intuition, a concept for them that included modes of mental imagery.

On cracking the problem, in 1925, of developing a new atomic physics, Werner Heisenberg recalled:

At first, I was deeply alarmed. I had the feeling that, through the surface of atomic phenomena, I was looking at a strangely beautiful interior, and felt almost giddy at the thought that I now had to probe this wealth of mathematical structures nature had so generously spread out before me.

Heisenberg's rapture is over the new quantum mechanics, whose mathematical structure he had realized through considerations based on the conservation law of energy, another guideline for scientific creativity.

In his characteristically emphatic way, the American physicist Richard Feynman described his immediate reaction to a new theory he developed in 1957 with his colleague Murray Gell-Mann: "There was a moment when I knew how nature worked. . . . It had elegance and beauty. The goddamn thing was gleaming." On aesthetic grounds Feynman and Gell-Mann were certain of its validity, despite experimental evidence to the contrary.

Not every great scientist achieves such moments of ecstasy and certainty. As we discussed in Chapter 9, von Helmholtz reported that his work sometimes took him "weeks or months," resulting in a "sharp attack of migraine." In Chapter 2 we saw that while almost everyone considered a wave representation for light and a particle one for objects to be aesthetic, Einstein did not; he judged that wave and particle side by side was *not* aesthetic. He suggested reconsidering the situation in which an accelerating electron emits light as one in which a *particulate* source emits *particles* of light. The case of electromagnetic induction in Chapter 8 is similar; everyone except Einstein considered there to be no asymmetries in its interpretation of two different causes for the same effect, which is the current generated in electromagnetic induction. In both cases Einstein's judgment of aesthetics was nonvisual and concerned redundancies in representation: Why have particle *and* wave; why have two explanations for a *single* measured effect (the generated electrical current)?

Another example of aesthetic choice was Schrödinger's disagreement with Heisenberg, which concerned a visual aesthetic in addition to Schrödinger's personal and bluntly stated preference for continuity with classical physics. As we discussed in Chapter 2, in Schrödinger's case it was distaste for Heisenberg's theory, which seemed to be completely at odds with what had always been a fundamental assumption for theory construction—continuity in nature along with visual representation.

We can only conclude that even if there were rules for assessing beauty, deviations from them would be among the defining hallmarks of creativity, as Einstein's aesthetic proclivities in 1905 attest. Some theories of beauty equate beauty with taste. In science, taste is decided by consensus, based on reasons taken as objective, in a scientific community. Certain theories of aesthetics assert that not to see beauty, in the face of such a consensus, is to lack taste. We might place Schrödinger in this category. In 1926, he called Heisenberg's theory repulsive because it conflicted with the traditional conception of what a theory ought to be. Consensus was for the particle theory of atomic physics, known as quantum mechanics. Einstein, Schrödinger, and others in the minority continued to opt for a wave representation of atomic physics, known as wave mechanics. Yet no one accused this minority group of lack of taste, nor did anyone accuse Einstein in 1905 of lack of taste in seeing asymmetries and not beauty in the theory of light and electromagnetism accepted at the time. In summary, although there is a consensus judgment in the scientific community, scientists are willing to change if the circumstances warrant it, although change can come slowly.

Science has general rules of aesthetics—of taste, if you wish—which can even lay claim to being universal. For example, some rules of aesthetics concern symmetries of nature. Mathematicians mean something very definite when they mention "aesthetics" or "beauty." If the mathematical function in question should have something to do with physics, then further important results emerge because a stunning theorem in physics is that every symmetry is related to a law of nature. For example, mirror symmetry is connected with the law that phenomena occur in exactly the same manner when viewed in a mirror. Yet mirror symmetry, or parity, is violated in certain processes involving elementary particles. This finding was surprising because until 1957 all physicists assumed that mirror symmetry is so fundamental to the macroscopic world that it ought to be maintained in the atomic world. The violation of mirror symmetry led to new areas of investigation, providing insight into unrealized symmetries of nature: In science, broken symmetries often reveal new ones. Asymmetries can be of the essence, in physics just as in art.

Sometimes a work characterized by perfect symmetry with no indication of imbalance or human-scale imperfection can seem cold, pristine, somewhat lacking in feeling, although perhaps majestic. The physicist and science writer Philip Morrison describes a similar case in science:

> A soap bubble is beautiful. [Yet it] has a kind of simplicity, a coldness, which bars it from the category of great beauty. In fact, the very reflections and color changes which make it something other than a perfect sphere enhance its beauty.

Nature and art do not abhor asymmetries, which can energize the artist's canvas. Consider the startling effects of the skewed faces in Picasso's groundbreaking *Les Demoiselles d'Avignon*, 1907 (Figure 1).

Today elementary particle physicists and cosmologists assume that our physical world is actually one of broken symmetries. Minute fractions of a second after the Big Bang some 10 to 15 billion years ago, there was only one force in an instant of purest symmetry, an instant of oneness. A slight imbalance or asymmetry between matter and antimatter began it all. Gradually, the four forces we are aware of today emerged—the weak, strong, gravitational, and electromagnetic—with distinctions among the various elementary particles. Among the deepest problems in modern physics is how this symmetry was broken and how to work backward to that one force: the quest for the Holy Grail of physics, the grand unified theory possessing pristine but profound symmetries.

So, just like artists, scientists have methods for taking account of symmetry considerations. Although aesthetics in art is more diffuse, it shares with science the motif of shifting between symmetry and asym-

FIGURE 1.

Pablo Picasso, *Les Demoiselles d'Avignon*, 1907. The Museum of Modern Art, New York. Acquired through the Lillie P. Bliss Bequest. Photograph © 1996 The Museum of Modern Art, New York.

metry. These shifts are essential for both disciplines. In art they helped to give rise to Cubism, and in science to the exciting discoveries in physics opened up by such asymmetries as parity nonconservation. Aesthetic taste is as important in science as in art. Consequently, in both disciplines the search for new aesthetics has the highest creative importance. Let us pursue the issue of the sometimes highly individual nature of aesthetic tastes in art and science.

ONE PERSON'S AESTHETICS AND INTUITION
MAY NOT BE ANOTHER'S

Just as in art, what is aesthetic in a scientific theory to one person might not be to another. We have already seen this with Einstein's light quantum and the asymmetries he noticed in electromagnetic theory. The aesthetic assessment of a work of art or science is an issue unto itself. Philosophers of art and art critics have written at great length on what defines a work of art. Examples include Monroe Beardsley's "aesthetic experience" emerging from contemplation of a work of art; Edward Bullough's "aesthetic attitude" meant to explain what the experience of art is about; the institutional theory of art, which is an umbrella term encompassing Arthur Danto's "artworld," meaning essentially that if a work is exhibited in an art museum or gallery (the artworld), then it is a work of art, be it a Brillo box, a painting by Picasso, or by a chimpanzee; and Ludwig Wittgenstein, whose writings on philosophy of language, particularly his classic analysis of the set of conditions for defining games, are used to argue that seeking any definition of art is beside the point.

The problem of what constitutes a work of art can be traced back to the work of a single man in 1917, Marcel Duchamp's sculpture entitled *Fountain*, which is a urinal. With good reason Duchamp called this type of art "Readymade," and it became the prototype for the "Conceptual Art" of the 1960s, as well as many of today's art movements. Art critics agree that after Duchamp's Readymade, art was never the same again. There were no more canons for aesthetics agreed to by consensus. *Almost* anything goes.

The situation differs in the sciences in at least two main points. First, there are no science critics who play a role similar to art critics—although there has never been a shortage of informed (and uninformed) outsiders criticizing the scientific enterprise. And there is no group of outsiders who by consensus can seriously declare a body of knowledge to be science, although certain Postmodernists attempt to do so, particularly with astrology and esoteric beliefs held by primitive tribes. Scientists critique the work of other scientists. They do not critique the work of humanists, unless it happens to pertain to science. The second difference is that there are canons of rationality and objectivity agreed to by consensus in the scientific community for assessing scientific theories. Among them are experimental agreement, predictive powers, degree of unification of disciplines, rigor in foundations, and agreement with conservation laws. Yet just as in any creative enterprise, rules of procedure are not always strictly adhered to and are sometimes set aside. For example, in the face of disconfirming data, adherents of a theory may choose either to ignore these data, to change the theory, or to hang on until new experiments are done, as we discussed in Chapter 3. The world of actual scientific practice is far too complex to be forced into any cultural or philosophical framework offering methods for assessment.

Just like artists, scientists believe that nature is beautiful. Artists attempt to interpret nature's beauty by crafting sculptured forms or applying colors to surfaces. The scientist uses mathematics to penetrate into a more intrinsic aesthetic, one beyond appearances in the atomic and subatomic realms. This aesthetic may have no direct visual representation based in our senses, because it can refer to an attribute of an elementary particle that does not correlate to anything we have ever visualized or even imagined, for example, its simultaneous wave and particle existence. Consequently, scientific representations are not "naturalistic" in the sense that this term is defined with reference to our world. But then our intuition with respect to this world has undergone dramatic transformations.

As we have argued in previous chapters, science is a means of extending our concept of intuition into worlds that ever-better approximate physical reality. This situation parallels the one in art where, for example, twentieth-century artists including the Cubists and Abstract Expressionists such as Mark Rothko have attempted to represent unnaturalistically a world beyond anything imagined with our sense perceptions. Unnaturalistic, that is, with respect to whether the meaning of this term is developed from phenomena we have actually witnessed.

Another element essential to aesthetics concerns style. Just as writers and artists have styles, so do scientists. An important ingredient in Einstein's style of doing theoretical physics was to seek out redundancies and then interpret them as asymmetries, as he did in his research on light and electromagnetic induction. The special theory of relativity is, in large part, a consequence of Einstein's aesthetic preference. In retrospect, in the light quantum and relativity papers, Einstein expressed strong aesthetic desires that, backed up by mathematics, produced stunning results.

A similarly strong aesthetic preference surfaced in 1923 in Bohr's suggestion to avoid the light quantum and to retain the traditional dual representation of *continuous* light waves interacting with *discrete* matter (Chapter 2). This aesthetic offended Einstein's minimalist one from 1905 based on a single representation for light and matter. To provide a foundation for his dualism, Bohr formulated new versions of his atomic theory with no visual imagery of the atom. Atomic physics slipped into an abyss in which the atom's electrons became unvisualizable. What is so remarkable is that Bohr's surrender of visualization was precipitated only in part by experimental data, but more in response to his aesthetic choice in favor of continuity and discontinuity existing side by side.

Heisenberg, on the other hand, opted in 1925 for Einstein's aesthetic of an exclusively particulate version of quantum mechanics, which he believed best reflected the essential discontinuities in atomic physics. In order to avoid falling into traps from the old Bohr theory, Heisenberg continued to renounce any visual imagery at all: Heisenberg's particles were unvisualizable (Chapter 2). Yet physicists yearned for the return of some sort of visual imagery. We glean from their physics papers a sense of despair because they were adrift with no anchor to the familiar world of sense perceptions.

Heisenberg's aesthetic preference "repelled" Schrödinger's, which was for all atomic matter to be represented as waves. Heisenberg, in turn, was "disgusted" with Schrödinger's aesthetic, regarding it as "trash." What emerged was an aesthetic that included both wave and particle representations and differed fundamentally from the ones Schrödinger and Heisenberg fought over so bitterly from 1925 to 1927. Neither Schrödinger nor Heisenberg, nor anyone else, has ever "seen" atoms or electrons behave like particles or waves. As we discussed in Chapter 2, these terms must be understood differently from the way they are in the world of sense perceptions. How this redefinition of terms like "wave" and "particle" was to be carried out was supposed to be dictated by Bohr's complementarity principle.

Bohr meant complementarity to be a means for folding the wave and particle aesthetics of classical physics into quantum mechanics through restrictions on these terms in the atomic domain, along with defining what an atomic entity is: The totality of an atomic entity's wave and particle properties constitutes the entire thing. Particle properties and wave properties complement each other, like yin and yang, but in any experimental setup the electron can reveal only one of its sides. The wave and particle aspects of an atomic entity are complementary but mutually exclusive.

The aesthetics of Schrödinger and Heisenberg reflect the philosopher George Santayana's understanding that beauty can be perceived in an inward sense. This parallels the poet Guillaume Apollinaire's attempt in 1911 to identify trends in Cubism in a way that evoked a spirit of art moving in entirely new directions of thought. He wrote of a "Scientific Cubism" as the "art of painting new structures out of elements borrowed not from the reality of sight, but from the reality of insight." This is the insight artists and scientists share and which we can describe as well with the term "intuition." It is the intuition of "seeing" a proof at a glance, as Poincaré put it; of "seeing" redundancies in explanation, in Einstein's case; of realizing that the mathematical framework of a new theory is consistent with the conservation law of energy, in Heisenberg's case; or that your new theory has "elegance and beauty," in Feynman's words.

Sharp differences of opinion between scientists on aesthetic viewpoints have occurred also among artists. Take, for example, Piet Mondrian and Jackson Pollock, two undeniably abstract artists. For Mondrian the essential starting point for nonfigurative art is our world, because, as he wrote in 1937,

> Nonfigurative art demands an attempt [at the] destruction of particular form and the *construction* of a rhythm of mutual relations, of mutual forms of free lines. [emphasis in original]

Pollock, at this time, was in a surrealistic stage based on his fascination with Jungian and Amerindian mythology. By 1946 he had stepped over the line into abstract art using the drip techniques with which he is associated today. In stark contrast to Mondrian, in the narration to a biographical film made in 1951, Pollack says that "My painting is direct. The method of painting is the natural growth out of a need. I want to express my feelings rather than to illustrate them." While Mondrian tried to strip away the perceptual world, right from the start Pollock leaves it behind and paints from emotions.

The purity and intensity with which Mondrian held to his view of seeking reality in terms of horizontal and vertical lines are reminiscent of the scientific disputes we have discussed. The reason is that Mondrian had his own view of what constituted good art. We know this because in 1925 Mondrian left the art movement *De Stijl* [The Style] because another major member, Theo van Doesburg, advocated the use of diagonals! For Mondrian this was too much to bear since it went entirely against the movement's original intent, which was emphasis on horizontal and vertical lines.

SCIENCE AS AESTHETICS

In arguing for a sun-centered universe, Copernicus made an astounding use of aesthetics in science. After all, no astronomical data existed to support his theory for over two hundred years. A sample of Copernicus's argument from his book of 1543, *De revolutionibus*, is as follows:

> In the middle of all sits Sun enthroned. In this most beautiful temple could we place this luminary in any better position from which we can illuminate the whole at once? . . . So the Sun sits as upon a royal throne ruling his children the planets which circle round him.

Copernicus used aesthetics for scientific data.

This approach was similar for Galileo, whose initial attraction to Copernicus's work was the elegance and symmetry of a sun-centered universe. By "elegance" Galileo meant that in a sun-centered universe the planets are placed in orbits that are related to how fast the planet moves around the central sun. No such relationship exists in an Earth-centered system. As it was for Copernicus, for Galileo the symmetry of a heliocentric universe resides in its better displaying the hand of God because the sun is the manifestation of light, warmth, and fertility.

Earlier in this chapter we mentioned Richard Feynman's immediate reaction to a new theory he developed in 1957 with his colleague at the California Institute of Technology, Murray Gell-Mann: "It had elegance and beauty." What notions of elegance and beauty did Feynman have in mind? Feynman's notion of elegance came from a mathematical formalism he had been honing since his university days and which had served as a basis for Feynman diagrams. We can conjecture that to Feynman, the beauty of the Feynman–Gell-Mann theory concerns its universality, by which Feynman meant the possibility of its covering every possible particle's involvement in the weak interactions. The link between unification and aesthetics is not surprising, because unification is also one of the guidelines for discovering scientific theories, as we discussed in Chapter 9. In the face of conflicting experimental data, Feynman and Gell-Mann offered these aesthetic notions in favor of their own theory. It turned out that the experiments were wrong (Chapter 3). So good were the theory's foundations that a decade later Salam and Weinberg folded them into their electroweak theory.

As a guide for elegance and beauty in theorizing, Feynman had in mind the British physicist P. A. M. Dirac's discovery in 1928 of the proper equation to describe the electron. Dirac sought an equation whose beauty could be measured by consistency with guidelines set down by the special theory of relativity. An unexpected consequence of Dirac's equation was the stunning prediction in 1931 of antimatter, a prediction that essentially was a consequence of one physicist's notion of a mathematical aesthetic. Dirac's positron, the antiparticle to the electron, was discovered in 1932. This confirmation of Dirac's approach lends support for belief in a mathematically structured universe.

Dirac's prescription for theorizing according to beauty does not always work. Take, for example, Heinrich Hertz who, in his reformulation of Maxwell's electromagnetic theory in 1890, suggested a way to make the theory completely symmetric between electricity and magnetism. Since there is a term in Maxwell's equations for electric charges to act as the source of electric fields, then why not have a magnetic charge as a source for the magnetic field. Hertz proposed adding such a term to predict magnetic charges or monopoles. But Lorentz's and later Poincaré's subsequent analyses of Hertz's theory revealed defects. Some of the problems were traceable to the extra term Hertz suggested, which they deleted. Sometimes asymmetries are only apparent. This was the case with Hertz's theory, because Lorentz's theory, which is a reformulation of Hertz's without magnetic monopoles, is consistent with the principle of relativity, which brings into play symmetries in space-time that turn out to be more important than Hertz's supposed symmetry between electricity and magnetism.*

*Magnetic monopoles are predicted by modern-day grand unified theories. These entities are of an entirely different sort than the ones assumed by Hertz. As yet, none have been found. The astrophysicist Alan Guth replied to this situation by proposing the notion of inflation according to which, in time intervals of the order of 10^{-35} seconds, the very early universe underwent several enormous changes in size. In this way, any monopoles present prior to inflation would become so sparse in comparison to the size of the universe as to be insignificant. Had Guth formulated his hypothesis of inflation only to explain away the lack of evidence for magnetic monopoles, it would have been an ad hoc hypothesis. It does more, however, because inflation explains adequately density inhomogeneities in the universe, among other problems.

In 1924 Einstein discovered permutation *symmetry*, which is based on the indistinguishability of elementary particles: All elementary particles "look" the same, and so these things are not at all the distinguishable billiard balls of Boltzmann and Maxwell.* We can imagine how astonished physicists were at this counterintuitive theoretical result which turned out to be the only way to understand such experimental data as why there are two distinct series of spectral lines for helium.

By 1927 severe representational problems beset the new atomic theory. Scientists needed a new manner in which to represent nature. This need was supposed to have been alleviated by Bohr's Copenhagen interpretation, which relegates any prequantum space-time description to the realm of naive idealizations. Bohr reached this conclusion through reasoning based on the complementarity principle, according to which in the atomic realm there was no longer the customary connection from the progress of an electron in space and time with Newtonian causality. This can be traced to Heisenberg's uncertainty principle, which denies even in-principle accurate measurements in a single experiment of velocity and position, and even deeper to the wave/particle duality of light and matter.

To many physicists, Bohr's complementarity was a breakthrough. It is also a clear-cut instance in which art actually influenced physics in the twentieth century.

CUBISM AND QUANTUM MECHANICS

It has been a puzzle in the history of scientific ideas how Bohr conceived of the complementarity view. I believe that a piece of the mosaic is to be found in his interest in art, particularly Cubist art. In the 1930s Bohr moved into a mansion owned by the Carlsberg Foundation and had carte blanche to furnish it. We might expect, therefore, to see in his

---

*Indistinguishability has startling occurrences in nature. Consider the helium atom, which has two electrons. According to quantum mechanics, if the two electrons were not indistinguishable, then the helium atom would be unstable.

study a painting by one of the acknowledged masters of Cubism, a Braque, a Gris, or a Picasso. Instead Bohr exhibited a Cubist painting by Jean Metzinger. Bohr's choice indicates a special interest in Cubism, and perhaps a clue to yet another path to complementarity—that is, assuming that Bohr knew about Metzinger prior to 1927. Let's assume Bohr had, and so we'll try to find what interested him in Metzinger.

Most art historians consider Metzinger to have been a minor Cubist painter, but everyone agrees that he was a major theorist of Cubism. In 1912, Metzinger and Albert Gleizes published a systematic exposition of Cubist methods in their widely read book *Du Cubisme*. Metzinger, particularly, had been greatly influenced by Maurice Princet, an insurance actuary and member of Picasso's inner circle who gave informal lectures on non-Euclidean geometry. A Cubist painting, write Gleizes and Metzinger, represents a scene as if the observer is "moving around an object [in order to] seize it from several successive appearances. . . . " Cubists achieve this motif through the interpenetration of figure and space in order to free the artist from a single perspective point in favor of multiple viewpoints.

I think this was what impressed Bohr about Cubism. Mogens Andersen, a Danish artist and friend of Bohr, recollected Bohr's pleasure in giving "form to thoughts to an audience at first unable to see anything in [Metzinger's] painting. They came with a preconceived idea of what art should be." Such had been the case in 1913, when atomic physicists had a preconceived visual image of the atom. By 1925 atomic physicists had come to realize the inadequacy of visual perception, as had the Cubists. In 1927 Bohr offered a motif for the world of the atom with striking parallels to the motif of multiple perspectives: According to complementarity, the atomic entity has two sides—wave and particle. Depending on how you look at it, that is, what experimental arrangement is used, that is what it is. While this suffices as a possible explanation of measurements, what about a visual representation of the atomic processes themselves? As we have discussed, Bohr had concluded that this was not possible. Some physicists, particularly Heisenberg, were not pleased with this consequence of complementarity.

Physics awaited the onset of its new mode of representation, a situation in which it sought a Giotto, a Cézanne, and a Picasso, all rolled into one. The breakthrough came in work initiated by Heisenberg in 1932 toward a theory of the nuclear force based on a visualizable metaphor.

PHYSICISTS RE-REPRESENT

At this juncture we can tie together much of what we have said in previous chapters about intuition and visualization under the more general concept of *representation*. Whereas the representation of the atom as a minuscule solar system could not be maintained (see Figure 2(a), to be compared with Chapter 2, Figure 7), the more abstract representation in Figure 2(b) still holds for atomic physics. In 1925 another representation of the material in Figure 2(c) appeared in what Heisenberg and his colleague in Copenhagen Hendrik Kramers referred to as a "term diagram," which is generated from the mathematics of the last throes of Bohr's dying atomic theory.

In Chapter 7 we discussed how Heisenberg's first move toward a theory of nuclear physics contained the seeds of a visual representation of atomic phenomena (see Figure 10 in Chapter 7). Paramount in analyzing this work was the concept of intuition coupled with visual imagery. This required physicists to distinguish between visualization [*Anschauung*] and visualizability [*Anschaulichkeit*]. Visualization pertains to visual imagery that we abstract from phenomena that we have actually witnessed in the world of perceptions. Visualizability is the visual imagery to which we are led by the mathematics of a physical theory. Heisenberg's early results on nuclear physics, pursued by Fermi and Yukawa, culminated in the Feynman diagrams, which made their appearance in 1949 (see Figure 13 of Chapter 7). With hindsight, Heisenberg wrote that the "term diagrams were like Feynman diagrams nowadays" because they were suggested by the mathematics of scattering processes treated within the old Bohr intuition. Figure 2(d) is a Feynman diagram that replaces the term diagram in Figure 2(c).

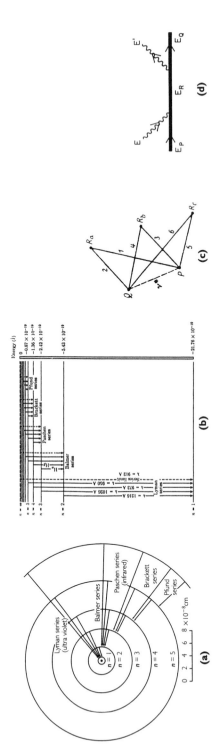

FIGURE 2.

Representations of the atom and its interactions with light. (a) A more detailed version of the hydrogen atom in Bohr's theory as depicted in Figure 7 in Chapter 1. The number n is the principal quantum number and serves to tag the atomic electron's permitted orbits. Lyman, Balmer, etc., are names for the series of spectral lines the atom emits when its electron drops from higher to lower orbits. (b) Another way of representing the Bohr atom. The horizontal lines are energy levels corresponding to permitted orbits, but are more general. The representation in (b) survived the demise of Bohr's atomic theory and remained essential to atomic theory. (c) Another manner of visually representing some of the information in (b). It is taken from a 1925 paper of Hendrik Kramers and Heisenberg, published shortly before Heisenberg formulated the quantum mechanics. Kramers and Heisenberg referred to the diagram in (c) as a "term diagram," in which R_a, R_b, R_c, Q, and P are energy levels in an atom struck by light. The incident light causes the atom to make transitions from a state P to a state Q via intermediate states R. The energy difference between the states P and Q is $h\nu^*$, where the frequency of the incident light is much greater than ν^* in order to promote the atom to its excited states. (d) A Feynman diagram for the processes in (a)–(c), for the case in which they were all caused by the interaction of atoms with light. In (d), $E(E')$ is the energy of the incident (scattered) light, E_P and E_Q are the energies of the atom's initial and final states, and E_R is the energy of possible intermediate states. The atom's trajectory in space-time is taken to be horizontal.

At this point we may ask why, in the late 1930s, no one drew any diagrams for particles being exchanged? After all, a higher plateau in problem solving is attained once visual representations are offered. The literature abounded with suggestive phrases, particularly when some scientists discussed the Coulomb force between two electrons as being mediated by a photon. In fact, Feynman-like diagrams appeared as a didactic aid in Gregor Wentzel's widely read text of 1943, *Introduction to the Quantum Theory of Fields* (Figure 3).

Besides the fact that most physicists were occupied otherwise in 1943, Wentzel's diagrams never caught on because there was no mathematics to back them up. They seemed to be precisely what Wentzel meant them as—didactic aids. Nevertheless, what could have motivated Wentzel to draw such diagrams, pictures that played an essential role in twentieth-century physics? Could it have been the result of "seeing" a deeper structure in Fermi's "intuitive" diagrams from Figure 11 of Chapter 7? Let's imagine that Wentzel's visual thinking proceeded as follows:

1. As Fermi did in 1934, *describe* β-decay as the transformation of a neutron (n) into a proton (p), an electron (e⁻), and neutrino (ν), that is, $n \rightarrow p + e^- + \nu$ (Figure 4(a)).

2. Eliminate the "imaginary" or analogical vertical arrow *depicting* a transition between neutron and proton as energy levels in Figure 4(a).

3. With step 1. in mind, replace the vertical arrow in Figure 4(a) with a horizontal one *depicting* exchange of a μ-meson. The result is the transformation of representation in Figure 4 (b).

FIGURE 3.

Wentzel's "schemata" for depicting β-decay as a "two-stage process" intermediated by a μ-meson. These drawings are taken from his 1943 book, *Introduction to the Quantum Theory of Fields*.

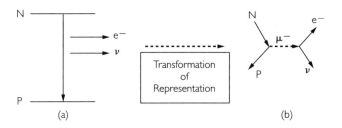

FIGURE 4.

Could it have been Wentzel's "seeing" the deep structure in (a) that led him to draw the "Feynman-like" diagram in (b)?

Support for this conjecture can be found in a paper Wentzel published in 1936, "On the theory of β-transformations and the nuclear force," where we find Wentzel's diagram in Figure 5. Clearly, this figure is a development of Fermi's two-level "intuitive" theory of β-decay.

The importance of visual representations to physicists is underscored by an intriguing event in the history of science. Coincident with Feynman's publication of his diagrammatic method in 1949, another formulation of quantum electrodynamics appeared, the one of Schwinger and Tomonoga. Schwinger's formulation was mathematically elegant and difficult to use, yet it had the aura of rigor that Feynman's did not. When, later in 1949, Freeman Dyson proved the equivalence of Feynman's and Schwinger's formulations, just about every physicist switched to Feynman's visual methods.

As an example of Feynman diagrams, in Chapter 7 we discussed the Coulomb interaction, or repulsion, between two electrons. We juxtaposed the visual imagery in that chapter's Figure 13(a) with the one in Figure 13(b): While the former is *imposed* upon physics, the latter is generated by it. Feynman diagrams are descendants of depictions and methods from Heisenberg's 1932 nuclear theory. They offer a means to transform the concept of naturalistic representation into one offering a glimpse of a world beyond the intuition of Galilean–Newtonian and rel-

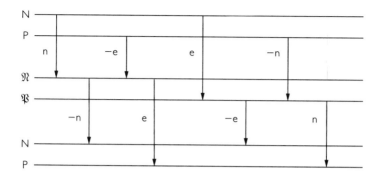

FIGURE 5.

Depicting β-decay as a "cascadelike" process analogous to the transition between energy levels in Fermi's "intuitive" theory of
β-decay. N(P) denotes a neutron (proton), 𝔑(𝔓) a "hypothetical
stationary state" of a neutron (proton), n(−n) a neutrino (antineutrino), and e(−e) an electron (positron). The illustration is taken
from Wentzel's 1936 paper, "On the theory of β-transformations
and the nuclear force."

ativity physics. They offer the proper *visualizability* of atomic physics,
which, alas, we can render only with the usual figure and ground distinction. These diagrams are presently the most abstract way of glimps-
ing an invisible world. In 1950 Heisenberg welcomed Feynman diagrams as the "intuitive [*anschaulich*] contents" of the new theory of
quantum mechanics. Once again, for Heisenberg, theory decided what
is intuitive, or *visualizable*.

FEYNMAN DIAGRAMS, REPRESENTATION, AND GESTALT PSYCHOLOGY

That two forms of quantum electrodynamics emerged is not unusual
from the standpoint of the underdetermination thesis. Neither is it unusual that one of them makes an ontological claim concerning entities
unique to it—namely, the intermediate or "virtual" photon in Figure

13(b) of Chapter 7. In Feynman's representation of quantum electrodynamics, any real particle can also be a virtual one. But virtual particles are not detected in the same way as real particles, because they are involved in processes that violate energy conservation. No doubt you cringed at the last statement, as well you should after all that has been said about the untouchable status of energy conservation. This situation can be understood only within the quantum theory, according to which it is possible for there to be fluctuations from the norm. Quantum theory gives new and radical meaning to the situations we encountered in Chapter 4 which illustrated that anything we can conceive of happening can occur. Situations such as an ink drop suddenly coalescing after spreading out in a liquid are statistically minute. The detailed properties of atoms, molecules, and elementary particles, on the other hand, depend on deviations from the norm, or fluctuations, and chief among them are virtual particles.

How virtual particles appear and disappear can be understood from Heisenberg's uncertainty principle for energy and time. Just like for position and momentum, the product of the errors in any simultaneous determination of energy and time is given by Planck's constant. So, over extremely short time intervals, an elementary particle with a certain amount of energy can be created and then disappear. An example is the "virtual" photon in Figure 13(b) of Chapter 7. This situation is analogous to your borrowing money from the bank and then repaying it so quickly that effectively no one ever notices. The background of space-time seethes with the creation and annihilation of elementary particles. Among the quantum phenomena that are *understandable* in terms of virtual particles is a minute shift in two of the energy levels in the hydrogen atom, referred to as the *Lamb shift*. Either version of quantum electrodynamics matches experimental accuracy, which is precise to about eight decimal places. Feynman's version brings out the physical aspects of the Lamb shift in ways transferable to understanding other phenomena such as nuclear forces and chemical bonding. Furthermore, as we saw in Chapter 7,

Feynman's version of quantum electrodynamics led to the successful electroweak theory. It turns out that formulating the electroweak theory requires certain virtual particles, such as Z^0. On this basis, the theory predicted that these particles occur freely in nature. Subsequently they were discovered. These, and other successful predictions, lead us to "infer to the best explanation" that virtual particles can be accorded a reality status, as is required by Feynman's version of quantum electrodynamics.

At this point it is appropriate to assemble some of the figures already presented of atomic and elementary particle phenomena as a panorama of transformations of representation from the Bohr atom through Feynman diagrams (see Figure 6). We note that each representation based on imagery imposed from the world of perceptions—ordinary intuition—has a perspective point. On the other hand, the visual imagery *generated* by the mathematics of physical theories has none, like the situation in Cubist art.

According to the Gestalt laws of organization, which Chapter 8 discussed and which are illustrated in Figure 15 of that chapter, Figures 2(a) and 6(a) possess the highest degree of symmetry of all the insets in Figures 2 and 6. Yet, the deepest representations are the Feynman diagrams in the pairs 2(d) and 6(j). These figures exhibit good continuation at the vertices, which gives the impression of lines flowing into one another. These Feynman diagrams also happen to possess a high degree of symmetry because there are two particles coming in and two going out. This lends the diagrams the Gestalt property of proximity. Not unreasonably, therefore, physicists have come to think in Feynman diagrams.

THE DEEPER STRUCTURE OF DATA

An important project in the Renaissance was the calculation of ballistics tables. It is one thing to know that cannon balls eventually hit the

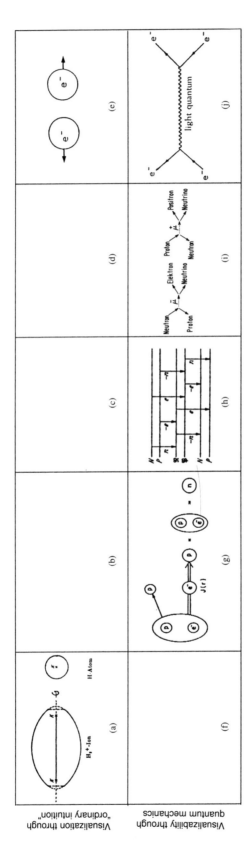

FIGURE 6.

The two rows of frames show how quantum theory distinguishes between visualization and visualizability. Frames (a) through (e) contain visualizations according to pictures constructed from objects and phenomena actually perceived. Frames (f) through (j) depict visualizability according to quantum mechanics. Thus, frame (f) is empty, as are frames (b), (c), and (d). The changing notions of physical reality that began with Heisenberg's remarkable extension of the exchange force from molecular physics to nuclear physics are in frames (g) through (j). Comparison of frames (e) and (j) illustrates the startling contrast between physical reality according to the world of perceptions and the atomic domain.

ground. The question is *where*. The correct answer includes the proper representation of its trajectory. First of all, only relatively recently have trajectories actually been observed with tracer shells. But even watching modern warfare in the comfort of our living room, it is difficult to discern whether a tracer shell's trajectory is circular or not.

Figure 4 of Chapter 1 is an example of an artist attempting to render occurrences beyond appearance. The artist is trying to understand the "deep structure" in a cannon ball's trajectory without ever having actually seen a trajectory. The assumed common-sense task in Aristotelian physics of visually representing the trajectory of an arrow or a cannon ball lasted a millennium before Galileo realized how to do it. This transition is shown in Figures 4 and 5 in Chapter 1. A rendering of a cannon ball's trajectory true to Aristotle's original prescription of a sharp difference between unnatural and natural motions would have no curved turnover. Variations on this theme were proposed by the sixteenth-century scientist Niccolò Tartaglia, who rightly concluded that the cannon ball's trajectory is curved (Figure 4). The problem was that Tartaglia could not calculate the curve.

Abstraction was basic to Galileo's correct solution; in this case, abstraction to a possible world in which there are vacuums, which turned out to be our world. Galileo calculated the cannon ball's trajectory to be a parabolic arc (Figure 5 of Chapter 1). We now "see," or represent, trajectories as parabolic arcs. We can add this to our list of examples where developments in science transform our "common-sense" intuition and visual imagery, too.

When we look at Figures 4 and 5 of Chapter 1 we notice a stunning change. Figure 4 is a fanciful rendering replete with tranquil landscape. There is no distinction between living and dead matter, and the trajectory is imposed on the projectile. By stark contrast, Galileo's drawing in Figure 5 is the sort of line drawing we see in modern physics texts, stripped of all relation to worldly surroundings. It is calculated from a mathematical formalism that deals only with dead matter, and it displays its result on a coordinate axis. Only then could the notion of time make its way explicitly into physics. This is the central difference between Aristotelian and Galilean world views.

Parallel to the way that Galileo's theory of physics leads to the "deep structure" of projectile motion, Feynman diagrams provide the "deep structure" of a world beyond appearances, the world of elementary particles. They offer a representation of nature from available data, for example, bubble chamber photographs.

Consider the famous bubble chamber picture in Figure 7(a) of a major discovery made in 1973 which went far to substantiate the electroweak theory. Just as in the experimental setup of Chapter 2 on double-slit diffraction, nature offers data (bubble chamber photograph) caused by the collision between two elementary particles. Details are inessential for our purpose, which is to argue that Feynman diagrams offer a glimpse into the deeper structure of Figure 7(a). The schematic in Figure 7(b) better identifies the event in the bubble chamber photograph in Figure 7(a), which is the scattering of two elementary particles: a muon antineutrino from an electron.*

Because of the complexity of the experimental setup, the bubble chamber photograph in Figure 7(a) is many layers removed from the "raw" primordial process of two elementary particles colliding. Taking as a guide the Feynman diagrams of how electrons interact by exchanging a light quantum (see Figure 13 of Chapter 7), Weinberg and Salam formulated the electroweak theory, which predicted the event in Figure 7(a).

Since the electroweak theory explains so many facets of the weak and electromagnetic processes, we infer to the best explanation that interactions between photons, neutrinos, and electrons can all be explained as a result of their properties being connected by the electroweak theory, which is predicated on certain symmetry groups and relations among coupling constants (recall Figure 14 of Chapter 7). In-

---

*Although data in Figure 7(a) are necessarily theory laden, the theories involved are basic quantum mechanics, electromagnetic theory, and the engineering science involved in construction of particle accelerators. So, data analysis does not concern the theory to be tested, thereby avoiding circularity.

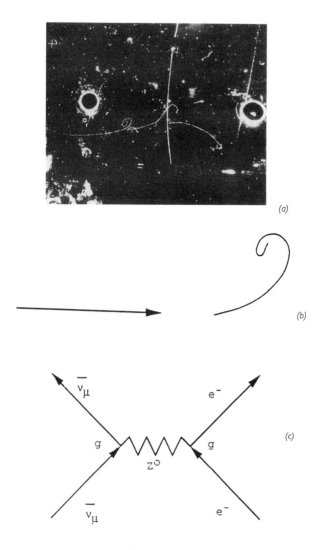

FIGURE 7.

(a). The first bubble chamber photograph of the scattering of a muon anti-neutrino ($\overline{\nu}_\mu$) from an electron (e^-). (b). We start to extract information from Figure 7(a) with Figure 7(b) which indicates that the muon anti-neutrino entered the bubble chamber from the left and struck an electron which moved in a trajectory that was curved by an externally imposed magnetic field. (c) The "deep structure" in Figures 7(a) and 7(b) according to the electroweak theory. Instead of two electrons interacting by exchanging a light quantum (Figure 8(b)), according to the electroweak theory an anti-neutrino ($\overline{\nu}_\mu$) and an electron (e^-) interact by exchanging a Z^0 particle, where g is the charge (coupling constant) for the electroweak force.

ferring to the best explanation is the process of reasoning we introduced in Chapter 2 and is a mainstay of scientific realism. This, in turn, entails the conclusion that the Feynman diagram in Figure 7(c) is a glimpse into the deep structure of the *real* world of particle physics.

In the original formulation of the electroweak theory, empirical data played an even lesser role than in special relativity. For Weinberg and Salam, aesthetics, representation, and the assumption of unification of physical theories were paramount. Woven into these guidelines for creativity was the proper visual imagery for the electroweak theory. As Chapter 7 explored, Weinberg and Salam assumed that unification of the electromagnetic and weak forces could be accomplished by a depictive analogy with quantum electrodynamics (Chapter 7's Figure 14).

This is another instance where visual representations are crucial for scientific discovery and understanding physical reality, in addition to their usefulness for calculational purposes. We have noted this already for the case of two electrons interacting with one another. These two situations are summarized in Figure 8. The visual representation of two electrons repelling one another in Figure 8(a) and of the Bohr atom in Figure 2(a) are pleasing to the eye not only because they are abstracted from phenomena that we actually witnessed, but also because they are set as self contained objects as "figure" set in a qualitatively different space, that serves as "ground."

Visual representations have become transformed by discoveries in science and, in turn, have transformed scientific theories. They offer a glimpse of an invisible world in which entities are simultaneously wave and particle, and so cannot even be imagined. Entities in this domain are desubstantialized, as we have come to understand this concept.

Certain abstract artists make the same claim for their paintings. Like Pollock, Mark Rothko did not take his themes from objects of our world. Instead a mixture of painting from feelings and emotions, in conjunction with a native Russian mysticism, led Rothko to write that he meant the mix of cool and warm colors in his later paintings to draw

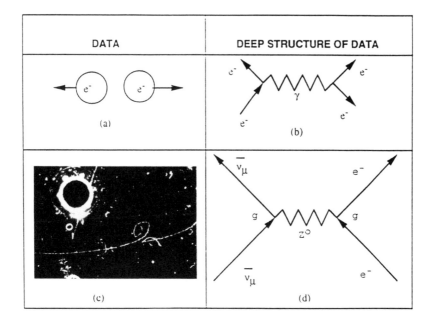

FIGURE 8.

Figures (a) and (c) exhibit data. (a) refers to the situation where two electrons are depicted as two like-charged macroscopic spheres that move apart because like charges repel. The arrows indicate their receding from one another. (b) provides a glimpse into this process. (c) The deep structure in the bubble chamber photograph from Figure (c) is given by the Feynman diagram in (d).

the observer into the "interior . . . illusory space" of the canvas to provide an "ideal" rather than an "achieved state," where intuition somehow bridges the gap. Rothko tried to create a world, an ambiance, approximating the unconscious. He sought to accomplish this with huge canvases displaying subtly vibrating large strips of colors, one flowing into the other:

> To paint a small picture is to place yourself outside experience. . . . However you paint the larger picture, you are in

it. . . . A large picture is an immediate transaction: it takes you into it.

In Rothko and the Abstract Expressionists, any vestiges of perceptual representations vanished. That the images of Abstract Expressionism have to be rendered as figure and ground is the only remaining reflection of our perceptual system. This situation in physics occurred almost at the same time as the discovery of Feynman diagrams. Science came a long way from the solar system image of the atom to an understanding of nature as it is *beyond* visual appearances. With little understatement we can say that the representation in a Feynman diagram is an advance in visual imagery akin to a jump from the art of Giotto's predecessors to Rothko. This fascinating turn of events completes our exploration of how visual representations increased in abstraction, particularly since the late nineteenth and early twentieth centuries.

We next focus on how a similar high degree of abstraction emerged in painting during the same time frame, and we will look at its relation to scientific developments. In this way we will be able to further explore the concepts of aesthetics, representation, and progress in art and science and so understand better why the older styles of painting were no longer good enough for artists such as Picasso.

ART IN THE TWENTIETH CENTURY

At the turn of the century, Paris had become a melting pot for artists. The Paris mystique emanates from its continuous physical transformation for over a thousand years, radiating outwards from the Ile de la Cité, and accomplished with exquisite architecture and conscious cultivation of the arts and sciences. The changes of weather and light over the Seine, with its ornate bridges, are breathtaking. Except for the decade or so of turmoil during the French Revolution, France's intellectual infrastructure remained relatively stable compared with those elsewhere in Europe. In this way Paris maintained its status as the locus of intellectual activity in France, and by consensus the world. The social, political, and economic changes taking place elsewhere in Europe to some extent seethed below the surface in France: Class structure was dissolving as reflected in the optimism of Impressionist art, and France was just recovering from the anarchy of 1890 and after that the near revolution caused by the Dreyfus affair. All of these topics were ripe for discussion in cafés on the Left Bank.

Avant-garde was the term loosely applied to the various art movements at the beginning of the twentieth century. It was taken to be synonymous with "experimental." The notion of experimentation was intentionally used because artists were being influenced, or at least inspired, by startling new scientific and technological developments such as x-ray imaging, which seemed to eliminate any difference between inside and outside; cinematography, with its multiple-frame exposures; new theories of color; and radical changes in space, time, and speed with the introduction of telephones, automobiles, and airplanes. Young French artists were discovering primitivism through studying African and Polynesian art. If you didn't hear about these developments at your local café, then you probably did through the newspaper. Ideas were everywhere and so was the desire for change. First artists and scientists, and then musicians and writers, became the jewels of the Third Republic formed in 1871 which, by the onset of the twentieth century, was a fundamentally stable and vital society with its own stamp of character.

Like the scientists, who walked head high and optimistic into the twentieth century, so did the artists. Scientists had overcome the *fin de siècle* mood with such new discoveries as radioactivity, x-rays, the electron, the noble gases, the promise of a theory of everything in Lorentz's theory of the electron, and Ernst Mach's positive message. Artists crashed headlong into the new century, having discovered the primacy of light and color over objects as achieved with dazzling success in Im-

pressionism, Post-Impressionism, and Pointillism, to name but three of
the current art movements. Artists even had their own philosopher,
Henri Bergson, with his vague lyrical rhetoric about a mysterious sort of
intuition that permits transference of the mind into the very interior of
things in order to be at one with their uniqueness. Bergson invited
everyone to journey beyond objectivism and boredom into the enigmatic realm of *élan vital*.

Change was in the air. The twentieth century was beginning to
evolve its trademark of increased abstraction. Coincidentally with the
avant-garde art movements, *the* trademark of the twentieth century
emerged in Bern, Switzerland, where a 26-year-old patent clerk just realized a result he had overlooked in his recent publication on space and
time: $E = mc^2$.

PICASSO, GEOMETRY, AND CUBISM: RELATIONS BETWEEN DEVELOPMENTS IN CUBISM AND SCIENCE

In 1907 the intensity of intellectual activity in Paris was riveting. In cafés
on the Left Bank, past, present, and future artistic and intellectual styles
were debated heatedly. The atmosphere was highly charged. The lightning stroke was a Spanish artist at the vortex of all currents, Pablo Picasso.
Over his almost 92-year-life span, this most prolific of artists constantly
changed styles, often inventing his own. Such was the mystique of what
seemed to be Picasso's apparently inexhaustible and supernatural creative
energy that on April 8, 1973, just six months before his 92nd birthday, he
managed to do something no one ever expected of him. He died.

Among the interplay of artistic and social factors that affected the
26-year-old Picasso in 1907 were Cézanne's play with forms, Edouard
Manet, and primitive African and Polynesian art, to which Henri Matisse
introduced him. From an interview with Picasso sometime between 1910
and 1911, the poet Guillaume Apollinaire quotes him as saying that see-

ing African art was his "greatest artistic revelation." According to the art historian and artist John Golding, what struck Picasso was that

> tribal art is more conceptual, much less conditioned by visual appearances [than is Western art]. The tribal sculptor tends to depict what he knows about his subject rather than what he sees. Or, to put it differently, he tends to express his *idea* of it.

Consequently, African pieces are often characterized by their great simplification with shortened limbs, enlarged navel with stomach enlarged and protruding, and angular features with simple forms for the trunk and head. To add to these inputs, we have Picasso as a master practitioner of so many contemporaneous art movements.

In his early days in Paris, Picasso mimicked all of the popular artists, producing canvases that played on their styles like verbal puns. In the first volume of his magnificent biography of Picasso, John Richardson has a chapter aptly entitled "Plundering the past," in which he emphasizes another influence on Picasso, namely, Manet. Picasso considered Manet to be the first modern artist. Picasso discovered Manet's paintings at the famous *Salon d'Automne* of 1905, before falling under the spell of Cézanne in 1906. Richardson discusses how Picasso was entranced by the "nonchalance of [Manet's] compositions and the way his figures stare back at the beholder." Incidentally, the distinctive "stare" in many of Manet's portraits can be correlated with photographs of the subjects. These techniques of Manet would surface two years later in Picasso's *Demoiselles*. Other sources of great help to Picasso were El Greco, of course, as well as Nicholas Poussin and Paul Gauguin, whose style never really fit in anywhere. Just as Bach "borrowed" from so much of Italian Baroque music, Picasso did the same with just about every artist. This is not meant to be derogatory, but rather to emphasize the continuity and complexity of the course of art history, just as in science history. Just as Bach surpassed everyone composing in the Italian Baroque style, so did Picasso with all artists that were his contemporaries and many of those from the past he had admired.

We must not forget, as John Golding reminds us, the influence of "social preoccupations [and] by literature." Most likely a good deal of this input came via Picasso's close friendship, beginning in 1904, with the flamboyant poet Apollinaire.* "Guillaume Apollinaire" is the pseudonym taken by the Pole Wilhelm de Kostowitzky. All of this input combined to influence Picasso's *Les Demoiselles d'Avignon* in 1907. Shortly after Apollinaire introduced Picasso to Braque, *Demoiselles* cleared the way for Cubism, which emerged fully in 1909 in the hands of Picasso and Braque.† As in science, art does not progress by revolutions. Progress in art is also gradual, with other interesting similarities and differences, some of which this chapter will explore.

If one wishes to impose on art history a critical event, then the *Demoiselles* is certainly a candidate. It is "simultaneously representational and anti-naturalistic," as Golding aptly describes it. The *Demoiselles* is the borderline between old and new. It is strikingly transitional in its sweep from the masklike Egyptian face of the demoiselle on the far left to the savagely distorted and primitive faces on the far right. It is an ingenious folding together of Paul Cézanne's geometrical forms and the conceptual quality of primitive art. It is Picasso's own style, in which he could paint as he *thought* things are. Like Cézanne, Picasso presents different viewpoints fused together on a single canvas and pushes all volumes up against the picture plane. This is the effect in, for example, Cézanne's *The Kitchen Table*, 1888–1890 (Figure 9), as well as in the transitions in his renderings of Mont Sainte-Victoire starting from 1885 to 1887, when he was still in his Impressionist period, through to 1904 when he painted it from his studio in Les Lauves. Cézanne carefully planned his conception of multiple perspectives on a single canvas. In

---

*John Richardson observed, "Poets exerted an especially formative influence [on Picasso]." Just as Max Jacob was Picasso's poet in residence during his earliest years in Paris, Apollinaire assumed that role with Cubism, having studied and practiced the Impressionist and Post-Impressionist styles.

†The name Cubism was coined in 1908 by Henri Matisse, in response to the geometrical aspect of some of Braque's paintings.

FIGURE 9.
Paul Cézanne's *The Kitchen Table*, 1888–1890.

Figure 10 is Erle Loran's famous diagram demonstrating how different parts of Cézanne's *The Kitchen Table* are in correct perspective when viewed from different positions.

But unlike Cézanne, Picasso's play with sharply defined geometrical forms is entirely original. Moving from left to right viewing Figure 1, one sees that the demoiselles' faces become increasingly distorted, with noses in profile even though they are presented full-faced. The transition is completed in the faces of the two demoiselles on the far right, which are geometrical caricatures replete with eyes off level. Linear perspective is gone, and form has been reduced to geometry. Cubism was an intensely intellectual adventure.

According to John Richardson, Picasso referred to *Demoiselles* as his "exorcism." Picasso considered it to be a rite of passage that opened the

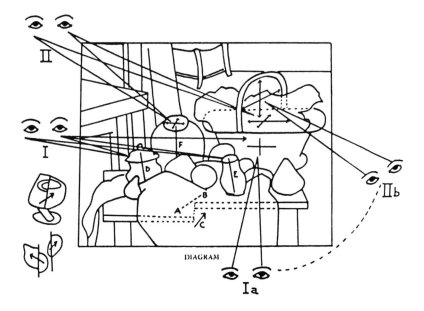

DIAGRAM

FIGURE 10.

Erle Loran's analysis of Cézanne's painting in Figure 9. Loran shows how the correct perspective is obtained for different positions in Figure 9. (From E. Loran, *Cézanne's Composition*, Univ. of California Press, 1943.)

floodgates of his artistic creativity. Never one to be modest, Picasso recalled that over the next decade he felt "as overworked" as God.

What Mozart was to Bach, Picasso was to Cézanne. As John Golding emphasizes, it was only in the "years following *Demoiselles* that Cézanne's reputation as the greatest and most influential figure in nineteenth-century art was firmly laid." Observations such as Golding's alert us to take *cum grano saltus* any historical scenario built on figures and/or events taken as revolutionary. Golding warns us not to use Cézanne as a means for oversimplifying the history of Cubism. Cézanne was no revolutionary. Just as Schrödinger tried to revive classical concepts in physics by suitably redefining them within the new quantum theory, Cézanne devoted himself to reviving, in Golding's words, "classical or traditional painting;

the means he used, although they were highly personal, were founded on the optical discoveries of the Impressionists."

On the other hand, the true Cubists, continues Golding, were consciously aware "of the fact they were the beginning of something completely new." True Cubists imagined Cézanne as bridging the old and the new: Cézanne's Impressionist colors and play with light defined figures that hearken back to classical art, suitably redefined. The early Cubism of Picasso and Braque, although figurative, radically geometrized Cézanne's world, while stripping it of color. Picasso's and Braque's complete break with Cézanne would not come until 1910, when a new phase of Cubism evolved which was nonfigurative. This situation is somewhat comparative to the complete break in about 1926 between quantum mechanics and Bohr's atomic theory. Max Born found the means to remove the last traces of Bohr's theory by reformulating the mathematics in Heisenberg's ingenious but clumsily written paper of June of 1925. In the case of Cubism, this break with the past came about as a result of Picasso's and Braque's exploring the nature of space.

PICASSO, BRAQUE, AND SPACE

The concerns of Picasso and Braque were similar to problems of interest to their contemporary in Bern, Albert Einstein. The three men were concerned with properties of space and how to represent it. All of them were concerned with how different observers represent space, and all of them were affected by the technology of their era, in Einstein's case, for example, how electrical dynamos worked. How did they "publish" their results? Einstein's relativity paper of 1905 is a jewel, perfect in its science and presentation. On the other hand, there are seminal scientific papers like Heisenberg's first one on quantum mechanics in June of 1925. Although clumsily executed, Heisenberg's paper contains exquisite ideas to be elaborated on and perfected, both by its author and by others more patient and poetic. Max Born played this role for Heisenberg. The *Demoi-selles* is closer to Heisenberg's landmark paper than Einstein's, with Braque playing the role of Born. Standing before the *Demoiselles,* one feels its pulsations and realizes its inevitability as a great yet flawed painting.

Among the problems the *Demoiselles* posed was the violent dislocations in the way that the figures of the five women are fractured and ruptured. There is a discordance, an internal inconsistency among its parts. What had interested Picasso in Cézanne was his notion that all forms in nature can be reduced to cylinders, cubes, spheres, and cones. But this is nothing new. Consult any elementary book on drawing then and now. What *is* unique is that Cézanne distorted these volumes in such a way as to maintain their integrity. He accomplished this by pushing them up against the picture plane. Picasso, on the other hand, pushed too hard and violently in the *Demoiselles,* which Braque considered to be a problem. Braque was horrified upon first seeing it. Picasso's geometrization squashed space and forms in ways that seemed to Braque much too unnatural. For Braque, space was literally an obsession. He believed that there is a quality in space that can be touched intuitively, which he called "tactile space." What drew him into *Demoiselles* was the possibility of materializing "this new [tactile] space which I sensed."

To clarify what Braque meant by a "tactile space" that could be touched "intuitively," we turn to John Golding. Braque had always been intrigued by the effect of three-dimensionality on a two-dimensional canvas. He seemed to feel the distance between objects on a canvas and even imagined himself moving around them. He wanted to paint these tactile spaces that he sensed intuitively. Braque was quoted as saying that tactile space fascinated him "so much, because that is what early Cubist painting was, a research into space."

In Picasso's squashed complex figures, Braque sensed a misuse of both visual and tactile space. Toward resolving this problem, Braque returned to the source, namely, to some of Cézanne's haunts in Provence. Taking a clue from the *Demoiselles*—with which he had come to terms— Braque simplified Cézanne's forms. He found that when pushing them against the picture plane, it was necessary for areas of "empty space" (negative space or tactile space) to be on equal footing with the painting's subject (positive space). The reason is that all "empty" or illusionistic space is squeezed out, all forms having pushed against the picture frame. Braque considered this process to be the "governing principle of Cubism [which] was the materialization of this new [tactile] space which I sensed."

Through one of the most intense collaborations in the history of art, Braque and Picasso developed Cubism between 1907 and 1909.* Their Cubist aesthetic begins with a naturalistic image, which is abstracted geometrically with multiple perspective. Background or negative spaces are treated on a par with the positive space of the subject. This permits Picasso's geometrized forms to achieve their places without violence. The multiperspective character of Cubist art satisfied Braque and Picasso, who rejected the traditional linear perspective because "it is mechanical," as Braque put it, and so cannot capture objects completely. Like scientists, Braque and Picasso continued to experiment with space, which led them during 1910 to 1912 to invent collage in order to enrich the two-dimensionality of picture space.†

An excellent example of the original Cubism of Braque and Picasso is Braque's *Violin and Palette*, 1910 (Figure 11). Braque gives us the feeling that he has actually explored each of the objects in it from several different viewpoints, while coming to grips with tactile space. Objects and space are fused together into a spatial continuum built up from interpenetrating planes, that seem almost as if they can be slid in and out. As Golding writes of Braque's paintings from this period, they give the sensation of "almost unprecedented complexity."

---

*Picasso recalled, "*C'était comme si nous étions mariés*" ["It was like we were married."].

†Undoubtedly unintended by Braque and Picasso, Cubism played a key role in World War I and in the military thereafter. The French army officer credited with the invention of camouflage, Guirand de Scévola, was quoted as saying: "In order to totally deform objects, I employed the means Cubists used to represent them." A variation on splashing structures and gun emplacements with earth colors was devised for ships: painting the sides of ships with geometric patterns in varying colors to confuse observers as to the ship's size and direction of travel.

FIGURE 11.
Georges Braque, *Violin and Palette*, 1910

CUBISM AND RELATIVITY

In 1907 Picasso stood at the focal point of artistic, literary, and social currents of Paris. What about science? The art historian Linda Dalrymple Henderson has argued in her book *The Fourth Dimension and Non-Euclidean Geometry in Modern Art* that advances in science had little effect, if any, on Picasso in 1907, nor during the invention of Cubism with Braque during 1907 through 1909. She writes:

> The sources of Cubism are to be found within art itself, primarily in African sculpture and the paintings of Cézanne.

Can Henderson's conjecture be maintained, despite her meticulous research? Some scientific input into Picasso's circle, the *bande à Picasso*, occurred in 1904 when Maurice Princet, an insurance actuary, became a member. We know that he gave informal lectures on non-Euclidean geometries based on Henri Poincaré's best-seller *Science and Hypothesis*. As we discussed in earlier chapters, Poincaré's writings were immensely popular and to artists in France he was particularly well received owing to his emphasis on the intuitive aspects of thought. On the basis of archival evidence, however, Picasso's biographer John Richardson concludes that "Picasso and Braque emphatically denied taking any interest in Princet's ideas."* Yet certain of Braque's terminology such as "visual space" and "tactile space" are in the chapter in Poincaré's *Science and Hypothesis* that discusses the origins of geometry. What are we to make of this? I can only say that we cannot discount the *Zeitgeist*. Change was in the air. There were any number of ways in which Picasso could have heard by 1907 about, for example, Poincaré's writings from sources other than the somewhat comical Princet. And we must not forget stunning advances in technology such as x-rays and cinematography, which we discussed earlier. Although, as far as we know, there is no supporting archival evidence, I find myself disagreeing with Henderson's finding: The roots of Cubism are not totally within art. Rather there is a network of input from many different fields of study which include science and technology. That science and technology affected how Cubism developed is indisputable.

---

*Princet seems to have been considered by the *bande à Picasso* rather as a disagreeable though somewhat comical figure. Gertrude Stein recalled that in 1904 Picasso had an affair with Princet's mistress, who reveled in being notoriously unfaithful, much preferring lusty young Spanish painters, particularly Picasso. Picasso did permit Princet to augment his salary as an actuary through selling some of Picasso's very early paintings.

FIGURE 12.
Marcel Duchamp, *Nude Descending a Staircase,* 1912.

Cubist art represents subject matter simultaneously from different perspectives. Relativistic to be sure. In 1912 Apollinaire referred to this intrinsic property of Cubism as exhibiting a "fourth dimension." The three spatial dimensions give depth and perspective, while the fourth is motion in time. Sometimes, however, certain Cubist artists took the fourth dimension to be an additional and intuitive fourth spatial dimension that lends itself to the viewers' feeling as if they can walk around the painting in order to view it from all sides. This is akin to Braque's tactile space.

We know that Metzinger and Duchamp were somewhat familiar

422

FIGURE 13.

E. J. Marey, *Man Walking,* ca. 1882.

with relativity theory and that it influenced their practice of Cubism, as in Duchamp's *Nude Descending a Staircase,* 1912 (Figure 12). A contemporary critic described it as "an explosion in a shingle factory." No nude figure is evident. But what we do see is motion. Change of position with time is portrayed on a single canvas. There is an unmistakable influence of Etienne-Jules Marey, the French physiologist and inventor who was a pioneer in motion picture photography. In 1882 he produced the first reel of motion pictures shot from a single camera (Figure 13). Marey's motion pictures were considered as nothing short of spectacular and had an enormous effect on artists.

FIGURATIVE TO NONFIGURATIVE, OBJECTS DISAPPEAR

The fact that no clear-cut nude is recognizable on Duchamp's canvas signals that something else has occurred in art, namely, trends away from any connection with living forms or designs from nature as we perceive it, whatsoever: a true abstract art. After all, we can always pick out parts of bodies or whatnot in Picasso's Cubist renderings, although there are exceptions such as his 1910–1911 *Female Nude.* The year 1910 is important because in that year the flamboyant Russian artist Wassily Kandinsky produced the first nonfigurative painting, entitled *Improvisation.*

Duchamp and Kandinsky are sometimes set into an offshoot of Cubism called "Orphism," replete with its own poet and philosopher.* The ubiquitous Apollinaire sang the praises of this group of Cubists trying to break away from the figuration and dull colors of Cubism as it had developed between 1907 and 1909. Form and color replaced recognizable objects. The philosophical flag flown was—you guessed it— Bergson's. In addition to all of this, Kandinsky was steeped in Russian occult teachings, advocating a music of colors, lines, and forms created not from this world but "in a mysterious and secret way," as he writes in *Concerning the Spiritual in Art,* published in 1912.

The goal of Orphism was to replace recognizable forms with a fundamental dynamics of line and color. They claimed inspiration from developments in atomic physics and, of course, relativity, with its new concepts of space and time, particularly the equivalence between mass and energy. The mass–energy equivalence represented to them the complete deobjectification of the physical world. Independently, this abstraction appeared in Kasimir Malevich's *Suprematism,* in which the painter's vocabulary is limited to the rectangle, circle, triangle, and cross. Objects have disappeared.

The kineticism of Duchamp's *Nude Descending the Staircase* inspired a group of Italian artists and writers to advocate a radical art movement in celebration of modern technology, with emphasis on the violence and beauty of speed. In a manifesto-like statement, they advocated glorifica-

---

*The nonfigurative theme of Orphism has parallels with German Expressionism with roots in Van Gogh, the Belgian James Ensor, and the Norwegian Edvard Munch. It turns out that Van Gogh's style of painting was more heartily taken up in Germany than in his adopted country, France. With its extreme mood swings of politics, historical fragmentation owing to German federalism, and its attraction to mysterious philosophies such as Nature Philosophy, it is not surprising that in turn-of-the-century Germany, the mood was one of an antinaturalistic subjectivism. Munch's summation of his most well known work, "I hear the scream in nature," sums up German Expressionism.

FIGURE 14.
Luigi Russolo, *Dynamism of an Automobile*, 1913.

tion of war to cleanse the present and future of the past in order to "glorify the life of today, unceasingly and violently transformed by victorious science." They called their movement Futurism. After World War I, Futurism collapsed under the weight and inappropriateness of its rhetoric. Yet some of its art and sculpture are stunning in conception. Luigi Russolo's *Dynamism of an Automobile*, 1913 (Figure 14), for example, anticipates pictures in wind tunnels. Any resemblance between Futurist paintings and Marey's multiple exposures was intentional.*

*The horrific carnage of World War I led to realist movements that took center stage away from pre-war nonfigurative or Expressionist genres. Among them were Dada and Surrealism. The thrust of Dada, which made its appearance in 1917, was that if society had lost its rationalism then artists too should forsake any attempts to find hidden meaning and order. Surrealism tried to fill the gap when Dada collapsed under the weight of literary obscurantism. Surrealism's message is one of despair, of the rationality that comes with nightmare. The melting watches and decaying life forms in Dali's *The Persistence of Memory*, 1931, is the paradigmatic example of Surrealism. Like Futurism, Dadaism and Surrealism were, at their core, literary movements.

ART THEORY AND SCIENCE THEORY

The aesthetic of the original Cubism of Braque and Picasso was to break down objects into geometrical forms. Despite the "geometrization" of nature in this type of art, objects can still be discerned. Around 1911, Braque and Picasso began to reverse this procedure, with collage, a technique invented by Picasso and applied ingeniously by Braque. In collage fragments of material are pasted onto the canvas. Besides the shock value of pieces of chair glued on a canvas, Picasso and Braque considered collage another means for probing space by lifting it out of the canvas itself. This new bottom-up phase, with its often colorful collages, satisfied the urge of many artists to break away from the original Cubism with its bland colors, high geometrization of figures, and apparent loss of emotive expression.

Among them was the Spanish painter Juan Gris, who made the useful distinction between what he called an analytical Cubism and a synthetic Cubism. Roughly speaking, analytical Cubism is the top-down analysis of nature, and synthetic is the reverse. Analytic Cubism was the original Cubism of Braque and Picasso in which figures are geometrized.

In 1913 Gris turned against synthetic Cubism, preferring a nonfigurative end result. The version of synthetic Cubism he preferred was one in which the painter begins and ends with geometry. Gris's approach was highly intellectual, as he described his work in 1921:

> I work with elements of the intellect, with the imagination. . . .
> Cézanne turns a bottle into a cylinder, but I begin with a
> cylinder and create an individual of a special type: I make a
> bottle—a particular bottle—out of a cylinder.

Seeking purity in forms and colors, Gris proposed a synthetic Cubism that crossed the line from figurative to nonfigurative art. Gris's writings exhibit his serious reading of the works of Einstein and Poincaré during the war. He explores a point of view that in art there are invariant laws of nature that his *"system"* can capture. With this system, which he never brought to completion, Gris hoped to find a way to achieve harmonious Cubist compositions in purely abstract, that is, geometrical terms. Although Gris never achieved a high degree of abstraction in own art as did Braque and Picasso, his eloquent essays suggesting such an exotic program affected Mondrian, among others such as Kazimir Malevich, who took it to its extreme: Art should not refer at all to objects of the visible world. Instead, objects should be deobjectified by reducing them to their bare component lines.

Piet Mondrian lived in Paris from 1910 to 1938 and so moved in the company of Picasso, Gris, and other Cubists. He sought a Platonic idealism in contrast to Cubism's figurativeness. This desire has roots in Mondrian's austere Calvinistic upbringing and adherence to the mystical purity of Theosophy. In 1917 he was one of the founding members of a Dutch offshoot *De Stijl* [The Style] of the abstract Cubism advocated by Gris. The combination of his well-written profound ideas and what is often described as saintly presence resulted in Mondrian's great influence on subsequent abstract movements, architectural style, and typography.

This deobjectification of nature was occurring simultaneously in atomic physics, where in 1923 the visual representation of the atom as a solar system was rejected in favor of a mathematical representation. In this sense Mondrian's art is particularly interesting keeping in mind parallel developments in atomic physics.

Like Gris, Mondrian believed that the artist's quest is to discover invariant laws of nature, as he wrote in 1937: For there are "made" laws, "discovered" laws, but also laws—a truth for all time." Art and science should proceed hand in hand toward unraveling these laws. For Mondrian the key to it all is investigating the properties of vertical and horizontal lines using pure colors. He deems these structures as basic to a dynamic equilibrium that the artist strives for rather than a static one. This means "painting can become much more real, much less subjective, much more objective," he writes in 1943. Mondrian believed he had accomplished this in such works as *Composition of Lines*, 1917 (Figure 15).

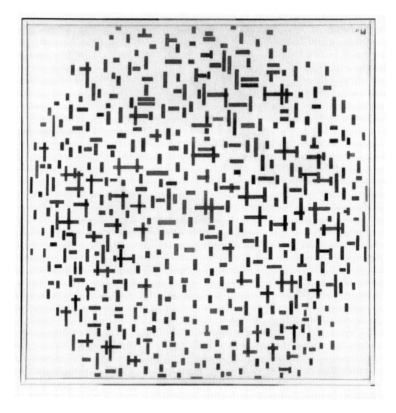

FIGURE 15.
Piet Mondrian, *Composition of Lines,* 1917.

The emergence of Heisenberg's nuclear physics in 1932, with its influence on diagrammatic representations such as Fermi's in 1934 and then Feynman's in 1949, parallels the deobjectification of nature much in the way Mondrian and other artists sought. According to Feynman diagrams, processes in the atomic world are represented with lines, as in a Mondrian painting, but with many of them oblique.

Mondrian's restriction to the horizontal and the vertical is required by his aesthetic, grounded in the definition of dynamic equilibrium, which is close to the one in Gestalt psychology. His art echoes as well findings of neurophysiologists such as Hubel and Wiesel who discovered that neurons in the primary visual cortex respond only to straight lines at particular orientations, as we discussed in Chapter 8.

This is the essence of early vision in which properties of objects are analyzed from a primitive sketch comprised of straight lines.

What is indisputably essential here is that, either consciously or unconsciously, many artists composed their works along guidelines of Gestalt psychology. Just as asymmetry can enhance a work of art or science, paintings rendered contrary to Gestalt principles of good organization should also be appealing. This was the case for Picasso's *Three Musicians*, as we noted in Chapter 8 (Figure 16).

This is reasonable, because the Gestalt principles of organization comprise essentially an aesthetic. There is a rather interesting circle of reasoning here. Extracting information from the neural image depends on analyzing straight lines, which is accomplished with the aid of innate Gestalt principles. Consequently, it should come as no surprise that at a time when artists and psychologists just happen to be focusing on straight lines, they reached similar conclusions. The more intellectual artists such as Paul Klee and Mondrian subsequently wrote at some length on "good" forms, but this was after their own studies in Gestalt psychology.

Thus far in this chapter we have compared and contrasted what artists and scientists in the late nineteenth and early twentieth centuries meant by aesthetics and representation of nature. Taking this into account, we will now widen our discussion of creativity in Chapter 9 to include science and art.

CREATIVITY IN ART AND SCIENCE

Joachim Gasquet, a young poet with whom Cézanne spoke at some length in the years 1896 to 1900, recalled Cézanne's saying the following:

> Painting is damned difficult. . . . You always think you've got it, but you haven't. . . . God knows how the old masters go through those acres of work. . . . As for me, I exhaust myself, work myself to death trying to cover fifty centimeters of canvas. . . . No matter . . . that's life. . . . I want to die painting.

Who can better sum up the torture and singlemindedness of the creative struggle in art and science than Picasso, who is to twentieth-century art what Einstein is to twentieth-century science: "The important thing is to create. Nothing else matters; creation is all." That's all well and good, but as we have seen, the creativity involved in a *Demoiselles d'Avignon* or in three physics papers every eight weeks in 1905 is many-faceted. The prepared mind is of the essence.

Behind these quotes, which one can multiply ad infinitum, there lies a very basic difference between creativity in art and creativity in science. Artists bare their hearts and souls to the world in a product that is intensely personal, while modern scientists are constrained to hide their hopes, dreams, aspirations, and angst in their personal correspondence and unpublished manuscripts. Artists and scientists alike put their reputations on the line each time they place one of their products in the public domain. But when we read Einstein's 1905 relativity paper, we learn nothing at all about his domestic problems or state of mind. In contrast, Vincent Van Gogh's *Starry Night* is almost autobiographical. It has an air about it of instant creation, of an orgasm of tortured beauty transferred to canvas in one passionate instant in which Van Gogh portrayed an almost nonfigurative world crying out with incredible loneliness.

On the other hand, like most works of art, Picasso's *Demoiselles* is closer to a scientific paper. The canvas is calculated: Picasso did not set it all down in one burst. Nor did Cézanne with his carefully calculated and executed multiperspective paintings. Like in a scientific paper, the artist invites the viewer to interpret the piece at hand. Scientific papers and works of art are on a par at that level. Yet a scientific paper has a limited set of acceptable interpretations, while works of art can have many.

Artists and scientists work according to procedures or rule-based systems. In art the formal system can be articulated in a descriptive manner, or in a visual language. In science the rule-based system is comprised of mathematics and certain physical principles assumed to be inviolate and notions of aesthetics, as Chapter 9 discussed under guidelines for creativity. Thought experiments spring from this hard-won structure.

The question is: How can new knowledge be created from already existing knowledge? How can the conclusion go beyond the premises? In part, the reply to this age-old problem is that some artist or scientist realizes a new aesthetic, which he or she assumes provides a better representation of nature. For Picasso it was play with geometrical figures representing faces and bodies all askew. This vision led to the Cubist aesthetic. Einstein realized thought experiments that revealed asymmetries, which he removed by proposing a particle nature of light in one instance, and a new approach to space and time in another. Einstein's style of creativity is one of extreme minimalism—preference for a *single* representation. Poincaré combined methods of geometry and group theory with the theory of differential equations in startling new ways. Essential to both of them were their own concepts of intuition. In 1925 Heisenberg opted for a minimalist aesthetic based on particles, which repelled Schrödinger in 1926 who chose the opposite, waves. In the meantime, Bohr realized the need for an aesthetic based on waves *and* particles. By this time Einstein had changed his mind at least twice: In 1905 he opted for a minimalist particle aesthetic; in 1909 for a wave/particle duality; and then by 1917 a minimalist aesthetic based on waves only. Creative thinkers must be flexible enough even to change conceptual frameworks.

Reflecting on his creative thinking, Einstein wrote that visual imagery occurred first and words followed. Einstein's statement about his mode of creative thinking bears a strong resemblance to the artist's. Joan Miró described his creative thinking thus:

> I begin painting and as I paint the picture begins to assert itself, or suggest itself, under my brush. . . . The first stage is

free, unconscious . . . the second stage is carefully calculated.

Picasso introspected along the same lines in 1935:

"A picture is not thought out and settled beforehand. . . . An
idea is a starting point and nothing more. . . . What I think
about a great deal, I find I have always had complete in my
mind.

Similarly Rothko writes in 1947 in a vein similar to Poincaré's:

Neither the action nor the actors can be anticipated, or described in advance. They begin as an unknown adventure in
an unknown space. It is at the moment of completion that in
a flash of recognition, they are seen to have the quantity and
function which was intended. Ideas and plans that existed in
the mind at the start were simply the doorway through
which one left the world in which they occur.

Art and science at their most fundamental are adventures into the
unknown.

The introspections of Einstein, Miró, Picasso, and Rothko bear a
striking similarity to Poincaré's emphasis on unconscious processing of information. They support network thought, exemplifying the lack of
boundaries in unconscious parallel processing of information, a conclusion
supported also by the results of controlled psychological experiments.

In order to study scientists' creativity, we delve into their notebooks. With what we found in Chapter 9, we are not surprised that page
after page of Poincaré's worksheets are like the one in Figure 2 of Chapter 9: not a line crossed out, not a figure drawn. Although Poincaré was
essentially not a visual thinker, for pedagogic purposes and in certain research problems he turned often to visual imagery in order to make a
point. In contrast to Poincaré, Einstein was a visual thinker. Figure 16
contains a diagram from a letter Einstein wrote on May 7, 1952, in which

FIGURE 16.

In a letter of May 7, 1952, Einstein wrote to his old friend Maurice Solovine, "I probably expressed myself badly [at attempts to explain in words to Solovine his philosophy of science]. I view such matters schematically thus." The figure is Einstein's sketch from this letter. (Permission granted by the Albert Einstein Archives, the Hebrew University of Jerusalem, Israel.)

he said in effect that a picture is worth a thousand words. In science there are descriptive and depictive problem solvers. For physicists visual thinking is the creative mode.

We can try to glimpse how artists seek new ways to represent nature by studying their sketchbooks. Like Einstein and Poincaré, Picasso was intensely interested in the creative process. John Richardson writes of Picasso that "Anything that might cast light on the mystery of his creative process—a mystery he was always trying and failing to fathom—intrigued him." One of the most famous searches for the proper representation of nature is in Picasso's sketchbooks for his classic 1937 painting *Guernica*. Picasso is quoted as saying:

> Paintings are but research and experiment. I never do a painting as a work of art. All of them are researches. I search constantly and there is a logical sequence in all this research. That is why I number them. It's an experiment in time. I number them and date them. Perhaps one day someone will be grateful.

And indeed we are, in no small measure thanks to Rudolf Arnheim's analysis. Figures 17, 18, and 19 show the first few sketches by Pi-

FIGURE 17.

Pablo Picasso, composition study for *Guernica,* dated "May 1 37 (I),"
pencil on blue paper, 10⅝ × 8¼ in, 1937. Picasso commences his
research on proper balance of point and counterpoint among faces,
bodies, and subthemes for *Guernica.* His experiments continue in
Figures 18 and 19.

FIGURE 18.

Pablo Picasso, composition study for *Guernica,* dated "May 1 37 (II),"
pencil on blue paper, 10⅝ × 8¼ in, 1937.

FIGURE 19.

Pablo Picasso, composition study for *Guernica*, dated "May 1 37 (III),"
pencil on blue paper, 10⅝ × 8¼ in, 1937.

casso of the finished mural. They constitute one day's work! Picasso
plays with rules of composition such as balance of point and counterpoint among faces, bodies, and subthemes. Arnheim writes:

> At the time sketch 1 was made, the total concept consisted
> essentially of four fairly simple and relatively independent el-
> ements: the upright figure of the bull, the upside down
> corpse of the horse, the horizontal pointer of the light-bear-
> ing woman, and the inarticulate spread of bodies on the
> ground.

Picasso works according to an a priori set of rules governing representation. Then, like creative scientists, he breaks away, going beyond
them. At the nascent moment of creativity, barriers dissolve between
disciplines.

SEARCHING FOR REALITY

The situation in science regarding aesthetics is not at all straightforward. Just as in art, the opinion of what is aesthetic is not shared by everyone (Heisenberg versus Schrödinger), nor is it immediately noticeable to everyone (Einstein and the light quantum). And just as in art, a sense of aesthetics can trigger creativity (Einstein and the light quantum, redundancies in electromagnetic induction, Bohr and atomic physics, Schrödinger and wave mechanics, Dirac and Feynman with new theories of elementary particles). We have come across a case where science generates visual imagery (Feynman diagrams that are far deeper than many products of the present state of computer-generated art).

We have found that a common pursuit of artists and scientists is to seek a *representation* of worlds both visible and invisible. Twentieth-century abstract artists attempt to represent unnaturalistically a world beyond anything imagined with our sense perceptions. "Unnaturalistic," that is, with respect to whether the meaning of this term is taken from phenomena we have actually witnessed in the world of sense perceptions. Parallel developments in science have revealed that what is at first sight unnaturalistic is, in fact, naturalistic with reference to the new meaning of this term, a meaning coincident with how our concept of what is intuitive has been irreversibly altered when extended to worlds beyond perceptions.

We have touched upon the notions of progress in art and have noted similarities and differences compared to that of scientific progress. One similarity is the gradualness of advance due to accumulation of knowledge. One difference is that while all of us, and particularly artists, view for pleasure and study for content and style, paintings by the old masters, just because Picasso came along, we don't discard or dismiss Michelangelo and Rembrandt. We make room for new representations. Not so in science, where, despite the gradualism of scientific change, once a "new" theory is accepted, scientists often relegate past and superseded theories to the dustbin of history. The concept of progress cannot be applied exactly in the same way to both art and science. One can argue for the notion of an irresistible direction for scientific progress, as in Figure 18 of Chapter 7, on the basis of an underlying view of scientific realism in which scientific theories are a way of understanding a reality beyond appearances in which there are real electrons, protons, neutrons, etc., and absolute truth, too. History tells us that scientific theories succeed one another gradually and in such a way that fertile concepts survive, whereas often the incubating theories do not and so, in time, are usually discarded. We express our admiration for Aristotle's physics, but scientists no longer study it and certainly don't use it. Similarly for Maxwell's original version of his electromagnetic theory, Lorentz's original electron theory, and Bohr's atomic theory, despite the deep physics they involve. Unfortunately, even such a scientific masterpiece as Einstein's 1905 relativity paper is hardly ever looked at by physicists today, despite the profound philosophy and physics it contains and presents in ways often superior to what one finds in modern expositions.

An exception to this discussion of the dissimilarity of progress in art and science are artists such as Gris and Mondrian who believe that like science, art set into a proper mathematical framework can somehow join in the exploration for what constitutes physical reality.

Figure 20 summarizes the parallelisms between art and science we have explored in this chapter. The straight time line for scientific development is meant to display persistent progress. The year 1897 is selected to begin the time line for science because in that year the electron was discovered and scientists began to explore systems they could not deal with directly. Abstraction soon followed because they found that concepts from the world of sense perceptions were increasingly unsuitable for use in the domain of electrons and atoms. On the other hand, the time line for art covers only developments in abstraction, which omits interesting returns to naturalism.

It is important to bear in mind that despite the forward motion of science, scientists sometimes dip back into the past to explore over-

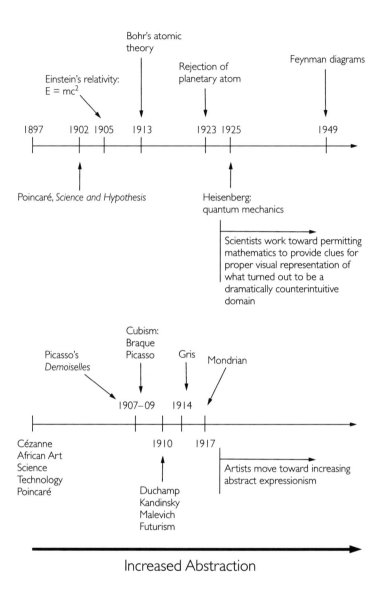

FIGURE 20.

This figure displays the parallelisms between science and art explored in this chapter.

looked treasures. Chaos theory is such an example, where good old Newtonian mechanics reformulated by Poincaré is studied anew in order to advance science, including quantum mechanics. As we discussed in Chapter 4, we can imagine that just as the second law of thermodynamics appeared to end the possibility of a completely deterministic classical physics, the dream certainly came to an end with the advent of quantum mechanics. Ironically, the final nail driven into the coffin of the Newtonian clockwork universe came from within Newtonian mechanics itself—chaos theory, which already existed in more than nascent form in results Poincaré obtained in 1890. Poincaré discovered that within the apparent orderly motion of the planets there lurks instability and eventual chaotic behavior. We have now become aware that this sort of behavior exists within most parts of nature.

Relations between art and science are less clear-cut today than they were during the Renaissance, when there were artists who moved with equal ease in the world of science, such as Albrecht Dürer and Leonardo da Vinci. With the publication of Newton's *Principia* in 1687, and the onset of the Age of Rationalism, science was deemed the only body of knowledge capable of revealing absolute truth. Art was relegated to the realm of frivolous entertainment. Since the early nineteenth century, this view has been changing. The strivings of artists at the turn of the twentieth century and statements like Mondrian's indicate that the link is still healthy. This is underscored in a lecture given at the California Institute of Technology in 1994 by the playwright Tom Stoppard, for whom science is central to his productions. Stoppard said,

> I think we've tended to create and talk about a false dichotomy, and I think that we acknowledge that it is a false one most of the time we're talking about it. Science and art are nowadays beyond being *like* each other. Sometimes they seem to *be* each other.

CONCLUSION: THE NEW SCIENCES

We have found that intuition plays a central role in scientific research and that scientific intuition is an extension of common sense intuition. This extension is accomplished through thought experiments, and our new intuition is then formally defined by a scientific theory. Galileo, for example, imagined a world in which there is a vacuum and in which, therefore, all objects fall with the same acceleration regardless of their weight. This abstraction contradicts our common sense intuition, but it was necessary for formulating a consistent theory of motion. Further thought experiments by Einstein demonstrated the need to extend our intuition into a world in which space and time are relative quantities. In turn, this counterintuitive situation became intuitive. In each case a scientific theory was the means for understanding worlds beyond sense perceptions.

This situation was brought home in a particularly striking way in atomic physics, where intuition was revealed to be synonymous with visualization. The demise of Bohr's solar system atom in 1923 and the subsequent formulations of atomic physics with no visual representation were troubling to physicists. Mathematics alone became the guide. The resulting tribulations and successes of quantum physics focused essentially on further sweeping redefinitions of intuition. Exploring this point we found that visual representation of a world unimaginable with our perceptual apparatus was made possible with diagrams generated by the mathematics of quantum physics: Feynman diagrams. These diagrams are not merely methods for facilitating calculations. They provide a glimpse into physical reality.

Understanding how intuition is transformed is essential to understanding scientific progress. We noted a societal element in how science progresses and we studied priority battles. These concerned the use of empirical data in choosing one theory over another, and the implementation of aesthetic criteria. We observed that scientists sometimes use reasoning with no solid basis whatsoever to argue for a theory's acceptance. Sometimes, in the end, they are correct.

Along with abstractions in intuition came transformations in the notions of visual representation and visual imagery. What emerged from this analysis is that researchers in the history of scientific ideas and cognitive science have useful things to say to each other. This line of investigation took us into scientific creativity where the power of unconscious parallel processing of information emerged as a central part of creative thought.

Artists as well as scientists seek aesthetic representations of nature, and at the nascent moment of creativity, boundaries between art and science dissolve. We saw that both disciplines tended toward increasingly abstract representations during the same era, the late nineteenth and twentieth centuries. Developments in science and technology influenced art's progress from Post-Impressionism to Cubism and then modern Abstract Expressionism. Several key artists expressed their goal as seeking invariant laws of nature, just as the scientists were doing.

Not unexpectedly, along with transformations in intuition and visual imagery came changes in the concept of scientific reality. This concept went from one that emphasized visual representations abstracted from phenomena we have actually witnessed, to one based on representations generated by the mathematics of physical theory. This is a fundamental shift. The notion that there are physically real entities, even if they can be measured only extremely indirectly, has historically been accompanied by the assumption that science can deliver absolute truths as to the nature of physical reality, in addition to revealing universal laws of nature.

Only along these realist lines can we understand science in a consistent and informative manner that views it as more than merely a fact-gathering enterprise with a hidden agenda based in the quest for power. We found that scientific progress parallels the construction of knowledge itself. From myriads of input data, structures emerge in a step-wise fashion, with one rung emerging continuously from the one below, which was thrown into turmoil by either new data or new theoretical realizations. Visually we may imagine this occurring in the manner of an upwardly growing spiral. After a while the curved sides are removed and just the rungs remain. Each rung is a scientific theory providing access to a possible world that is a more exact version of the one open to sense perceptions. Science offers a staircase to physical reality.

The astonishing emergence of an organized level from a disorganized lower one resembles phenomena in complexity theory.* One of

---

*As its name implies, complexity theory deals with how complex systems far-from-equilibrium eventually achieve equilibrium states. Examples of complex systems range from fluids in turbulent states, chemical reactions going to-and-fro between different states, pendula whose bobs are pulled so far away from the vertical that they can go round 360°, to ant colonies in unpredictable environments and brain patterns.

this field's pioneers, the Belgian scientist Ilya Prigogine, boldly describes disequilibrium as being the "source of order." This bears directly on the view presented here of the mind as an open system, never completely in equilibrium with its surroundings, seeking ever-deeper understanding. An important result of complexity theory is that systems far from equilibrium eventually adjust to their environments in a variety of ways. The outcome is given the obvious name "self-organization." We are speaking about teleology here, the causality advocated by Aristotle and supposedly supplanted by the relentless determinism of Newton's laws.

What is most astonishing about complexity theory is its goal: a single set of fundamental concepts to describe the behavior of living *and* physical systems. Complexity theory aims for nothing less than unification of physics and biology. The formulation of such a theory at this time accords with our view of scientific progress. Was not the pronouncement of Galileo and Descartes to *discriminate* between living and dead matter a stepping stone to complexity theory? Because only after having in hand laws governing the motion of dead matter (physical systems) could we even dream of such a subject as complexity theory.

Complexity theory supports the view of scientific progress we see emerging from the history of scientific thought: somehow, complex systems far from equilibrium achieve self-organization. Consequently, complexity theory may well contain clues for a deeper understanding of the assimilative/accommodative process. The key issue is inherent in the word *somehow*, which is shorthand for the dynamics driving systems to equilibrium. Whatever it is, the laws of this dynamics will have to supplement the known laws of physics if complexity theory is to cover biological systems, too. Newton's laws deal only with dead matter and so include a causal law inapplicable to living systems. The study of the dynamics behind self-organization is a frontier problem in complexity theory. Concerning scientific progress, with such allied issues as scientific creativity and its parallels with art, we look to new and fresh approaches for insights into science and mind that we cannot even conceive at present. What we can be certain of, however, is that the ultimate instrument in art and science is the human mind, whose imagination is decisive for reading nature.

Are there limits to science? An often used metaphor for scientific exploration is that it is like peeling an onion. This is not without interest because the implication is that there is some hard core that we can reach some time in the future. What awaits us there cannot even be imagined, nor can we but wonder about the methods needed to reach it. It is a fallacy of reasoning to project far into the future on the basis of what we know of the present. Who could have guessed even fifty years ago what science would be like today, with its dependence on computers not only as number crunchers but as creative tools? As for the form science will take, complexity theory tells us we are in for some surprises.

Looking back over our journey from ancient Babylonia to the present, we discern today a trend toward new sciences that incorporate insights from many millennia ago. Scientific progress occurs because ideas emerge, one after the other, about events visible and invisible to our perception. What chutzpah in trying to understand such phenomena! For this purpose a subject was formulated to supplement insights from literature, poetry, music and art, and to explain in ways they could not the operation of the world about us. It became known as science.

Photo Credits

Figure 5, p. 16: Biblioteca Nazionale Centrale, Florence

Figure 2, p. 39; Figure 6, p. 52; Figure 1, p. 137; and Figure 2, p. 398: From: Holton G. and Roller D.: *Foundations of Modern Physical Science*, 1958; reprinted with permission of Addison-Wesley

Figure 3, p. 40: From Halliday D. and Resnick R.: *Physics, Parts I and II*, New York, John Wiley, 1967, p. 1070

Figure 5, p. 42: From Wood, B.A. and Oldham, F.: *Thomas Young, Natural Philosopher, 1773–1829*; reprinted with permission of Cambridge University Press

Figure 1, p. 75: W. Kaufman, Göttinger Nachrichten, 1901

Figure 2, p. 75: W. Kaufman, Göttinger Nachrichten, 1902

Figure 5, p. 126: From Miller, A.I. (Ed.): *62 Years of Uncertainty*, New York, Plenum, 1990

Figure 3, p. 163: © CERN

Figure 1, p. 184: © Niedersächsische Staats und Universitätsbibliothek

Figure 1, p. 222: From Niven W.D.: *The Scientific Papers of James Clerk Maxwell*, New York, 1965; reprinted with permission from Dover Publications

Figure 1, p. 267: Reprinted with permission of Erlbaum Publishing

Figure 2, p. 268: From Kosslyn S.M.: *Image and Mind*, 1980, pp. 43 and 44, Harvard University Press

Figure 6, p. 283: From *Proc. Roy. Soc. Lond. Ser. B*, 207; D. Marr and E. Hildreth, Theory of Edge Detection, 1980

Figure 7, p. 286: From Stillings N.A. et al.: *Cognitive Science: An Introduction*, MIT Press, 1987

Figure 9, p. 291: From Barlow H., Blakemore C., and Weston-Smith M. (Eds.), *Images and Understanding*; reprinted with permission of Cambridge University Press

Figure 18, p. 306: From Kosslyn S.M.: *Image and Brain*, MIT Press, 1994, p. 300

Figure 19, p. 310: Figure from *Descartes' Error*, 1994; Putnam

Figure 2, p. 347, Figure 3, p. 348, and Figure 4, p. 349: Courtesy of the Poincaré Archives

Figure 7(a), p. 407: From Hasert F.J. et al.: Muon-Neutrino Electron Scattering, *Phys. Lett. B 46*, 1973; Physikalisches Institut der Technischen Hochschule, Aachen, Germany

Figure 9, p. 415: Paul Cézanne, *The Kitchen Table*, 1888–1890, Courtesy Bridgeman Art Library

Figure 11, p. 420: Georges Braque, *Violin and Palette*, 1910. Photography by Robert E. Mates © The Solomon R. Guggenheim Foundation, New York (FN 54.1212)

Figure 12, p. 422: Marcel Duchamp: *Nude Descending a Staircase*, 1912. Courtesy Philadelphia Museum of Art: The Louise and Walter Arensberg Collection. Photo by Graydon Wood, 1994

Figure 13, p. 423: E.J. Marey: *Man Walking*, ca. 1882. Courtesy Bibliothèque Nationale de Paris

Figure 14, p. 425: Luigi Russolo: *Dynamism of an Automobile*, 1911. Oil on canvas, 104 × 140 cm. Courtesy Musée National d'Art Moderne Centre Georges Pompidou, Paris. Gift: Sonia Delaunay

Figure 15, p. 428: Piet Mondrian, *Composition with Lines*, 1917. Courtesy Kröller-Müller Museum

Figures 17 and 18, p. 434 and Figure 19, p. 435: Compositions for *Guernica*. Courtesy of The Bridgeman Art Library. These figures as well as *Three Musicians* (p. 301) and *Les Demoiselles d'Avignon* (p. 387) are reprinted with permission of ARS.

BIBLIOGRAPHY

Wherever possible, within the text I have indicated the source of a quotation. This bibliography contains these sources in addition to a selection of writings into which the interested reader can delve further. Archival sources that have been useful to me are the Poincaré Archives, at the University of Nancy, France, and the *Archives for the History of Quantum Physics,* on deposit at the American Institute of Physics in College Park, Maryland, the American Philosophical Society in Philadelphia, the University of California, Berkeley, and at the Niels Bohr Institute in Copenhagen.

Anderson, J. R., *Cognitive Psychology and Its Implications* (2nd ed., New York: Freeman, 1985).

Anderson, M., "An impression," in S. Rozental (ed.), *Niels Bohr: His Life and Work as Seen by His Friends and Colleagues* (New York: Wiley, 1967).

Apollinaire, G., "Les peintres Cubistes [The cubist painters]," in H. B. Chipp (ed.), *Theories of Modern Art* (Berkeley: Univ. of Calif. Press, 1968).

Arnheim, R., *The Genesis of a Painting: Picasso's Guernica* (Berkeley: Univ. of Calif. Press, 1962).

Arnheim, R., *Visual Thinking* (Berkeley: Univ. of Calif. Press, 1971).

Aspect, A. and P. Granger, "Wave–particle duality: A case study." in A. I. Miller (ed.), *Sixty-Two Years of Uncertainty: Historical, Philosophical, and Physical Inquiries into the Foundations of Quantum Mechanics* (London: Plenum, 1990).

Barlow, H., C. Blakemore, and M. Weston-Smith, *Images and Understanding* (Cambridge, U.K.: Cambridge Univ. Press, 1990).

Barrow, J. D., *Theories of Everything: The Quest for Ultimate Explanation* (Oxford: Oxford Univ. Press, 1991).

Barrow, J. D., *Pi in the Sky: Counting, Thinking, and Being* (Oxford: Oxford Univ. Press, 1992).

Bell, J. S., *Speakable and Unspeakable in Quantum Mechanics* (Cambridge: Cambridge Univ. Press, 1987).

Bergson, H., *Durée et Simultanéité* (Paris: Presses Universitaires de France, 1968).

Berkeley, G., *Principles, Dialogues and Philosophical Correspondence* (New York: Bobbs-Merrill, 1965).

Black, J. T., *Ernst Mach: His Life, Work, and Influence* (Berkeley: Univ. of Calif. Press, 1972).

Black, M., "More about Metaphor," in A. Ortoney (ed.), *Metaphor and Thought*, 2nd ed. (Cambridge: Cambridge Univ. Press, 1993).

Block, N., *Readings in Philosophy of Psychology* (2 vols., Cambridge: Harvard Univ. Press, 1981).

Bohr, N. "On the constitution of atoms and molecules," *Philosophical Magazine*, **26**, 1–25 (1913). Reprinted with introductory material by L. Rosenfeld in L. Rosenfeld (ed.), *On the Constitution of Atoms and Molecules* (New York: Benjamin, 1963).

Bohr, N., *The Theory of Spectra and Atomic Constitution* (Cambridge: Cambridge Univ. Press, 1924).

Bohr, N., "The quantum postulate and the recent development of atomic theory, *Nature (Supplement)*, 580–590 (1928).

Born, M., "Quantentheorie und Störungsrechnung [Quantum theory and perturbation calculations]," *Die Naturwissenschaften*, **27**, 537–550 (1923).

Boyd, R., P. Gasper, and J. D. Trout (eds.), *The Philosophy of Science* (Cambridge: MIT Press, 1991).

Boyd, R., "Metaphor and theory change: What is 'metaphor' a metaphor for?" in A. Ortony (ed.), *Metaphor and Thought* (2nd ed., Cambridge: Cambridge Univ. Press, 1993).

Brecht, B., "Life of Galileo," in J. Willett and R. Manheim (eds.), *Collected Plays of Bertolt Brecht: Volume 5* (London: Metheuen, 1980).

Brock, W. H., *The Fontana History of Chemistry* (London: Fontana Press, 1992).

Brush, S. G., "Thermodynamics in history," *The Graduate Journal*, **VII**, 477–525 (1967).

Brush, S. G., *The Kind of Motion We Call Heat: A History of the Kinetic Theory of Gases in the 19th Century* (2 vols., Amsterdam: North Holland, 1976).

Buchwald, J. Z., *The Rise of the Wave Theory of Light: Optical Theory and Experiment in the Early Nineteenth Century* (Chicago: Univ. of Chicago Press, 1989).

Carmicheal, L., H. P. Hogan, and A. A. Walters, "An experimental study of the effect of language on the reproduction of visually perceived form," *Journal of Experimental Psychology*, **15**, 73–86 (1932).

Cartwright, N., *How the Laws of Physics Lie* (Oxford: Oxford Univ. Press, 1983).

Cassidy, D., *Uncertainty: The Life and Science of Werner Heisenberg* (New York: Freeman, 1992).

Cassirer, E., *Determinism and Indeterminism in Modern Physics* (New Haven: Yale Univ. Press, 1956).

Chalmers, A., *Science and Its Fabrication* (Minneapolis: Univ. of Minnesota Press, 1991).

Chipp, H. B. (ed.), *Theories of Modern Art* (Berkeley: Univ. of Calif. Press, 1968).

Clagett, M., *Greek Science in Antiquity* (London: Collier-Macmillan, 1955).

Claparède, E., E. Fehr, and T. Flournoy, "L'Enquête sur la méthode de travail du mathématician [Inquiry on the mathematician's method of work]," *L'Enseignement Mathématique*, **4**, 208–211 (1902); **6**, 376, 481 (1904); **7**, 387–395, 473–478 (1905); **8**, 43–48, 217–225, 293–310, 463–495 (1906); **9**, 123–135, 204–217, 306–312, 473–479 (1907); **10**, 152–172 (1908).

Collins, A. and E. E. Smith (eds.), *Readings in Cognitive Science: A Perspective from Psychology and Artificial Intelligence* (San Mateo, Calif.: Morgan Kaufmann Publishers, Inc., 1988).

Cushing, J., *Quantum Mechanics: Historical Contingency and the Copenhagen Hegemony* (Chicago: Univ. of Chicago Press, 1994).

Damasio, A. R., *Descartes' Error: Emotion, Reason and the Human Brain* (New York: Putnam, 1994).

Danto, A., *Transfiguration of the Commonplace: A Philosophy of Art* (Cambridge: Harvard Univ. Press, 1981).

Davidson, D., "On the very idea of a conceptual scheme," *Proceedings of the American Philosophical Association*, **47**, 5–20 (1973–74).

Davies, P. (ed.), *The New Physics* (Cambridge: Cambridge Univ. Press, 1989).

Davies, P., *The Last Three Minutes* (London: Weidenfeld & Nicholson, 1994).

Devitt, M. and K. Sterelny, *Language and Reality: An Introduction to the Philosophy of Language* (Cambridge: MIT Press, 1990).

Drake, S., *Galileo Studies: Personality, Tradition, and Revolution* (Ann Arbor: Univ. of Michigan Press, 1970).

Drake, S. and I. E. Drabkin (eds.), *Mechanics in Sixteenth Century Italy* (Madison: Univ. of Wisconsin Press, 1969).

Duhem, P., *The Aim and Structure of Physical Theory*, translated by P. P. Wiener (New York: Athenium, 1962).

Dyson, F. W., A. S. Eddington, and C. Davidson, "A determination of the deflection of light by the sun's gravitational field, from observations made at the total eclipse of May 29, 1919," *Philosophical Transactions of the Royal Society,* **220A**, 291–333 (1920).

Einstein, A., "Über einen die Erzeugung und Verwandlung des Lichtes betreffenden heuristischen Gesichtpunkt, *Annalen der Physik,* **17**, 132–148 (1905); translated by A. B. Arons and M. B. Peppard, as "On a heursitic viewpoint concerning the production and transformation of light," in *American Journal of Physics,* **33**, 367–374 (1965).

Einstein, A., "Die von der molekularkinetischen Theorie der Wärme geforderte Bewegung von in ruhenden Flüssigkeiten suspendierten Teilchen," *Annalen der Physik,* **17**, 549–560 (1905); translated by A. D. Cowper as "On the movement of small particles suspended in a stationary liquid demanded by the molecular-kinetic theory of heat," in R. Fürth (ed.), *Albert Einstein: Investigations on the Theory of Brownian Movement* (New York: Dover, 1956), pp. 1–18.

Einstein, A., "Zur Elektrodynamik bewegter Körper," *Annalen der Physik,* **17**, 891–921 (1905); translated by A. I. Miller as "On the electrodynamics of moving bodies," in A. I. Miller, *Albert Einstein's Special Theory of Relativity: Emergence (1905) and Early Interpretation (1905–1911)* (Reading, Mass.: Addison-Wesley, 1981).

Einstein, A., "Relativitätsprinzip und die aus demselben gezogenen Folgerungen [On the relativity principle and the consequences that follow from it]," *Jahrbuch der Radioaktivität und Elektronik,* **4**, 411–462 (1907).

Einstein, A., *Essays in Science* (New York: Philosophical Library, 1934).

Einstein, A., B. Podolsky, and N. Rosen, "Can quantum-mechanical description of physical reality be considered complete?" *Physical Review,* **47**, 777–780 (1935).

Einstein, A., "Autobiographical notes," in P. A. Schilpp (ed.), *Albert Einstein: Philosopher-Scientist* (Evanston: The Library of Living Philosophers, 1949).

Einstein, A., *Lettres à Maurice Solovine* (Paris: Gauthiers-Villars, 1956).

Einstein, A., *Ideas and Opinions* (New York: Bonanza Books, n.d.).

Einstein, A., *Out of My Later Years* (Totawa, N.J.: Littlefield, Adams and Co., 1967).

Fauvel, J. (ed.), *Let Newton Be!* (Oxford: Oxford Univ. Press, 1980).

Fermi, E., "Versuch einer Theorie der β-Strahlen [Research on a theory of β-Rays]," *Zeitschrift für Physik*, **88**, 161–177 (1934).

Ferris, T., *Coming of Age in the Milky Way* (New York: Morrow, 1988).

Feyerabend, P. K., *Against Method* (London: New Left Books, 1975).

Feynman, R. P. and M. Gell-Mann. "Theory of the Fermi interaction," *Physical Review*, **109**, 193–198 (1958).

Feynman, R. P., *The Character of Physical Law* (Cambridge: MIT Press, 1965).

Fine, A., *The Shaky Game: Einstein, Realism and the Quantum Theory* (Chicago: Univ. of Chicago Press, 1986).

Fox, R., *The Caloric Theory of Gases from Lavoisier to Regnault* (Oxford: Oxford Univ. Press, 1971).

Frank, P., *Einstein: Sein Leben und seine Zeit* (New York: Knopf, 1947), translated by G. Rosen, and edited and revised by S. Kusaka as *Einstein: His Life and Times* (New York: Knopf, 1952).

van Fraassen, B., *The Scientific Image* (Oxford: Oxford Univ. Press, 1980).

Friedman, M., *Kant and the Exact Sciences* (Cambridge: Harvard Univ. Press, 1992).

Galilei, Galileo, *The Starry Messenger*, in S. Drake (ed.) and translator, *Discoveries and Opinions of Galileo* (Garden City: Doubleday, 1967).

Galilei, Galileo, *Dialogue Concerning the Two Chief World Systems*, translated by S. Drake (Berkeley: Univ. of Calif. Press, 1967).

Galilei, Galileo, *Two New Sciences*, translated by S. Drake (Madison: Univ. of Wisconsin Press, 1974).

Galison, P., *How Experiments End* (Chicago: Univ. of Chicago Press, 1987).

Galison, P., "Aufbau/Bauhaus: Logical positivism and architectural modernism," in *Critical Inquiry*, **16**, 710–752 (1990).

Gardner, H., *Creating Minds* (New York: Basic Books, 1993).

Gell-Mann, M., *The Quark and the Jaguar: Adventures in the Simple and the Complex* (New York: Freeman and Company, 1994).

Gillispie, C. C., *The Edge of Objectivity: An Essay in the History of Scientific Ideas* (2nd ed., Princeton: Princeton Univ. Press, 1990).

Gingerich, O., "The Galileo affair," *Scientific American*, 133–143 (August 1982).

Glass, A. L., K. J. Holyoak, and J. L. Santa, *Cognition* (Reading, Mass., Addison-Wesley, 1979).

Gleick, J., *Genius: The Life and Science of Richard Feynman* (New York: Pantheon, 1992).

Golding, J., *Cubism: A History and an Analysis, 1907–1914* (London: Faber and Faber, 1959).

Golding, J., "Cubism," in N. Stangos (ed.), *Concepts of Modern Art* (London: Thames and Hudson, 1991).

Gombrich, E. H., *Art and Illusion: A Study in the Psychology of Pictorial Representation* (Princeton: Princeton Univ. Press, 1961).

Gombrich, E. H., J. Hochberg, and M. Black, *Art, Perception and Reality* (Baltimore: Johns Hopkins Press, 1972).

Goodman, N., *Languages of Art* (Indianapolis: Hackett, 1976).

Gooding, D., *Experiment and the Making of Meaning: Human Agency in Scientific Observation and Experiment* (Dordrecht: Kluewer, 1990).

Gould, S. J., *Time's Arrow, Time's Cycle: Myth and Metaphor in the Discovery of Geological Time* (Cambridge: Harvard Univ. Press, 1987).

Gray, J., *Ideas of Space* (Oxford: Oxford Univ. Press, 1989).

Gray, J., *Linear Differential Equations and Group Theory from Riemann to Poincaré* (Boston: Birkhäuser, 1986).

Gregory, R. L., *Mind in Science: A History of Explanations in Psychology and Physics* (London: Weidenfeld and Nicholson, 1981).

Gross, P. R. and N. Levitt, *Higher Superstition: The Academic Left and Its Quarrels with Science* (Baltimore: Johns Hopkins Press, 1994).

Gruber, H. E., "On the relation between 'Aha experiences' and the construction of ideas," *History of Science*, **19**, 41–59 (1981).

Guth, A. and P. Steinhardt, "The inflationary universe," in P. Davies (ed.), *The New Physics* (Cambridge: Cambridge Univ. Press, 1989).

Hacking, I., *Representing and Intervening: Introductory Topics in the Philosophy of Natural Science* (Cambridge: Cambridge Univ. Press, 1983).

Hadamard, J., *The Psychology of Invention in the Mathematical Field* (New York: Dover, 1954).

Hanfling, O. (ed.), *Philosophical Aesthetics: An Introduction* (Oxford: Blackwell, 1992).

Hanson, N. R., *Patterns of Discovery* (Cambridge: Cambridge Univ. Press, 1969).

Harman, G., "Inference to the best explanation," *Philosophical Review*, **74**, 88–95 (1965).

Hayles, K. H., *Chaos Bound: Orderly Disorder in Contemporary Literature and Science* (Ithaca: Cornell Univ. Press, 1990).

Hawking, S., *A Brief History of Time: From the Big Bang to Black Holes* (New York: Bantam, 1988).

Heilbron, J., and T. S. Kuhn, "The genesis of the Bohr atom," *Historical Studies in the Physical Sciences*, **1**, 211–290 (1969).

Heilbron, J., *The Dilemmas of an Upright Man: Max Planck As Spokesman for German Science* (Berkeley: Univ. of Calif. Press, 1986).

Heisenberg, W., "Über den Baue der Atomkerne. I [On the structure of the atomic nucleus. I]," *Zeitschrift für Physik*, **77**, 1–11 (1932).

Heisenberg, W., Über die in der Theorie der Elementarteilchen auftretende universelle Länge," *Zeitschrift für Physik*, **32**, 20–33 (1938); translated as "The universal length appearing in the theory of elementary particles," in A. I. Miller, *Early Quantum Electrodynamics: A Source Book* (Cambridge: Cambridge Univ. Press, 1994).

Heisenberg, W., "Über quantentheoretische Umdeutung kinematischer und mechanischer Beziehungen," *Zeitschrift für Physik*, **33**, 879–893 (1925); translated as "On the quantum theoretical interpretation of kinematical and mechanical relations," in B. W. van der Waerden (ed.), *Sources of Quantum Mechanics* (New York: Dover, 1968).

Heisenberg, W., "Zur Quantenmechanik [On quantum mechanics]," *Die Naturwissenschaften,* **14**, 899–904 (1926).

Heisenberg, W., "Über den anschaulichen Inhalt der quantentheoretischen Kinematik und Mechanik [On the intuitive content of the quantum theoretical kinematics and mechanics]," *Zeitschrift für Physik,* **43**, 172–198 (1927).

Heisenberg, W., *Der Teil und das Ganze: Gespräche im Umkreis der Atomphysik* (Munich: Piper, 1969), translated by A. J. Pomerans as *Physics and Beyond: Encounters and Conversations* (New York: Harper, 1971).

Heisenberg, W., "Was ist ein Elementarteilchen [What is an elementary particle]," *Die Naturwissenschaften,* **63**, 1–7 (1976).

von Helmholtz, H., "An autobiographical sketch," in R. Kahl (ed.), *Selected Writings of Hermann von Helmholtz* (Middletown, Conn.: Wesleyan Univ. Press, 1971).

Henderson, L. D., *The Fourth Dimension and Non-Euclidean Geometry in Modern Art* (Princeton: Princeton Univ. Press, 1983).

Hendry, J., *The Creation of Quantum Mechanics and the Bohr–Pauli Dialog* (Boston: Reidel, 1984).

Hermann, A., *The Genesis of Quantum Theory (1899–1913),* translated by C. W. Nash (Cambridge: MIT Press, 1971).

Hertz, H., *Electric Waves,* translated by D. E. Jones (New York: Dover, 1962).

Hertz, H., *The Principles of Mechanics,* translated by D. E. Jones (New York: Dover, 1956).

Hildesheimer, W., *Mozart,* translated by M. Fabor (New York: Vintage Books, 1983).

Holmes, F. L., "Research trails and the creative spirit: Can historical case studies integrate the short and long time scales of creative activity?," in *Creativity Research Journal,* (to be published) (1996).

Holton, G., *Thematic Origins of Scientific Thought: Kepler to Einstein* (Cambridge: Harvard Univ. Press, 1973).

Holton, G., *Scientific Imagination: Case Studies* (Cambridge: Cambridge Univ. Press, 1978).

Holton, G., *Science and Anti-Science* (Cambridge: Harvard Univ. Press, 1993).

Holton, G., *Einstein, History, and Other Passions* (Woodbury, N.Y.: AIP Press, 1995).

Howard, M. (ed.), *The Impressionists by Themselves* (London: Conran Octopus Limited, 1991).

Hubel, D. H., *Eye, Brain and Vision* (New York: Scientific American Library, 1988).

Hulten, P. (ed.), *Futurismo & Futurismi* (Milan: Fabbri, 1986).

Ishai, A. and D. Sagi, "Common mechanisms of visual imagery and perception," *Science*, **268**, 1772–1774 (1995).

Jammer, M., *Conceptual Development of Quantum Mechanics* (New York: McGraw-Hill, 1967).

James, W., *The Principles of Psychology* (New York: Dover, 1950).

Johnson-Laird, P. N., *Mental Models* (Cambridge: Harvard Univ. Press, 1983).

Kant, I., *Critique of Pure Reason*, translated by N. K. Smith (New York: St. Martin's Press, 1929).

Kant, I., *Prolegomena to Any Future Metaphysics*, translated by P. Carus (La Salle: Open Court, 1967).

Kemp, M., *The Science of Art: Optical Themes in Western Art from Brunelleschi to Seurat* (New Haven: Yale Univ. Press, 1990).

Kern, S., *The Culture of Time and Space* (Cambridge: Harvard Univ. Press, 1983).

Klein, M. J., "Max Planck and the beginnings of the quantum theory," *Archives for History of Exact Sciences*, **1**, 459–479 (1962).

Klein, M. J., A. J. Kox, R. Jürgen, and R. Schulmann (eds.), *Albert Einstein: Volume 3: The Swiss Years: Writings, 1909–1911* (Princeton: Princeton Univ. Press, 1993).

Klein, M. J., A. J. Kox, and R. Schulmann (eds.), *Albert Einstein: Volume 5: The Swiss Years: Correspondence , 1902–1914* (Princeton: Princeton Univ. Press, 1993).

Koestler, A., *The Sleepwalkers* (London: Hutchinson, 1959).

Kosslyn, S. M., *Image and Mind* (Cambridge: Harvard Univ. Press, 1981).

Kosslyn, S. M., *Image and Brain* (Cambridge: MIT Press, 1994).

Koyré, A., *Metaphysics and Measurement: Essays in the Scientific Revolution* (Cambridge: Harvard Univ. Press, 1968).

Koyré, A., *From the Closed World to the Infinite Universe* (Baltimore: Johns Hopkins Press, 1957).

Kuhn, T. S., *The Copernican Revolution* (New York: Vintage Books, 1957).

Kuhn, T. S., *The Structure of Scientific Revolutions* (Chicago: Univ. of Chicago Press, 1962).

Kuhn, T. S., *The Essential Tension: Selected Studies in Scientific Tradition and Change* (Chicago: Univ. of Chicago Press, 1977).

Kulkarni, D. and H. A. Simon, "The processes of scientific discovery: The strategy of experimentation," *Cognitive Science*, **12**, 139–175 (1988).

Lakoff, G. and M. Johnson, *Metaphors We Live By* (Chicago: Univ. of Chicago Press, 1980).

Langley, P., H. A. Simon, G. L. Bradshaw, and J. M. Zytkow, *Scientific Discovery: Computational Explorations of the Creative Processes* (Cambridge: MIT Press, 1987).

Leplin, J. (ed.), *Scientific Realism* ((Berkeley: Univ. of Calif. Press, 1984).

Levenson, T., *Measure for Measure: A Musical History of Science* (New York: Simon and Schuster, 1994).

Liberman, A., *The Artist in His Studio* (New York: Random House, 1988).

Lindberg, D. C., *Theories of Vision from Al-Kindi to Kepler* (Chicago. Univ. of Chicago Press, 1976).

McGuiness, B. (ed.), *Ludwig Boltzmann: Theoretical Physics and Philosophical Problems*, translated by P. Foulkes (Boston: Reidel, 1974).

Mach, E., "Die Leitgedanken meiner naturwissenschaftlichen Erkenntnislehre und ihre Aufnahme durch die Zeitgenossen," *Physikalische Zeitschrift*, **11**, 599–606 (1910); reprinted as "The guiding principles of my scientific theory of knowledge and its reception by my contemporaries," in S. Toulmin (ed.), *Physical Reality* (New York: Harper Torchbooks, 1970).

Mach, E., *Science of Mechanics: A Critical and Historical Account of Its Development*, translated by T. McCormack (La Salle: Open Court, 1960).

Mach, E., "On thought experiments," in E. Mach, *Knowledge and Error: Sketches on the Psychology of Enquiry*, translated by T. McCormack (Boston: Reidel, 1976).

Mach, E., *Knowledge and Error: Sketches on the Psychology of Enquiry*, translated by T. McCormack (Boston: Reidel, 1976).

McMullin, E., "A case for scientific realism," in J. Leplin, (ed.), *Scientific Realism* (Berkeley: Univ. of Calif. Press, 1984).

Mandler, G., "Hypermnesia, incubation, and mind popping: On remembering without really trying," in C. Umiltà and M. Moscovitch, *Attention and Performance XV* (Cambridge: MIT Press, 1994).

Manuel, F. E., *A Portrait of Isaac Newton* (Cambridge: Harvard Univ. Press, 1968).

Marr, D., *Vision* (Cambridge: MIT Press, 1982).

Merz, J. T., *A History of European Scientific Thought in the Nineteenth Century* (4 vols., 1904–1912: New York, Dover, 1965). Volumes 3 and 4 are entitled *A History of European Thought in the Nineteenth Century.*

Metzinger, J. and A. Gleizes, *Du Cubisme* (Paris: Éditions Présence, 1912).

Metzler J. and R. Shepard, "Transformational studies of the internal representations of three dimensional objects," in R. L. Solso (ed.), *Theories of Cognitive Psychology: The Loyola Symposium* (Hillsdale, N.J.: Lawrence Erlbaum Associates, 1974).

Miller, A. I., "The myth of Gauss' experiment on the Euclidean nature of physical space," *Isis*, **63**, 345–348 (1972).

Miller, A. I., "Visualization lost and regained: The genesis of the quantum theory in the period 1913–1927," in J. Wechsler (ed.), *Aesthetics in Science* (Cambridge: MIT Press, 1978).

Miller, A. I., *Albert Einstein's Special Theory of Relativity: Emergence (1905) and Early Interpretation (1905–1911)* (Reading, Mass.: Addison-Wesley, 1981).

Miller, A. I., "Albert Einstein and Max Wertheimer: A Gestalt psychologist's view of the genesis of special relativity theory," in A. I.

Miller, *Imagery in Scientific Thought: Creating 20th-Century Physics* (Boston: Birkhäuser, 1984, Cambridge: MIT Press, 1986).

Miller, A. I., *Imagery in Scientific Thought: Creating 20th-Century Physics* (Boston: Birkhäuser, 1984, Cambridge: MIT Press, 1986).

Miller, A. I., *Frontiers of Physics: 1900–1911* (Boston: Birkhäuser, 1986).

Miller, A. I., "Symmetry and imagery in the physics of Bohr, Einstein and Heisenberg," in M. Doncel, A. Hermann, L. Michel, and A. Pais (eds.), *Symmetries in Physics (1600–1980)* (Barcelona: Servei de Publicacions, UAB, 1987).

Miller, A. I. (ed.), *Sixty-Two Years of Uncertainty: Historical, Philosophical and Physical Inquiries into the Foundations of Quantum Mechanics* (London: Plenum, 1990).

Miller, A. I., "Albert Einstein's 1906 *Jahrbuch* paper: The first step from SRT to GRT," in J. Eisenstadt and A. J. Kox (eds.), *Studies in the History of General Relativity* (Boston: Birkhäuser, 1992), pp. 319–335.

Miller, A. I., *Early Quantum Electrodynamics: A Source Book* (Cambridge: Cambridge Univ. Press, 1994).

Miller, A. I. and F. W. Bullock, "Neutral currents and the history of scientific ideas," *Studies in History and Philosophy of Modern Physics*, **6**, 895–931 (1994).

Miller, A. I., "Why Poincaré did not create special relativity in 1905," in J. L. Greffe, G. Heinzmann, and K. Lorenz (eds.), *Henri Poincaré: Science and Philosophy* (Berlin: Akademie Verlag, 1996).

Minkowski, H., "Space and time," translated by W. Perrett and G. B. Jeffery, in *The Principle of Relativity* (New York: Dover, 1923).

Mondrian, P., "Plastic art and pure plastic art," in H. B. Chipp (ed.), *Theories of Modern Art* (Berkeley: Univ. of Calif. Press, 1968).

Moore, W., *Schrödinger: Life and Thought* (Cambridge: Cambridge Univ. Press, 1989).

Morrison, P., "On broken symmetries," in J. Wechsler (ed.), *Aesthetics in Science* (Cambridge: MIT Press, 1978).

Munitz, M. K., *Theories of the Universe from Babylonian Myth to Modern Science* (New York: The Free Press, 1957).

Murray, D. J., *A History of Western Psychology* (Englewood Cliffs, N.J., Prentice-Hall, Inc., 1983).

Newton, I., *Mathematical Principles of Natural Philosophy*, translation by A. Motte, revised by F. Cajori (2 vols., Berkeley: Univ. of Calif. Press, 1966).

Newton-Smith, W. H., *The Rationality of Science* (London: Routledge & Kegan Paul, 1981).

Niven, W. D., *The Scientific Papers of James Clerk Maxwell* (2 vols., New York: Dover, 1965).

Ortony, A. (ed.), *Metaphor and Thought* (2nd ed., Cambridge: Cambridge Univ. Press, 1993).

Pais, A., *Subtle Is the Lord: The Science and the Life of Albert Einstein* (Oxford: Oxford Univ. Press, 1982).

Pais, A., *Inward Bound: On Matter and Forces in the Physical World* (Oxford: Oxford Univ. Press, 1986).

Pais, A., *Niels Bohr's Times, in Science, Philosophy and Polity* (Oxford: Oxford Univ. Press, 1991).

Pauli, W., *Wissenschaftlicher Briefwechsel mit Bohr, Einstein, Heisenberg, U.A.: Volume I: 1919–1929*, A. Hermann, K. von Meyenn, and V. F. Weisskopf (eds.) (Berlin: Springer-Verlag, 1979).

Pauli, W., *Wissenschaftlicher Briefwechsel mit Bohr, Einstein, Heisenberg, U.A.: Volume II: 1930–1939*, K. von Meyen (ed.) (Berlin: Springer-Verlag, 1985).

Pearce Williams, L., *Michael Faraday: A Biography* (New York: Simon and Shuster, 1971).

Pera, M., *The Discourses of Science*, translated by C. Botsford (Chicago: Univ. of Chicago Press, 1994).

Piaget, J., *Play, Dreams and Imitation in Childhood*, translated by G. Gattegno and F. M. Hodgson (New York: Norton, 1962).

Piaget, J., *The Child's Conception of Physical Causality* translated by M. Gabain (Totowa, N.J.: Littlefield, Adams and Co., 1969).

Piaget, J., *Genetic Epistemology*, translated by E. Duckworth (New York: Columbia Univ. Press, 1970).

Piaget, J., *The Principles of Genetic Epistemology*, translated by W. Mays (New York: Basic Books, 1972).

Pinker, S., *The Language Instinct: How the Mind Creates Language* (New York: William Morrow and Company, Inc., 1994).

Planck, M., "Die Einheit des physikalischen Weltbildes," *Physikalische Zeitschrift*, 10, 62–75 (1909); reprinted as "The unity of the physical world-picture," in S. Toulmin (ed.), *Physical Reality* (New York: Harper Torchbooks, 1970).

Planck, M., "Zur Machschen Theorie der physikalischen Erkenntnis," *Physikalische Zeitschrift*, 11, 1186–1190 (1910); reprinted as "On Mach's theory of physical knowledge," in S. Toulmin (ed.), *Physical Reality* (New York: Harper Torchbooks, 1970).

Poincaré, H., "Sur les hypothèses fondamentales de la géométrie," *Bulletin de la Société mathématique de France*, 15, 203–216 (1887).

Poincaré, H., *Science and Hypothesis* (New York: Dover, 1952), translator unknown (originally published by Flammarion in 1902).

Poincaré, H., Book review of "D. Hilbert, *The Foundations of Geometry*," *Bulletin of the American Mathematical Society*, 10, 1–23 (1903).

Poincaré, H., *Value of Science* (New York: Dover, 1958), translated by G. Halsted (originally published by Flammarion in 1905).

Poincaré, H., *Science and Method* (New York: Dover, n.d.), translated by F. Maitland (originally published by Flammarion in 1908).

Poincaré, H., *Mathematics and Science: Last Essays* (New York: Dover, 1963), translated by J. W. Bolduc (originally published by Flammarion in 1913).

Popper, K. R., *Conjectures and Refutations: The Growth of Scientific Knowledge* (London: Routledge & Kegan Paul, 1963).

Prigogene, I. and I. Stengers, *Order out of Chaos* (London: Heinemann, 1984).

Putnam, H., *Mind, Language and Reality: Philosophical Papers: Volume 2* (Cambridge, U.K.: Cambridge Univ. Press, 1975).

Putnam, H., "The 'corroboration' of theories," in R. Boyd, P. Gasper, and J. D. Trout (eds.), *The Philosophy of Science* (Cambridge: MIT Press, 1991).

Pylyshyn, Z. W., *Computation and Cognition* (Cambridge: MIT Press, 1986).

Quillian, M. R., "Semantic memory," in A. Collins and E. E. Smith (eds.), *Readings in Cognitive Science: A Perspective from Psychology and Artificial Intelligence* (San Mateo, Calif.: Morgan Kaufmann Publishers, Inc., 1988).

Radman, Z., (ed.), *From a Metaphorical Point of View: A Multidisciplinary Approach to the Cognitive Content of Metaphor* (New York: de Gruyter, 1995).

Rattansi, P. M. and J. M. McGuire, "Newton and the wisdom of the ancients," in J. Fauvel (ed.), *Let Newton Be!* (Oxford: Oxford Univ. Press, 184–201, 1980).

Redondi, P., *Galileo: Heretic*, translated by R. Rosenthal (Princeton: Princeton Univ. Press, 1987).

Richards, I. A., *The Philosophy of Rhetoric* (Oxford: Oxford Univ. Press, 1965).

Richardson, J., *A Life of Picasso: Volume I, 1881–1906* (New York: Random House, 1991).

Root-Bernstein, R., "On paradigms and revolutions in science and art: The challenge of interpretation," *Art Journal*, Summer, 109–118 (1984).

Rosenfeld, L., "Niels Bohr in the thirties," in S. Rozental (ed.), *Niels Bohr: His Life and Work as Seen by His Friends and Colleagues* (New York: Wiley, 1967).

Rozental, S., (ed.), *Niels Bohr: His Life and Work as Seen by His Friends and Colleagues* (New York: Wiley, 1967).

Rumelhart, D. E. and J. L. McClelland (eds.), *Parallel Distributed Processing: Explorations in the Microstructure of Cognition, Volume I: Foundations* (Cambridge: MIT Press, 1986).

Ryle, G., *The Concept of Mind* (New York: Harper, 1949).

Salam, A., "Weak and electromagnetic interaction," in N. Svartholm (ed.), *Elementary Particle Theory* (Stockholm: Almavist and Wiksell, 1968).

Sambursky, S., *The Physical World of the Greeks* (Princeton: Princeton Univ. Press, 1987).

Schaffner, K., *Nineteenth Century Aether Theories* (New York: Pergamon, 1972).

Schrödinger, E., "Über das Verhältnis der Heisenberg-Born-Jordan-schen Quantenmechanik zu der meinen," *Annalen der Physik*, **70**, 734–756 (1926); translated in part as "On the relationship of the Heisenberg-Born-Jordan quantum mechanics to mine," in G. Ludwig (ed.), *Wave Mechanics* (New York: Pergamon, 1968).

Schweber, S. S., *QED and the Men Who Made It: Dyson, Feynman, Schwinger and Tomonaga* (Princeton: Princeton Univ. Press, 1994).

Seelig, C., *Albert Einstein: Eine dokumentarische Biographie* (Zürich: Europa, 1954).

Settle, T., "An experiment in the history of science," *Science*, **133**, 19–23 (1961).

Smith, C. and N. Wise, *Energy and Empire: A Biographical Study of Lord Kelvin* (Cambridge: Cambridge Univ. Press, 1989).

Smith, N. K., *A commentary to Kant's "Critique of Pure Reason"* (New York: Humanities Press, 1962).

Snow, C. P., *The Two Cultures and A Second Look* (Cambridge: Cambridge Univ. Press, 1969).

Soskice, J. M. and R. Harré, "Metaphor in science," in Z. Radman (ed.), *From a Metaphorical Point of View: A Multidisciplinary Approach to the Cognitive Content of Metaphor* (New York: de Gruyter, 1995).

Stachel, J. (ed.), *Albert Einstein: Volume 1: the Early Years, 1879–1902* (Princeton: Princeton Univ. Press, 1987).

Stangos, N. (ed.), *Concepts of Modern Art* (London: Thames and Hudson, 1991).

Stewart, I., *Does God Play Dice?: The New Mathematics of Chaos* (London: Penguin, 1989).

Stillings, N. A., M. Feinstein, J. L. Garfield, E. L. Rissland, D. A. Rosenbaum, S. E. Weisler, and L. Baker-Ward, *Cognitive Science: An Introduction* (Cambridge: MIT Press, 1987).

Stoppard, T., "Playing with science," *California Institute of Technology: Engineering and Science*, 3–13 (Fall, 1994).

Swenson, L., *The Ethereal Aether* (Austin: Univ. of Texas, 1972).

Tootell, R. B. H., M. S. Silverman, E. Switkes, and R. L. De Valois, "Deoxyglucose analysis of retinotopic organization in primate striate cortex," *Science,* **218,** 902–904 (1982).

Toulmin, S. (ed.), *Physical Reality* (New York: Harper Torchbooks, 1970).

Toulouse, E., *Henri Poincaré* (Paris: Flammarion, 1910).

Turing, A., "Computing machinery and intelligence," *Mind,* **59,** (1950).

Tweney, R. D., "Reflections on scientific creativity," in *Creativity Research Journal* (to be published) (1996).

Ullman, S., "Visual routines," in A. Collins and E. E. Smith (eds.), *Readings in Cognitive Science: A Perspective from Psychology and Artificial Intelligence* (San Mateo, Calif.: Morgan Kaufmann Publishers, Inc., 1988).

Vitz, P. C. and A. B. Glimcher, *Modern Art and Modern Science: The Parallel Analysis of Vision* (New York: Praeger, 1984).

van der Waerden, B. W. (ed.), *Sources of Quantum Mechanics* (New York: Dover, 1968).

Watson, J. B., *The Ways of Behaviorism* (New York: Harper and Brothers, 1928).

Weinberg, S., *Gravitation and Cosmology: Principles and Applications of the General Theory of Relativity* (New York: Wiley, 1972).

Weinberg, S., *Dreams of a Final Theory* (London: Hutchinson Radius, 1993).

Wentzel, G., *Quantum Theory of Fields,* translated by C. Houtermans and J. M. Jauch (New York: Interscience, 1949).

Westfall, R. S., *Never at Rest: A Biography of Isaac Newton* (Cambridge: Cambridge Univ. Press, 1980).

White, L., *Dynamo and Virgin Reconsidered* (Cambridge: MIT Press, 1968).

Whorf, B., *Language, Thought and Reality* (Cambridge: MIT Press, 1964).

Wigner, E., "The unreasonable effectiveness of mathematics in the natural sciences," *Communications on Pure and Applied Mathematics,* **XIII,** 1–14 (1960).

Wood, A., *Thomas Young, Natural Philosopher, 1773–1829* (Cambridge: Cambridge Univ. Press, 1954).

Wotiz, J. H. and S. Rudofsky, "Kekulé's dream: Fact or fiction?" *Chemistry in Britain*, **20**, 720–723 (1954).

Zeki, S., *A Vision of the Brain* (Oxford: Blackwell, 1993).

INDEX